Advances in Intelligent Systems and Computing

Volume 781

Series editor

Janusz Kacprzyk, Polish Academy of Sciences, Warsaw, Poland
e-mail: kacprzyk@ibspan.waw.pl

The series "Advances in Intelligent Systems and Computing" contains publications on theory, applications, and design methods of Intelligent Systems and Intelligent Computing. Virtually all disciplines such as engineering, natural sciences, computer and information science, ICT, economics, business, e-commerce, environment, healthcare, life science are covered. The list of topics spans all the areas of modern intelligent systems and computing such as: computational intelligence, soft computing including neural networks, fuzzy systems, evolutionary computing and the fusion of these paradigms, social intelligence, ambient intelligence, computational neuroscience, artificial life, virtual worlds and society, cognitive science and systems, Perception and Vision, DNA and immune based systems, self-organizing and adaptive systems, e-Learning and teaching, human-centered and human-centric computing, recommender systems, intelligent control, robotics and mechatronics including human-machine teaming, knowledge-based paradigms, learning paradigms, machine ethics, intelligent data analysis, knowledge management, intelligent agents, intelligent decision making and support, intelligent network security, trust management, interactive entertainment, Web intelligence and multimedia.

The publications within "Advances in Intelligent Systems and Computing" are primarily proceedings of important conferences, symposia and congresses. They cover significant recent developments in the field, both of a foundational and applicable character. An important characteristic feature of the series is the short publication time and world-wide distribution. This permits a rapid and broad dissemination of research results.

More information about this series at http://www.springer.com/series/11156

Isabel L. Nunes
Editor

Advances in Human Factors and Systems Interaction

Proceedings of the AHFE 2018 International
Conference on Human Factors and Systems
Interaction, July 21–25, 2018,
Loews Sapphire Falls Resort at Universal Studios,
Orlando, Florida, USA

 Springer

Editor
Isabel L. Nunes
Faculty of Science and Technology
Universidade NOVA de Lisboa
Caparica, Portugal

ISSN 2194-5357   ISSN 2194-5365   (electronic)
Advances in Intelligent Systems and Computing
ISBN 978-3-319-94333-6   ISBN 978-3-319-94334-3   (eBook)
https://doi.org/10.1007/978-3-319-94334-3

Library of Congress Control Number: 2018947359

Printed on acid-free paper

This Springer imprint is published by the registered company Springer International Publishing AG
part of Springer Nature
The registered company address is: Gewerbestrasse 11, 6330 Cham, Switzerland

Advances in Human Factors
and Ergonomics 2018

AHFE 2018 Series Editors

Tareq Z. Ahram, Florida, USA
Waldemar Karwowski, Florida, USA

9th International Conference on Applied Human Factors and Ergonomics and the Affiliated Conferences

Proceedings of the AHFE 2018 International Conferences on Design for Inclusion, held on July 21–25, 2018, in Loews Sapphire Falls Resort at Universal Studios, Orlando, Florida, USA

Advances in Affective and Pleasurable Design	Shuichi Fukuda
Advances in Neuroergonomics and Cognitive Engineering	Hasan Ayaz and Lukasz Mazur
Advances in Design for Inclusion	Giuseppe Di Bucchianico
Advances in Ergonomics in Design	Francisco Rebelo and Marcelo M. Soares
Advances in Human Error, Reliability, Resilience, and Performance	Ronald L. Boring
Advances in Human Factors and Ergonomics in Healthcare and Medical Devices	Nancy J. Lightner
Advances in Human Factors in Simulation and Modeling	Daniel N. Cassenti
Advances in Human Factors and Systems Interaction	Isabel L. Nunes
Advances in Human Factors in Cybersecurity	Tareq Z. Ahram and Denise Nicholson
Advances in Human Factors, Business Management and Society	Jussi Ilari Kantola, Salman Nazir and Tibor Barath
Advances in Human Factors in Robots and Unmanned Systems	Jessie Chen
Advances in Human Factors in Training, Education, and Learning Sciences	Salman Nazir, Anna-Maria Teperi and Aleksandra Polak-Sopińska
Advances in Human Aspects of Transportation	Neville Stanton

(continued)

(continued)

Advances in Artificial Intelligence, Software and Systems Engineering	*Tareq Z. Ahram*
Advances in Human Factors, Sustainable Urban Planning and Infrastructure	*Jerzy Charytonowicz and Christianne Falcão*
Advances in Physical Ergonomics & Human Factors	*Ravindra S. Goonetilleke and Waldemar Karwowski*
Advances in Interdisciplinary Practice in Industrial Design	*WonJoon Chung and Cliff Sungsoo Shin*
Advances in Safety Management and Human Factors	*Pedro Miguel Ferreira Martins Arezes*
Advances in Social and Occupational Ergonomics	*Richard H. M. Goossens*
Advances in Manufacturing, Production Management and Process Control	*Waldemar Karwowski, Stefan Trzcielinski, Beata Mrugalska, Massimo Di Nicolantonio and Emilio Rossi*
Advances in Usability, User Experience and Assistive Technology	*Tareq Z. Ahram and Christianne Falcão*
Advances in Human Factors in Wearable Technologies and Game Design	*Tareq Z. Ahram*
Advances in Human Factors in Communication of Design	*Amic G. Ho*

Preface

Human Factors and Systems Interaction aims to address the main issues of concern within systems interface with a particular emphasis on the system lifecycle development and implementation of interfaces and the general implications of augmented and mixed reality with respect to human and technology interaction. Human Factors and Systems Interaction is, in the first instance, affected by the forces shaping the nature of future computing and systems development. The objective of this book is to provide equal consideration of the human along with the hardware and software in the technical and technical management processes for developing systems that will optimize total system performance and minimize total ownership costs. This book aims to explore and discuss innovative studies of technology and its application in system interfaces and welcomes research in progress, case studies, and poster demonstrations.

A total of four sections are presented in this book:

 I. Applications of Human Factors and System Interactions;
 II. Human Factors and System Interactions in Complex Systems;
 III. Applications in Healthcare and Patient Safety; and
 IV. Management of Productivity in Smart and Sustainable Manufacturing—Industry 4.0

Each section contains research papers that have been reviewed by members of the International Editorial Board. Our sincere thanks and appreciation to the board members as listed below:

Amy Alexander, USA
Musaed Alzeid Alnaser, Kuwait
Pedro Arezes, Portugal
Francesco Biondi, UK
Nina Berry, USA
James P. Bliss, USA
Filipa Carvalho, Portugal

Denis Coelho, Portugal
Alexandra Fernandes, Norway
Frank Flemisch, Germany
José Manuel Fonseca, Portugal
Kazuo Hatakeyama, Brazil
Christopher Lowe, UK
Ravi Mahamuni, India
Pamela McCauley, USA
Bonnie Novak, USA
Maria Papanikou, UK
Alexandra B. Proaps, USA
William Prugh, USA
Yves Rybarczyk, Ecuador
Michael W. Sawyer, USA
Mario Simoes-Marques, Portugal
Vesna Spasojevic Brkic, Serbia
Paolo Trucco, Italy

July 2018   Isabel L. Nunes

Contents

Applications of Human Factors and System Interactions

**Relationships Between Cognitive Workload and Physiological
Response Under Reliable and Unreliable Automation** 3
Jangwoon Park, Heejin Jeong, Jaehyun Park, and Byung Cheol Lee

**Analysis of Trust in Automation Survey Instruments
Using Semantic Network Analysis** 9
Heejin Jeong, Jangwoon Park, Jaehyun Park, Thanh Pham,
and Byung Cheol Lee

Development of an Intuitive, Visual Packaging Assistant 19
Benedikt Maettig, Friederike Hering, and Martin Doeltgen

Visual Capability Estimation Using Motor Action Pattern 26
Yanyu Lu, Lu Ding, and Shan Fu

**Pedestrian Perception of Autonomous Vehicles with External
Interacting Features** 33
Christopher R. Hudson, Shuchisnigdha Deb, Daniel W. Carruth,
John McGinley, and Darren Frey

**Information and Communication Technology (ICT) Impact
on Education and Achievement** 40
Mahnoor Dar, Fatima Masood, Muhammad Ahmed, Maryam Afzaal,
Asad Ali, Zarina Bibi, Imran Kabir, and Hafiz Usman Zia

**The Use of Task-Flow Observation to Map Users' Experience
and Interaction Touchpoints** 46
Adriano B. Renzi

**Serbian and Libyan Female Drivers' Anthropometric
Measurements in the Light of the Third Autonomy Level Vehicles** 56
Ahmed Essdai, Vesna Spasojevic Brkic, and Zorica Veljkovic

**#LookWhatIDidNotBuy: Mitigating Excessive Consumption
Through Mobile Social Media** 69
Pedro Campos, Luisa Soares, Sara Moniz,
and Arminda Guerra Lopes

The Antecedents of Intelligent Personal Assistants Adoption 76
Tihomir Orehovački, Darko Etinger, and Snježana Babić

**Research on Comparison Experiment of Humanized Interface
Design of Smart TV Based on User Experience** 88
Na Lin, Haimei Wu, Huimin Hu, and Wei Li

Human Factors and System Interactions in Complex Systems

Moving Forward with Autonomous Systems: Ethical Dilemmas 101
Aysen K. Taylor and Sarah Bouazzaoui

**Signal-Processing Transformation from Smartwatch to Arm
Movement Gestures** 109
Franca Rupprecht, Bo Heck, Bernd Hamann, and Achim Ebert

**How Can AI Help Reduce the Burden of Disaster
Management Decision-Making?** 122
Mário Simões-Marques and José R. Figueira

**Tackling Autonomous Driving Challenges – How the Design
of Autonomous Vehicles Is Mirroring Universal Design** 134
Susana Costa, Nelson Costa, Paulo Simões, Nuno Ribeiro,
and Pedro Arezes

**Assessment of Pilots Mental Fatigue Status with the Eye
Movement Features** 146
Liwei Zhang, Qianxiang Zhou, Qingsong Yin, and Zhongqi Liu

**Effect of Far Infrared Radiation Therapy on Improving
Microcirculation of the Diabetic Foot** 156
Chi-Wen Lung, Yung-Sheng Lin, Yih-Kuen Jan, Yu-Chou Lo,
Chien-Liang Chen, and Ben-Yi Liau

**Microinteractions of Forms in Web Based Systems Usability
and Eye Tracking Metrics Analysis** 164
Julia Falkowska, Barbara Kilijańska, Janusz Sobecki,
and Katarzyna Zerka

**Design and Realization of Shooting Training System
for Police Force** 175
Bo Shi

An Examination of Close Calls Reported Within the International
Association of Fire Chiefs Database 184
James P. Bliss and Lauren N. Tiller

Applications in Healthcare and Patient Safety

Analysis and Improvement of the Usability of a Tele-Rehabilitation
Platform for Hip Surgery Patients 197
Hennry Pilco, Sandra Sanchez-Gordon, Tania Calle-Jimenez,
Yves Rybarczyk, Janio Jadán, Santiago Villarreal, Wilmer Esparza,
Patricia Acosta-Vargas, César Guevara, and Isabel L. Nunes

Educational Resources Accessible
on the Tele-rehabilitation Platform 210
Patricia Acosta-Vargas, Wilmer Esparza, Yves Rybarczyk,
Mario González, Santiago Villarreal, Janio Jadán, César Guevara,
Sandra Sanchez-Gordon, Tania Calle-Jimenez, Jonathan Baldeon,
and Isabel L. Nunes

Design of an Architecture for Accessible Web Maps
for Visually Impaired Users 221
Tania Calle-Jimenez, Adrián Eguez-Sarzosa,
and Sergio Luján-Mora

Analysis and Improvement of the Web Accessibility
of a Tele-rehabilitation Platform for Hip Arthroplasty Patients 233
Tania Calle-Jimenez, Sandra Sanchez-Gordon, Yves Rybarczyk,
Janio Jadán, Santiago Villarreal, Wilmer Esparza, Patricia Acosta-Vargas,
César Guevara, and Isabel L. Nunes

Interaction with a Tele-Rehabilitation Platform Through a Natural
User Interface: A Case Study of Hip Arthroplasty Patients 246
Yves Rybarczyk, Santiago Villarreal, Mario González,
Patricia Acosta-Vargas, Danilo Esparza, Sandra Sanchez-Gordon,
Tania Calle-Jimenez, Janio Jadán, and Isabel L. Nunes

Comparison of Theory of Mind Tests in Augmented Reality
and 2D Environments for Children with Neurodevelopmental
Disorders 257
N. Tugbagul Altan Akin and Mehmet Gokturk

A Real-Time Algorithm for Movement Assessment Using Fuzzy
Logic of Hip Arthroplasty Patients............................. 265
César Guevara, Janio Jadán-Guerrero, Yves Rybarczyk,
Patricia Acosta-Vargas, Wilmer Esparza, Mario González,
Santiago Villarreal, Sandra Sanchez-Gordon, Tania Calle-Jimenez,
and Isabel L. Nunes

An Integrated System Combining Virtual Reality with a Glove
with Biosensors for Neuropathic Pain: A Concept Validation 274
Claudia Quaresma, Madalena Gomes, Heitor Cardoso, Nuno Ferreira,
Ricardo Vigário, Carla Quintão, and Micaela Fonseca

Development of a Low-Cost Eye Tracker – A Proof of Concept 285
Ricardo Vigário, Filipa Gamas, Pedro Morais, and Carla Quintão

Management of Productivity in Smart and Sustainable
Manufacturing - Industry 4.0

Socio-Technical Capability Assessment to Support Implementation
of Cyber-Physical Production Systems in Line with People
and Organization 299
Fabian Noehring, René Woestmann, Tobias Wienzek,
and Jochen Deuse

Competences and Competence Development in a Digitalized
World of Work 312
Walter Ganz, Bernd Dworschak, and Kathrin Schnalzer

Opportunities of Digitalization for Productivity Management 321
Tim Jeske, Marc-André Weber, Marlene Würfels, Frank Lennings,
and Sascha Stowasser

How Digital Assistance Systems Improve Work Productivity
in Assembly 332
Sven Hinrichsen and Sven Bendzioch

Human Work Design: Modern Approaches for Designing
Ergonomic and Productive Work in Times of Digital
Transformation – An International Perspective 343
Martin Benter and Peter Kuhlang

Evaluation and Systematic Analysis of Ergonomic and Work
Safety Methods and Tools for the Implementation in Lean
Production Systems 353
Uwe Dombrowski, Anne Reimer, and Tobias Stefanak

Indicators and Goals for Sustainable Production Planning
and Controlling from an Ergonomic Perspective 363
Maximilian Zarte, Agnes Pechmann, and Isabel L. Nunes

Environment-Integrated Human Machine Interface Framework
for Multimodal System Interaction on the Shopfloor 374
Katrin Schilling, Simon Storms, and Werner Herfs

**Digitization of Industrial Work Environments and the Emerging
Challenges of Human-Digitized System Collaborative Work
Organization Design** 384
Mohammed-Aminu Sanda

**Design of a Platform for Sustainable Production Planning
and Controlling from an User Centered Perspective** 396
Maximilian Zarte, Agnes Pechmann, and Isabel L. Nunes

Author Index 409

Applications of Human Factors
and System Interactions

Relationships Between Cognitive Workload and Physiological Response Under Reliable and Unreliable Automation

Jangwoon Park[1]($\boxtimes$), Heejin Jeong[2], Jaehyun Park[3],
and Byung Cheol Lee[1]

[1] Department of Engineering, Texas A&M University – Corpus Christi,
Corpus Christi, TX, USA
jangwoon.park@tamucc.edu
[2] Department of Industrial and Operations Engineering, University of Michigan,
Ann Arbor, MI, USA
heejinj@umich.edu
[3] Department of Industrial and Management Engineering,
Incheon National University, Incheon, South Korea
jaehpark@inu.ac.kr

Abstract. Although reliable automation can reduce a lot of the mental workload from humans, unreliable automation can increase cognitive workload and physiological changes. The objective of this study is to quantify the relationships between cognitive workload and electrodermal activities (EDA) in reliable and unreliable auto-proofreading tasks. Nineteen native English speakers participated in sentence correction tasks under a reliable or unreliable auto-proofreading support. During the tasks, the participants' EDA signals were measured by using Empatica E4 wrist band and cognitive workload were evaluated by NASA-TLX indices including mental demand, effort, performance, and frustration level with 21-point scale. Overall, significant Pearson's correlation coefficients were observed between the slope of EDA signal and mental demand ($r = 0.477$, p = 0.039), effort ($r = 0.428$, $p = 0.068$), performance ($r = -0.500$, $p = 0.029$), and frustration ($r = 0.474$, $p = 0.040$). Detailed analysis results are described in the paper. To our best knowledge, the linear relationships between physiological responses and cognitive workload are quantified in reliable and unreliable automation for the first time. The findings of this study can be applied to guide future research to understand human behavior in unreliable automation.

Keywords: Trust in automation · Cognitive workload
Physiological measurement · Human-automation interaction

1 Introduction

Today, we are enjoying the benefits of successful automation in terms of time efficiency [1], increasing work performance [2], and trust [3]. For example, the automation technology of the various automakers that introduced the self-driving vehicle feature

© Springer International Publishing AG, part of Springer Nature 2019
I. L. Nunes (Ed.): AHFE 2018, AISC 781, pp. 3–8, 2019.
https://doi.org/10.1007/978-3-319-94334-3_1

enabled the driver to arrive safely and accurately at the destination with less workload. On the other hand, as a simple automation used in the office environment, the auto spelling checks or auto proofreading function provided by Microsoft® Word or Grammarly®, we can dramatically reduce typos and grammatical error in writing documents. These successful automations allowed users to reduce cognitive workload as well as improve work performance during the tasks.

Reduced cognitive workload through automation can result from the accumulation of trust. This trust can be established when the automation performs its function correctly and consistently. For example, while current self-driving vehicle technology is evolving and increasingly popular, many drivers still prefer manual driving because of the need to establish trust in the trustworthiness for safety of self-driving.

To establish trust in automation, reliability of a system and the enough interaction time between user and system are required. If a user recognizes that an automation is not reliable, the user will immediately find an alternative manual method. In addition, if a task that is highly dependent on the automation of the system (e.g., a self-driving vehicle or a safety device in a nuclear power plant) has a problem with unreliable automation, a problem that could lead to irreversible safety accidents. Even though the system works smoothly, but the user does not have enough time to build the trust on the system, the user is more likely to prefer the manual mode or mixed manual and automated mode.

The objective of this study is to analyze the effects of system reliability and interaction time on cognitive workload in automation. In this paper, we present an auto-proofreading task as a simple example of an automation task. Participants' cognitive workload and trust on reliable and unreliable auto proofreading systems were assessed by using a subjective questionnaire and their electrodermal activities (EDA) signals were recorded throughout the experiment.

2   Methods

Participants. Nineteen English speakers (11 females, 8 males) were participated in the experiment. Most participants had at least two more years of experience of using Auto proofreading feature in Microsoft Word.

Experimental Program. For the experiment, a custom-build program was developed. The program provided four different auto-proofreading sessions (e.g., sessions A, B, C, and D; see more details below). Figure 1 shows the developed auto-proofreading program.

- Session A ran a reliable auto-proofreading condition that highlighted a grammatical error with an underline and did not provide any suggestion.
- Session B ran the same reliable auto-proofreading condition and provided a correct suggestion.
- Session C ran an unreliable auto-proofreading condition that highlighted a correct word with an underline and did not provide a suggestion.

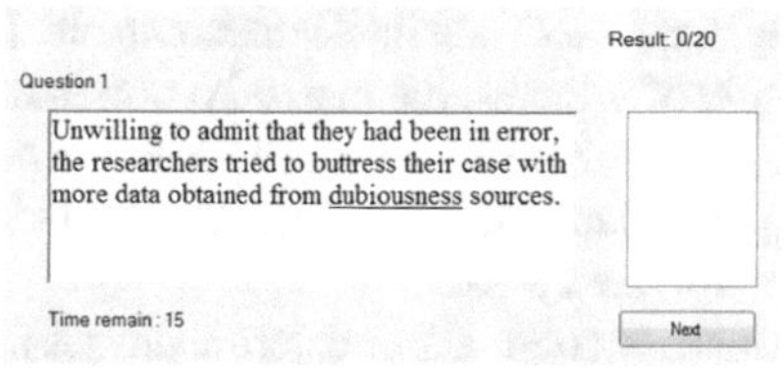

(a) Session A: Reliable without suggestion

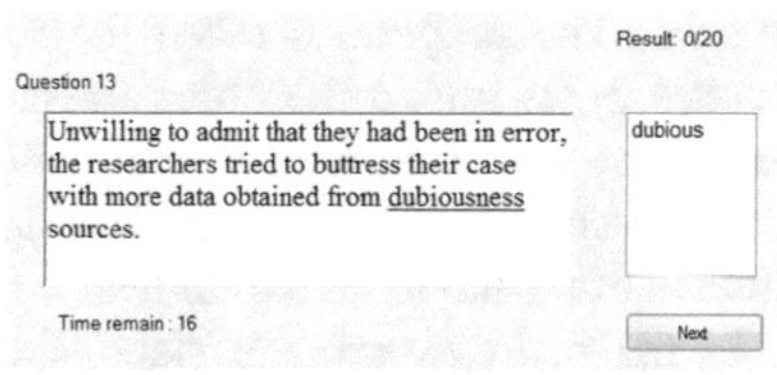

(b) Session B: Reliable with a correct suggestion

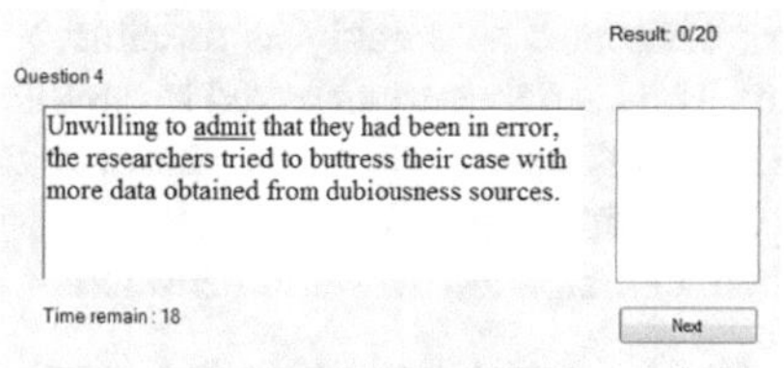

(c) Session C: Unreliable without suggestion

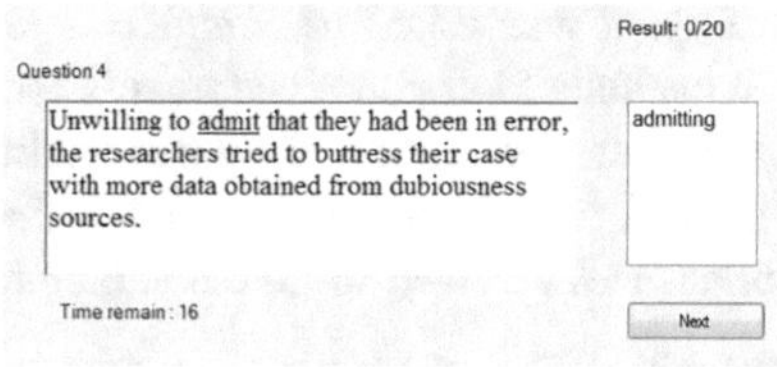

(d) Session D: Unreliable with an incorrect suggestion

Fig. 1. Four different auto-proofreading sessions provided by the custom-built program.

- Session D ran an unreliable auto-proofreading condition that highlighted a correct word with an underline and provided an incorrect suggestion.

The sentences for the proofreading tasks were carefully selected by considering their readability score (www.webpagefx.com/tools/read-able/) to standardize level of difficulty. Among the selected 34 sentences, four sentences were used for a training session; ten sentences were used for a manual proofreading session; and the rest of 20 sentences were used for each of the four sessions (sessions A, B, C, and D). More details about the program is described in Jeong et al. [4].

Experimental Procedure. The experiment was conducted in three stages (i.e., preparation, practice, and main experiment). In the preparation stage, the purpose and procedure of the experiment were explained to the participant, after which a written informed consent was obtained using procedures approved by the Texas A&M University Corpus Christi Institutional Review Board (human subjects research protocol #59-17). Each participant wore the Empatica E4 wristband on his/her wrist to measure EDA signal throughout the experiment. A two-minute rest period was conducted before starting the practice stage. During the experiment, a moderator recorded starting time of each proofreading task which allows us to synchronize the tasks with measured EDA signals.

In the practice stage, each participant conducted the training session to become familiar with the provided auto-proofreading system. During the training, the participant was asked to complete the proofreading on the selected four sentences as quickly and correctly as possible. After the participant corrected an error from the sentences, he/she was asked to click the "Next" button at the bottom right of the page, and then the next sentence would appear in the program. To increase the anxious levels during the proofreading tasks, each sentence must be corrected within 20 s. If the sentence was not completed within the 20 s, then the program would move automatically to the next sentence. The program showed the remaining time (sec) available to complete the sentence. Next, after a short break lasting a couple of minutes, the manual proofreading session for the 10 sentences was conducted, where there was no automated proof-reading system. After the manual session, the participant had a two-minute rest period before starting the main experiment which is automated session.

In the main experiment stage, the participant started a pre-determined session among the four sessions (A, B, C, and D) in a random order. Totally, the 19 participants were distributed to conduct sessions A (5), B (5), C (4), and D (5). In each session, the participant was asked to complete a set of five sentences as quickly as possible, and then evaluate their perceived trust in that system. The participant was asked to complete the 20 sentences, randomly separated into four sequential sets; subjective measures was evaluated at the end of each set. After completing the four sequential sets, the participant was allowed to take a short break. Their participation was compensated.

Measures. Five subjective measures (mental demand, effort, performance, frustration level, and trust) were evaluated as cognitive workload by a questionnaire with 21-point Likert scale (-10 is very low; 0 is neutral, 10 is very high) and one physiological measure (electrodermal activities, EDA) was measured by Empatica wristband (E4, Empatica Inc., USA) with a sampling rate of 4 Hz throughout the proofreading tasks. In order to reduce individual difference in EDA signals, each participant's EDA signal was standardized with z-score normalization.

3   Results

Electrodermal Activities in the Auto-Proofreading Tasks. Figure 2 shows the standardized EDA signal (roughly from -3 to $+3$) of subject #1 throughout the experiment. The four bold lines indicate the EDA signals in the four sequential sets of the auto-proofreading tasks. The dotted line is a linear regression model on the standardized EDA data for the four sequential sets that is showing a trend whether it is increasing or decreasing. In this case, subject #1's EDA signal had a decreasing pattern with negative regression coefficients (which is slope of a line; -0.00068184). Note that the subject #1 participated the session A (Reliable without suggestion) auto-proofreading task (see Table 1). Table 1 shows the averaged subjective ratings and calculated regression coefficients of the standardized EDA signals for each of the 19 participants.

In order to identify the relationships between subjective measures (mental demand, effort, performance, frustration level, and trust) and physiological response (regression coefficient of EDA signals), Pearson's correlation coefficients were calculated. Overall, significant Pearson's correlation coefficients were observed between the slope of EDA signal and mental demand ($r = 0.477$, $p = 0.039$), effort ($r = 0.428$, $p = 0.068$), performance ($r = -0.500$, $p = 0.029$), and frustration ($r = 0.474$, $p = 0.040$).

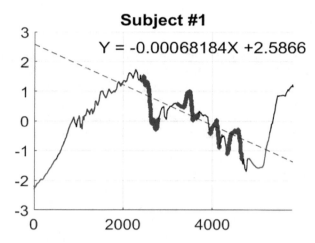

Fig. 2. Standardized EDA signal of subject #1 and a linear regression model for the four sequential sets.

Table 1. Averaged subjective ratings and regression coefficients of EDA signals for the 19 participants

Session	Subject#	Mental demand	Effort	Performance	Frustration level	Trust	Regression coefficient
A: Reliable without suggestion	1	−5.25	−5.25	1.50	−6.00	0.00	−0.000681840
	2	4.00	3.25	−1.50	0.75	9.50	0.000475140
	9	1.25	0.75	3.25	−3.00	8.00	0.000689500
	16	2.50	3.00	5.00	−2.25	−0.50	0.000405010
	18	6.25	5.00	2.25	4.75	0.00	−0.000175570
B: Reliable with a correct suggestion	8	10.00	7.50	8.50	−9.00	9.25	−0.000024129
	10	4.75	5.50	9.50	−1.50	9.75	−0.001313200
	12	−7.50	−7.50	9.50	−10.00	8.75	−0.001499500
	13	0.00	−1.25	8.50	0.00	9.50	0.000086513
	17	−9.25	−10.00	9.75	−10.00	9.75	−0.000582950
C: Unreliable without suggestion	3	10.00	10.00	−9.00	4.00	−2.50	0.000405130
	4	7.75	7.25	−9.25	0.75	−9.75	−0.000088698
	14	10.00	9.50	−2.00	0.25	−1.25	0.000829120
	15	2.25	2.00	2.75	−0.25	−6.75	0.000448690
D: Unreliable with an incorrect suggestion	5	4.75	5.50	2.75	3.75	2.00	0.000247410
	6	6.00	6.50	1.50	5.00	−1.25	−0.000292280
	7	3.25	3.25	2.25	4.25	−5.50	0.000378870
	11	−7.00	−7.00	5.25	3.25	−10.00	0.000181100
	19	8.50	3.50	−1.75	−0.25	−10.00	0.000680010

4   Discussion

The effects of system reliability on cognitive workload and physiological response have been quantified. In this study, an auto-proofreading system was selected as a simple automated system and the system was manipulated in terms of reliability (reliable vs. unreliable auto-proofreading system) and suggestion option (with suggestion vs. without suggestion) to identify the relationship between cognitive workload and physiological response. Main findings of this study can be listed as follow:

- First, significant correlations were quantified between cognitive workload and physiological response (EDA signal).
- Second, the suggestion feature can substantially elevate perceived trust if the auto-proofreading system worked reliably; On the other hand, if the system worked unreliably, the suggestion feature did not affect to the perceived trust. Third, the EDA signals were highly dependent on the system reliability.
- Lastly, the significant positive correlations between the perceived mental demand ($r = 0.477$, $p = 0.039$), mental effort ($r = 0.428$, $p = 0.068$), performance ($r = -0.500$, $p = 0.029$), and frustration ($r = 0.474$, $p = 0.040$) and the EDA signals in the auto-proofreading systems was quantified. Among the cognitive workload measures, the performance measure shows a slightly higher correlation coefficient with the EDA signals than others.
- The findings in this study indicates that if the cognitive workload has been increased due to unexpected unreliable automation supports, then that might affect to an increase of a slope of EDA signal.
- Further study is needed to identify the relationships between cognitive workload, physiological responses, and emotional status such as anxiety level in automation failure. Ultimately, a user's cognitive workload or emotional level on automation could be estimated by using the user's physiological response.

Acknowledgements. This work was supported by the National Research Foundation of Korea (NRF) grant funded by the Korea government (MSIP) (No. 2015R1C1A1A01054148). We thank our undergraduate students at Texas A&M University – Corpus Christi who contributed significantly to this work, Tri Vo who developed the custom-build auto-proofreading system for this study and Celeste Branstrom who led the data collection and literature review.

References

1. Crocell, W., Coury, B.: Status or recommendation: selecting the type of information for decision aiding. In: Proceedings of the Human Factors Society 34th Annual Meeting, p. 1524 (1990)
2. Dzindolet, M., Peterson, S., Pomranky, R., Pierce, L., Beck, H.: The role of trust in automation reliance. Int. J. Comput. Stud. **58**, 697–718 (2003)
3. Merritt, S.: Affective processes in human-automation interactions. Hum. Factors **53**(4), 356–370 (2011)
4. Jeong, H., Park, J., Park, J., Lee, B.C.: Inconsistent work performance in automation, can we measure trust in automation? Int. Robot. Autom. J. **3**(6) (2017)

Analysis of Trust in Automation Survey Instruments Using Semantic Network Analysis

Heejin Jeong[1], Jangwoon Park[2], Jaehyun Park[3], Thanh Pham[4],
and Byung Cheol Lee[2(✉)]

[1] Department of Industrial and Operations Engineering, University of Michigan,
1205 Beal Avenue, Ann Arbor, MI, USA
heejinj@umich.edu
[2] Department of Engineering, Texas A&M University – Corpus Christi,
6300 Ocean Dr., Corpus Christi, TX, USA
{jangwoon.park,byungcheol.lee}@tamucc.edu
[3] Department of Industrial and Management Engineering, Incheon National
University (INU), Academy-ro 119, Incheon, Republic of Korea
jaehpark@inu.ac.kr
[4] Department of Computing Science, Texas A&M University – Corpus Christi,
6300 Ocean Dr., Corpus Christi, TX, USA
tpham@islander.tamucc.edu

Abstract. This study analyzed existing survey instruments to provide an integrated list of keywords/constructs to measure the various perceptions of trust building in automation. While the trust between users and automated functions or systems has been an area of substantial research interest to understand the interactions between human and automation, research efforts to measure the trust to date have led to inconclusive and mixed outcomes. Of the existing scales for measuring trust in automation, inadequate development of constructs and the lack of reliability and validity have been identified as major causes for such outcomes. To develop a scale in a more objective and systematic approach, 86 keywords from existing 9 survey instruments were identified. The keyword network was developed based on the semantic textural similarity, and the network centrality analysis provided total 14 keywords with high centrality and degree matrics. The results can suggest some potential solutions about the lack of consistency and the wide array of constructs without adequate analytic justification in prior survey instruments. The outcomes will be utilized to develop a new integrated scale that can be generally applicable to a wide variety of automation adoption or, with slight modifications, in most trust in automation applications.

Keywords: Trust in automation · Natural language processing
Network analysis · Semantic textural similarity

1 Introduction

As technology develops, the interaction between human and automation has become increasingly common and important. Trust works as a backbone of the interaction and can affect how much people accept and rely upon increasingly automated systems [1].

© Springer International Publishing AG, part of Springer Nature 2019
I. L. Nunes (Ed.): AHFE 2018, AISC 781, pp. 9–18, 2019.
https://doi.org/10.1007/978-3-319-94334-3_2

While the interaction is a critical aspect to measure trust, the characteristics of inter-action are difficult to be defined. Prior research has examined the primary character-istics and attitudes of automation found inconsistent or ambiguous outcomes, and variance depends on distinct automation types or levels [2, 3]. Trust also plays a role in influencing users' strategies toward the use of automation [4]. A key factor shaping people's mindsets towards autonomous systems is their trust of the system; hence, understanding the factors that influence trust and measuring the degree of trust are vital to determine how the automation is smoothly adopted and effectively functioned. Insufficient or excessive trust seems to bring misuse and disuse of automation [5].

Prior survey instruments on trust in automation tended to be based on the existing interpersonal trust model, which tried to understand the user trust on automation by using simple adaptation of interpersonal survey instruments. The instruments, mainly based on the precedents of trust building process in certain specific automated systems, lack validity and reliability, and their applications were limited to build a universal trust scale to measure trust levels [6]. Survey questionnaires have been used to measure trust in human-machine system. Singh, Molloy, and Parasuraman measured complacency levels by examining attitudes towards automated teller machines [7]. Subjective rating scales to evaluate participants' perceptions of the reliability and trustworthiness of the automated systems was suggested [8]. Instead of measuring trust, the instruments have focused on the connection between the automation usage and reliability. Furthermore, there had been not enough field work done in trust on automation. Existing data collected via empirical approach was mainly focused on simulated or prototype sys-tems rather than practically operational systems. Measuring trust in automation is still one of the challenging issues facing the human – automation interaction.

This study investigated and analyzed existing trust in automation survey ques-tionnaires to provide potential solutions about the lack of consistency and the wide array of constructs without adequate analytic justification. Due to inadequate devel-opment of constructs and the lack of reliability and validity, research efforts have been considered as major causes for inconclusive outcomes to measure trust in automation. To develop an integrated survey scale and verify the convergent validity of the scale, keywords from existing survey instruments were identified and categorized. The similarity between collected keywords were computed by a natural language pro-cessing algorithm, and the network centrality analysis was conducted to identify the list of an integrated keywords of trust in automation. The results can suggest some potential solutions about the lack of consistency and the wide array of constructs without adequate analytic justification in prior survey instruments. The outcomes will be utilized to develop a new integrated scale that can be generally applicable to a wide variety of automation adoption or, with slight modifications, in most trust in automa-tion applications.

2   Background

Survey instrument has been a dominant tool to measure trust. Larzelere and Huston [9] began to use questionnaires to measure trust, in terms of honesty and benevolence. [10] suggested a structural model of trust and believed that certain aspects of trust may not be maintained at the same degrees and change with emotional states. As common constructs in general trust concept, integrity, ability, and benevolence, are considered [11, 12]. Lewis and Weigert asserted that integrity and ability are associated with the macro level trust in organizations or societies, while benevolence is suited for individual relationships [13]. In addition, vulnerability, uncertainty and risk are other constructs to describe trust [14]. While vulnerability is the state that an individual may induce harm, uncertainty and risk are referred to as the potential to cause vulnerability [15]. However, it is not easy to develop a well-designed psychometric instrument. The instruments containing these constructs were not successful to provide evidence to support their selection or modification of existing measures.

Recently several survey instruments to measure the trust in automation have been suggested. [16] provided an empirical survey instrument to assess trust. While this study attempted to identify the similarities and differences among general trust, trust between people, and trust between human and automated systems, the analyzed results had not confirmed the difference due to the assumption that three trusts are shared with the same construct dimensions. Also, since this study mainly focused on emotional aspects of trust related words, emotional dimensions of the trust are overly rated, and it lacks validation with actual automation. The characteristics of automation or systems were not covered enough. [17] suggested a psychometric instrument to measure human-computer trust (HCT). Based on existing interpersonal trust models, this instrument consists of cognition-based components and affect-based components such as user's emotional responses to the system. This study, however, did not provide full description of data collection process and task that used in the experiment and missed to conduct confirmatory factor analysis and other goodness of fir indexes, which leave the dimensionality uncertain. Chien, Semnani-Azad, Lewis, and Sycara conducted the first empirical study to validate trust in automation [18]. It sought to integrate existing instruments in which trust mediates reliance on automation across cultures and to understand fundamental principles and factors pertaining to trust in automation. However, the instrument needs more systematic analysis to identify the dimensionality of the data and factor loading.

2.1 Survey Instrument Development Process in Existing Trust in Automation Literature

Previous survey instruments of trust in automation can be categorized into construct-oriented and validation-oriented instruments. The construct-oriented instruments were built from the basis of automation and operator relationship and the instrument validation process were mainly conducted with human attitude elicitation or actual system application. For example, Jian et al. developed a unique three-phase experimental study comprising a word elicitation, a questionnaire and a paired comparison phases [16]. This study mainly focused on collecting the words that are well represented to trust and distrust and developed a survey scale based on underlying similarity among them. Though comprehensive validation tests were not accomplished, the scale frequently cited as a reliable tool to measure trust in automated systems. in other studies. HCT Human-Computer trust instrument also identifies 10 constructs that affect the trust levels using Nominal group technique and developed a scale incorporating a new construct, liking [17]. Merritt and Ilgen [19] and Chien et al. [18] gathered initial items from existing trust scales or combining various psychological attitude surveys.

Validation-oriented instruments are mainly grounded on interpersonal trust models. The trust can be understood as a partial component in the holistic technology adoption process and studies had been conducted in Information Science domain. Survey instrument developing process highly focused on the statistical validation on instrument items from pilot studies. The validation includes several fit statistics such as nonnormed fit index (NNFI), comparative fit index [20], root mean error of approximation (RMSEA), Chi square. Reliability can be assessed by coefficient alpha estimates, composite reliability, and AVE (Average Variance Extracted). Convergent validity is evaluated by the magnitude of correlation among item measures of a construct across multiple methods of measurement, and discriminant validity was also tested using Fornell and Larker's test [20–22].

In both instrument groups, factor analysis has been used as a major scale development approach. Basically, factor analysis is a statistical method to identify latent variables and dimensions with a wide survey response data. Though this approach has been extensively used in survey development process, it has some limitations. Factor analysis can be only as a solid method as the initial survey items are reliable. Collecting a comprehensive and legitimate set of the items is complicated, and the items may include less valid and reliable measures such as self-created items. In addition, if the items are too much scattered, the results from factor analysis would be difficult to reflect the data patterns or to reveal meaningful constructs. In addition, the selecting threshold of hidden factors is another issue.

3   Method

This study consists of three stages of analysis. First, we collected existing survey instruments from previous literatures. The first stage was to search relevant existing survey instruments and collect a pool of survey instrument items by classifying to fit the definition of trust in automation. Then, keywords were extracted from the pool and consolidated. The next stage in the process was semantic textural similarity analysis. Using the UMBC semantic similarity service, the similarity between keywords was computed. Finally, based on the similarity scores, network analysis was conducted to identify central keywords.

3.1   Survey Questionnaires Collection and Keyword Elicitation

The snowball method was used to collect proper trust in automation survey instruments. An initial collection of survey instruments was obtained from [18]. The study evaluated the existing and previous trust instruments and developed an inter-cultural trust in automation survey instrument with the application of general and specific automation. The study includes 8 survey instruments about trust instruments, and the collection was expanded to human factors, information systems, and Human Computer interaction articles in which trust in automation related survey instruments would likely be found. 228 questions were collected from nine survey instruments. After deleting redundant and ambiguous items, trust in automation survey pool has total 156 question items.

The keywords were identified from the survey pool. Four subject matter experts (SMEs) who have at least five years of academic research experience in human factors identified the keywords from the all items based on the similarity and difference among the keywords. Once the keywords were identified, four sets of keywords were compared and then re-evaluated to eliminate those which appeared redundant or ambiguous. The identified and consolidated number of keywords is 86. The value of inter-rater reliability analysis (Fleiss' Kappa) is 0.753.

3.2 Natural Language Processing Algorithm: Semantic Textural Similarity Analysis

This study used the University of Maryland Baltimore County (UMBC) Semantic Similarity Service system to compute the similarity between keywords [23]. The system provides semantic similarity between words/phrases based on using a thesaurus and statistics from a large word database [23]. The statistical method used in the system is based on Latent Semantic Analysis (LSA) and semantic relations extracted from WordNet [24]. Textural database used in this system is a web corpus from Stanford WebBase project which contains 100 million web pages from more than 50, 000 websites and a three billion words corpus of good quality English (Fig. 1).

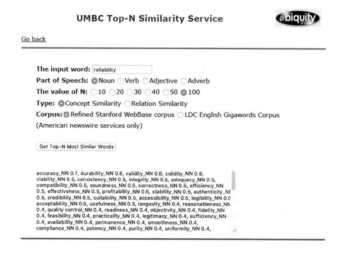

Fig. 1. University of Maryland Baltimore County (UMBC) Semantic Similarity Service [23]

Among three semantic textural similarity services, this study used the Top-N Similarity which provides top-n most similar words to an input word. This generated 100 similar words and their similarity values from 0.00 to 1.00 for each keyword from the consolidated 86 keywords. The words that are not included in the keyword pool and whose similarity values are less than 0.20 were removed from the results. An 86×86 semantic textural similarity matrix was developed.

3.3 Network Analysis and Centrality Index Computation

Network analysis was conducted to identify the relationship among keywords. Undirected weighted network diagram was developed, and degree centrality was measured to identify central nodes. Each keyword was taken as a node and the similarity value between two keywords were used as a weighted edge. Once the network diagram was developed, the next step involved calculating the network related metrics. Among the various possible metrics defined for an undirected weighted network, weighted degree, betweenness centrality, and closeness centrality indexes were computed. The degree of a focal node is a basic indicator of the network and indicates the number of adjacencies in a network. The closeness and betweenness centrality measures depend on the identification and length of the shortest paths among nodes in the network. The equations for calculating the different centrality measures is given below.

$$Degree(k_i) = C_D(i) = \sum_{j}^{N} x_{ij} \tag{1}$$

Where, i is the focal node, j represents s all other nodes, N is the total number of nodes, and x is the adjacency matrix, in which the cell x_{ij} id defined as 1 if node i is connected to node j, and 0 otherwise.

$$Betweenness\ centrality(C_B(i)) = \sum_{j<k} g_{jk}(i)/g_{jk} \tag{2}$$

Where, g_{jk} = the number of geodesics connecting jk, and
$g_{jk}(i)$ = the number of geodesics that actor i is on.

$$Closeness\ centrality(C_c(i)) = \left[\sum_{j=1}^{N} d(i,j) \right]^{-1} \tag{3}$$

Where d_{ij} = shortest distance from i to j.
For the present case, the generalized formula by Opsahl et al. [25] has been applied to calculate the mentioned metrics values using the tnet package in R.

4 Results

Figure 2 shows a network representation of the keywords from trust measuring survey question items based on semantic textural similarity indexes. The diagram demonstrates relationship between the keywords, and the font size of the nodes shows their relative importance within the network.

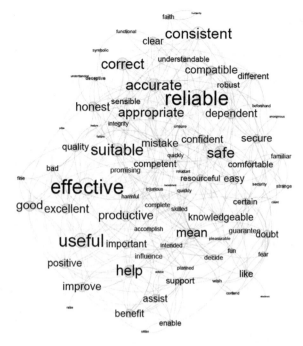

Fig. 2. Network diagram for keywords from trust measuring survey question items

Table 1 shows the metric values of weighted degree, betweenness centrality and closeness centrality. We selected 10 keywords per each metrics with the comparison purpose. While nine out of ten keywords are identical in closeness and betweenness centrality, only four keywords (appropriate, effective, reliable, and suitable) in weighted degree are overlapped with those in centrality indexes.

Table 1. Top 10 keywords in weighted degree, betweenness centrality, and closeness centrality.

Keyword	Degree	Keyword	Betweenness centrality	Keyword	Closeness centrality
Reliable	33	Effective	301.959	Effective	0.525
Effective	33	Safe	272.756	Safe	0.516
Useful	28	Reliable	253.053	Mean	0.506
Suitable	25	Mean	226.657	Reliable	0.503
Accurate	24	Accurate	211.458	Accurate	0.503
Safe	24	Dependent	211.347	Appropriate	0.494
Consistent	24	Mistake	190.949	Dependent	0.494
Help	23	Appropriate	174.630	Suitable	0.491
Correct	23	Useful	165.945	Mistake	0.488
Appropriate	21	Suitable	161.536	Productive	0.483

5 Discussion

As mentioned in the introduction, out of the various kinds of trust in automation survey instruments discussed, most of the existing instruments have used a subjective approach to elicit major constructs or survey questions. Therefore, it is meaningful to develop objective and systematic approach that can quantify the relationship and closeness among trust in automation survey questions.

The network diagram in Fig. 2 and centrality indexes in Table 1 provide a list of crucial keywords of trust in automation survey instruments. Generally, the network analysis identifies which nodes are more essential than others, and the central nodes have more ties and reach other nodes more quickly. They also can control the connections among other nodes. With these characteristics, three indexes of node centrality, degree, betweenness and closeness, offer quantitative criteria to sort out central keywords in trust in automation surveys [26]. Specifically, degree represent the number of nodes that a focal node is connected to and measures the involvement of the node in the network. It, however, lacks to understand a global structure of the network. Closeness centrality supports to overcome this limitation and offers distance information to other nodes from a focal node, which is valuable in measuring access resources of the nodes. Betweenness assess the degree to which a node controls the connection in the network, and it also aids to view the global network structure.

From the keyword list of three centrality indexes, total 14 keywords were collected. While 8 out of 10 words were identical in closeness and betweenness centrality indexes, five and six words in degree index were shown in closeness and betweenness centrality respectively. Thus, closeness and betweenness centrality resulted in a consistent list of keywords, but degree provides a quite different list. This outcome may indicate that trust in keyword network consists of multiple connected keyword groups, and local and overall centralities are differently described in the keyword network.

In this paper, we only presented top 10 keywords for each centrality indexes, and we have not a clear threshold to cut off the keyword list. A further study may be required to evaluate the validity of the selected keywords and define the criteria to select the number of keywords. Another limitation would be the possible application of other network metrics such as eigencentrality, clustering coefficient, and page rank. Additional metrics can bring more meaningful insights of the relationship between keywords. Moreover, new layout and visualization of the network diagram needs to be considered to optimize the readability of the graph.

6 Conclusion

This study suggests an innovative approach to investigate the survey constructs or keywords with natural language processing algorithm. Based on the semantic textural similarity and network analysis, total 14 keywords were sorted from 86 keywords of existing 9 trust in automation survey instruments. The results will be potentially used for developing an integrated survey questionnaire to assess trust in automation, and it provides a first step towards the development of a survey instrument based on artificial intelligence approach. Such methods bear the advantage over traditional survey development methods that they enable a more objective assessment of user attitudes. Though the results need to be validated by comparison with prior trust in automation surveys, this study suggest some potential to improve consistency and to bring adequate analytic justification in survey construct structure. The approach used in this study will be utilized to develop a new integrated scale that can be generally applicable to evaluate a wide variety of automation adoption or to assess trust in automation applications.

References

1. Sheridan, T.B.: Task allocation and supervisory control. In: Handbook of Human-Computer Interaction. 8 (1988)
2. Onnasch, L., Wickens, C.D., Li, H., Manzey, D.: Human performance consequences of stages and levels of automation: an integrated meta-analysis. Hum. Factors 56, 476–488 (2014)
3. Jeong, H., Park, J., Park, J., Lee, B.C.: Inconsistent work performance in automation, can we measure trust in automation. Int. Robot. Autom. J. 3, 00075 (2017)
4. Lee, J.D., Moray, N.: Trust, self-confidence, and operators' adaptation to automation. Int. J. Hum.-Comput. Stud. 40, 153–184 (1994)
5. Lee, J.D., See, K.A.: Trust in automation: designing for appropriate reliance. Hum. Factors J. Hum. Factors Ergon. Soc. 46, 50–80 (2004)
6. Chien, S.-Y., Lewis, M., Hergeth, S., Semnani-Azad, Z., Sycara, K.: Cross-country validation of a cultural scale in measuring trust in automation. Proc. Hum. Factors Ergon. Soc. Annu. Meet. 59, 686–690 (2015)
7. Singh, I.L., Molloy, R., Parasuraman, R.: Automation-induced "complacency": development of the complacency-potential rating scale. Int. J. Aviat. Psychol. 3, 111–122 (1993)
8. Muir, B.M., Moray, N.: Trust in automation. Part II. Experimental studies of trust and human intervention in a process control simulation. Ergonomics 39, 429–460 (1996)
9. Larzelere, R.E., Huston, T.L.: The dyadic trust scale: toward understanding interpersonal trust in close relationships. J. Marriage Fam. 595–604 (1980)
10. Rempel, J.K., Holmes, J.G., Zanna, M.P.: Trust in close relationships. J. Pers. Soc. Psychol. 49, 95 (1985)
11. Doney, P.M., Cannon, J.P.: Trust in buyer-seller relationships. J. Mark. 61, 35–51 (1997)
12. Ganesan, S.: Determinants of long-term orientation in buyer-seller relationships. J. Mark. 1–19 (1994)
13. Lewis, J.D., Weigert, A.: Trust as a social reality. Soc. Forces 63, 967–985 (1985)
14. Mayer, R.C., Davis, J.H., Schoorman, F.D.: An integrative model of organizational trust. Acad. Manag. Rev. 20, 709–734 (1995)
15. Friedman, B., Khan Jr., P.H., Howe, D.C.: Trust online. Commun. ACM 43, 34–40 (2000)
16. Jian, J.-Y., Bisantz, A.M., Drury, C.G.: Foundations for an empirically determined scale of trust in automated systems. Int. J. Cogn. Ergon. 4, 53–71 (2000)
17. Madsen, M., Gregor, S.: Measuring human-computer trust. In: 11th Australasian Conference on Information Systems, pp. 6–8. Citeseer (2000)
18. Chien, S.-Y., Semnani-Azad, Z., Lewis, M., Sycara, K.: Towards the development of an inter-cultural scale to measure trust in automation. In: International Conference on Cross-Cultural Design, pp. 35–46. Springer (2014)
19. Merritt, S.M., Ilgen, D.R.: Not all trust is created equal: dispositional and history-based trust in human-automation interactions. Hum. Factors 50, 194–210 (2008)

20. Lee, I., Choi, B., Kim, J., Hong, S.-J.: Culture-technology fit: effects of cultural characteristics on the post-adoption beliefs of mobile Internet users. Int. J. Electron. Commer. **11**, 11–51 (2007)
21. Hwang, Y., Lee, K.C.: Investigating the moderating role of uncertainty avoidance cultural values on multidimensional online trust. Inf. Manage. **49**, 171–176 (2012)
22. Mcknight, D.H., Carter, M., Thatcher, J.B., Clay, P.F.: Trust in a specific technology: an investigation of its components and measures. ACM Trans. Manag. Inf. Syst. TMIS **2**, 12 (2011)
23. Han, L., Kashyap, A.L., Finin, T., Mayfield, J., Weese, J.: UMBC_EBIQUITY-CORE: semantic textual similarity systems. In: * SEM@ NAACL-HLT, pp. 44–52 (2013)
24. Miller, G.A.: WordNet: a lexical database for English. Commun. ACM **38**, 39–41 (1995)
25. Opsahl, T., Agneessens, F., Skvoretz, J.: Node centrality in weighted networks: generalizing degree and shortest paths. Soc. Netw. **32**, 245–251 (2010)
26. Freeman, L.C.: A set of measures of centrality based on betweenness. Sociometry, 35–41 (1977)

Development of an Intuitive, Visual Packaging Assistant

Benedikt Maettig$^{(\boxtimes)}$, Friederike Hering$^{(\boxtimes)}$, and Martin Doeltgen$^{(\boxtimes)}$

Fraunhofer-Institute for Material Flow and Logistics IML,
Joseph-von-Fraunhofer-Str. 2-4, 44227 Dortmund, Germany
{benedikt.maettig, friederike.hering,
martin.doeltgen}@iml.fraunhofer.de

Abstract. High volume of information and rising dynamics require a flexible interaction of humans with systems in their work environment. To fulfill this and with regard to logistics demands, new solutions for human-machine-interaction are being developed at Fraunhofer IML. Packaging in e-commerce is characterized by heterogeneous products. Therefore, a solution is required that guides staff efficiently through the packaging process based on optimizing algorithms in the backend. The developed packaging assistant guides the staff through the process with the help of hardware and concurrently considers the characteristics of human cognition. The packaging information is provided through traditional user interfaces like displays or textual-based information as well as visual signals given through LEDs. Using visual signals, the positioning of goods inside a package is visualized and verified by a camera-based system. The purpose of this kind of system is to reduce the amount of information as well as the susceptibility to errors within the process.

Keywords: Human factors · Human-systems integration · Logistics
Intralogistics · E-commerce · Intuitive interface · Cognitive ergonomics
Packaging assistant · Smart packaging · Context-sensitive packaging

1 Changes in Business Due to the Technologic Developments and the Need for Cognitive-Ergonomic User Interfaces

The technology developments of recent times show an evolution according to Moore's Law [1]. This law states that technology developments become higher-performing as well as affordable as the time goes on after their initial invention and introduction to the commercial market. This facilitates the throughout deployment of these technologies within privates as well as commercial spheres. Nowadays, businesses and the execution of respective work processes rely on the use of technologic means. Especially the usage of machines are given a high importance as it lead to, among others, more robust product outcomes and better scalable business processes.

The demand to integrate humans into the work process efficiently is given and continuously growing due to the increasing degree of automation and digitization – especially in the field of production and logistics [2]. The nature of the logistics business is characterized by a high volume of information and ever-changing physical

© Springer International Publishing AG, part of Springer Nature 2019
I. L. Nunes (Ed.): AHFE 2018, AISC 781, pp. 19–25, 2019.
https://doi.org/10.1007/978-3-319-94334-3_3

product dimensions which stipulate the need for dynamic processes. Hence, it requires a flexible interaction of the employee with the systems in his surrounding work environment. New solutions for the human-machine-interaction need to be developed in order to fulfill this requirement. These concepts need to be driven by modern and intuitive methods for the visualization of the relevant information. Especially the tasks within operational logistics execution have certain requirements that demand specific technical solutions [3].

An important aspect when developing interfaces for human beings is the consideration of in how far a system is cognitive ergonomic and intuitive [4]. This means how easy a guidance is understandable by an employee as well as it can be used by the end user or majority of users [5]. It also implies that the system can be used by a human being without the need of profound explanations and the interaction with the system happens unconsciously. That means that the guidance falls back on the fundamental understanding a human being has inherent through processing stimulus form the outer world and the pre-existing links in the human brain. It is important to provide an efficient and simple guidance to the working staff in order to maintain a happy and healthy work force, among other aspects. Also, faultless work processes are enabled by well-working guidances for the human work force. Therefore, the design of cognitive ergonomic and intuitive interfaces according to standardized guidelines is meant to be pursued when designing new assistance systems. But apart from a theoretical approach of realization, these systems also need to be evaluated in practice to see in how far the system is actually improving work process in the sense they are meant to do.

2 Importance and State-of-the-Art in the Field of Packaging Assistance

In order to master the rising demands within e-commerce, robust packaging assistants need to be implemented. The background derives from the fact that packaging in e-commerce is characterized by a dynamic, heterogeneous product range and hence large amounts of different packaging variables such as dimensions, weight and priority. Especially the e-commerce business in the B2C sector is highly susceptible to packaging challenges because an order oftentimes consists of several products with one amount of each product. It therefore requires a large flexibility of the system in the end process that packs a respective order [6].

A packaging assistant is based on two major foundations. One is an efficient algorithm in the backend. This algorithm calculates the volume of the ordered goods that have to be packed in a cardboard box for delivery. Based on this calculation it generates an optimal packaging scheme for the smallest possible box size. The second part of the packaging assistant covers the way the calculated packaging scheme is presented to an operating human worker. The main goal of this presentation is an efficient packaging process and to prevent wrong packaging.

The aspect of efficient algorithms has been largely addressed and solved in many ways as different software solutions can be purchased on the market (e.g. PUZZLE[1]). These software solutions run on different operating systems. Furthermore, they address different packaging needs. The majority of these software systems that are available address optimized packaging demands of industrial solutions. Oftentimes, this kind of software is used in the production and automotive sector. There are a few solutions addressing the needs of the e-commerce sector. Several reasons account for the different needs. For one, higher revenues can be generated at an early stage in the supply chain and hence the needed funds for optimization tools such as implementing a packaging software are available. Another reason is simply that there was no need for it before but with the rising shift form stationary retail to online business the work processes in this field generate more money and hence require more efficient tools. An important part of the packaging assistant as presented in this paper is the availability of flexible and dynamic algorithms to cope with manifold product dimensions and requirements. In the future, throughout the use and generated knowledge through AI and Big Data it is assumed that the implemented algorithms will become as smart and dynamic as needed in the real world. The research for such algorithms is included in the research at Fraunhofer but just remains a topic in the background for this paper.

The main object of reflection is the way the packaging system considers how instructions can be given to a human being. In this respect the usability plays an important role. Currently, packaging in e-commerce relies mainly on the knowledge and expertise of the employee – hence it is a not very digitized process. But with the rising challenges in the e-commerce business dynamic and digital processes need to be implemented to cope with the daily business. In the current packaging process, the only guidance on behalf of the backend system is given by telling the employee the products of an order throughout the simple interface of an ERP software displayed on a monitor. Eventually, the ERP software includes a volume calculation of the goods to be shipped with the respective order. With this basis, the employee has to decide which of the available box sizes is suitable for the shipment. In order to reduce the complexity of this decision many businesses have less than five box sizes available between which an employee has to decide. If a box has empty space inside the process usually requires that it is filled up with filling material such as air cushion foil. But with regard to saving money this is a rather unsustainable business practice. On one hand it is using up materials such as paper and plastic for the mere purpose of keeping the products inside of the box in place. Furthermore, it is a waste of loading space inside a truck. Therefore, the perfect fit between products and box is an important improvement. But having more box sizes available enhances the complexity of an employee to make a decision and during his working hours there is no time for a simple trail and error course of actions.

With regard to the e-commerce business, the logistics sector mainly provides technical solutions for the storing, warehousing and picking of goods. Efficient packaging assistances are not available. This field has merely risen as a theoretical part of research. As of yet, there are different ideas available but none of these is commercially

---

[1] PUZZLE packaging optimization software, http://demo.mypuzzle.de/.

available. These research ideas include the use of smart cameras which for example can detect motions [7]. However, the research is in its early beginning and no robust end product has been thoroughly built yet.

3   New Solution for an Intuitive Packaging Assistant for the E-commerce Business

In this context, a team at the Fraunhofer IML in Dortmund developed a low cost solution for an intuitive packaging assistant which helps to create an optimized packaging process in the field of e-commerce.

With this background, a solution was developed to meet the demands. The solution enables the consideration of dynamic processes with a large amount of variables. The solution is a packaging assistant which aims to guide the packaging staff and thus to optimize the packaging process of orders packaging. It is based on an dynamic optimizing algorithm in the computer backend that is communicated to the staff through a modern and intuitive interface. Concerning this issue, the developed packaging assistant considers the essential characteristics of the human cognition. The set-up is a packaging desk customary in the trade that was enhanced by modern hardware that guides a staff member through the process. The packaging assistants' hardware is comprised of a standard computer screen, a camera system and a hybrid array of LEDs and light depend resistors (see Fig. 1).

A computer touch screen is attached to the packaging desk. The hardware is a traditional user interfaces with a display that shows short process information. The screen gives instructions to the end-user of what to do, for example, it can first indicate what packaging material is needed and in the proceeding of the packaging process indicate which product to take one after another.

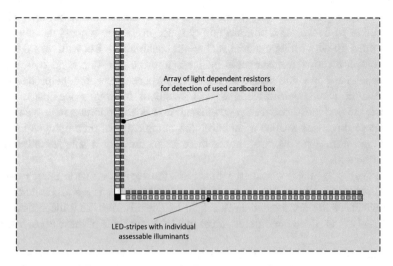

Fig. 1. Technical configuration of the developed arrays (topview).

In order to provide the relevant packaging information, the packaging assistant uses visual signals represented by LED-stripes that can show different colors. These LED stripes are attached to several edges of the packaging desk and can enclose the minimum as well as maximum dimensions of utilized cardboard boxes. The LED colors that can be represented are linked to meanings that are quickly perceived by humans because of their culturally universal meaning [8]. Hence, the color provides a simple, cognitive guidance to the end-user. Green colors indicate the correct product has been selected and the packaging process is faultless. Red light indicates a problem which has been induced by the user which can be the outcome of different actions. Within the process a well-fitting box for the current order that has to be packed is suggested by the system by illuminating the LEDs with the corresponding x, y and z sizes. After the user placed the corresponding cardboard box on the table, the sensors within the system validate the selected box. If the size differs the user would be informed and the system calculates a new possible packaging scheme based on the new box.

In the next process steps the placement of the products within the box should be done. Through the length and location of the visual LED signals, the user will be informed where the current product has to be positioned within the box (see Fig. 2).

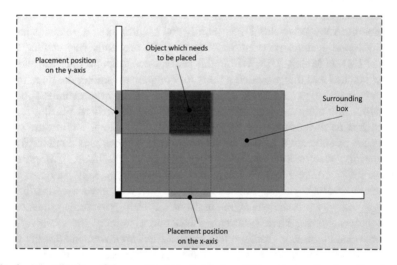

Fig. 2. Visualization of the positioning of objects within a surrounding box (topview).

A camera system is furthermore part of the packaging assistant. The camera examines the packaging process and detects faults such as for example a misplacement of a product. Depending on the choice of camera as well as implementation with the backend system, the cameras can also function as a scanning unit. The information it generates is interpreted by the backend system's algorithms and communicated to the end-user with the help of the LED lights as well as the computer touch screen.

All in all, the staff member is guided through the process with help of this assistant. The purpose of this kind of information provisioning is to reduce the amount of information and to reduce the susceptibility to errors within the packaging process in e-commerce environments. Focusing on only representing the fundamental set of information, hence providing the information specifically context-sensitive, it minimizes the cognitive effort to avoid any misunderstanding. This concept and finding can be adapted to other use cases within production and logistics and opens new perspectives in the human-machine-interaction. But especially to the field of e-commerce is represents a suitable and needs assistance.

Considering the solution further another advantage of the assistance system can be revealed. The system works by light and information that is visualized through a computer screen. Noise information is not intended to be used for this system. Therefore, it also provides a good guidance for workers who have a hearing impairment or are suspect to any other impairment.

4   Evaluation of the New Packaging Assistant and Next Steps

The described hardware and software system has been implemented in a proof-of-concept solution at the Fraunhofer IML. In March 2018 it has been tested for the first time among human being within a study as well as it has been presented to the public as a trade fair exhibit.

To generate knowledge about the usability and usefulness of different aspects of the newly developed system, a scientific study with 23 probands was conducted at the Fraunhofer IML in March 2018. Within this study, three different interfaces were given to the test persons and this provided a basis for comparison among the three interfaces. Each interface instructed a human worker to pick an order which entails choosing the right product one after another, scanning the product as well as placing it within a cardboard box as foreseen by the packaging assistance system. During the study, the measurement of subjective as well as objective data sources was conducted with the help of sensors, observations as well as questionnaires (e.g. NASA-TLX [9, 10]). The results cannot be communicated as they have not been completely analyzed as of yet. As a general result, it can be stated that the test persons gave a generally positive feedback to the way that the packaging assistance system provided a guidance as well as good process times have occurred. Furthermore, among the observations and throughout open comments from the test persons, first improvements for the design of hardware and software for the packaging assistance system could be derived which will be implemented on behalf of Fraunhofer IML for the future development of the packaging assistance system.

Furthermore, the proof-of-concept solution has been presented to a larger public as an exhibit on the trade fair LogiMAT 2018 in Germany. Trade fair visitors were able to use and test the system. The general feedback during the trade fair was very positive as well as a mutual interest by several companies for deploying and testing the packaging assistance solution has been received.

The Fraunhofer IML will continue to improve the prototype of the packaging assistance system in respect to the hardware as well as the software. The next steps of the project will be to implement the revealed aspects for improvements into the system and continually implement further scientific revelations. This will happen in regard to usability guidelines in order to ensure a system that is ergonomic in respect to physical as well as cognitive aspects.

An important part of research for a robust packaging assistant solution remains the dynamic implementation and data processing from the received data to the output for the employee. That means that every order with its specific characteristics requires as new calculation with regard to various aspects, such as for example the packaging material and especially the best arrangement of products within a certain packaging scheme.

References

1. Schaller, R.R.: Moore's law: past, present and future. IEEE Spectr. **34**(6), 52–59 (1997)
2. Reif, R., Walch, D.: Augmented & virtual reality applications in the field of logistics. Vis. Comput. **24**, 987–994 (2008)
3. Grosse, E.H., Glock, C.H., Neumann, W.P.: Human factors in order picking: a content analysis of the literature. Int. J. Prod. Res. **55**, 1260–1276 (2017). Taylor & Francis
4. Rinkenauer, G., Kretschmer, V., Kreutzfeldt, M.: Kognitive Ergonomie in der Intralogistik. In: Future Challenges in Logistics and Supply Chain Management, vol. 2. Fraunhofer IML, Dortmund (2017)
5. ISO 9241-210:2011-01, Ergonomics of human-system interaction - Part 210: Human-centred design for interactive systems. Beuth, Berlin (2011)
6. Bayles, D.L., Bhatia, H.: E-Commerce Logistics & Fulfillment: Delivering the Goods. Prentice Hall PTR, Upper Saddle River (2000)
7. Bannat, A.: An assistance system for digital worker support in industrial production. Technical University of Munich, Munich (2014)
8. Heimgärtner R.: Intercultural user interface design – culture-centered HCI design – cross-cultural user interface design: different terminology or different approaches? In: Marcus, A. (ed.) Design, User Experience, and Usability. Health, Learning, Playing, Cultural, and Cross-Cultural User Experience. Lecture Notes in Computer Science, vol. 8013. Springer, Heidelberg (2013)
9. Hart, S.G., Staveland, L.E.: Development of NASA-TLX (Task Load Index): results of empirical and theoretical research. In: Hancock, P.A., Meshkati, N. (eds.) Human Mental Workload, pp. 139–183. Elsevier, Amsterdam (1988)
10. Hart, S.G.: NASA-Task Load Index (NASA-TLX); 20 Years Later. Sage Publications, Los Angeles (2006)

Visual Capability Estimation Using Motor Action Pattern

Yanyu Lu[1], Lu Ding[2], and Shan Fu[1(✉)]

[1] School of Electronic Information and Electrical Engineering,
Shanghai Jiao Tong University, No. 800 Dongchuan Road,
Minhang District, Shanghai, China
sfu@sjtu.edu.cn
[2] School of Aeronutics and Astronautics, Shanghai Jiao Tong University,
No. 800 Dongchuan Road, Minhang District, Shanghai, China

Abstract. Visual channel is a very import input for human recognition during complex tasks. The critical challenge is to tell exactly how well the visual capability of a subjects in real time dynamically. In this research, we assume that the visual capability of a subject varies according to the real task situation, and that the performance of the subject on the task is a very important measurement for estimation of the visual capability of the subject. We use the motor action pattern as the indicator to investigate the level of visual capability of the subjects in relation to the characteristics of the visual information, the nature of the task and state of the subject in a simulated task environment. The research found there was a strong indication that variation of the visual capability was related to the nature of the task and the state of the subject. The motor action pattern was good indicator for visual capability.

Keywords: Visual capability · Motor action pattern · Recognition state

1 Introduction

Visual channel is very important for information input. However, it has been a long-time challenge to tell exactly how well the visual capability of a subject is in real time dynamically. As a quite common experience that people watch but not see, it suggested that eye-tracking device told very accurately the place where a subject was looking at but was not able to tell if the subject did take and understand the visual information. It is a critical problem of civil aviation since there are a lot of information on the flight desk interface requires timely access to pilots and respond appropriately. Statistics shows that more than 60% of modern aircraft accidents are caused by human factors. In this situation, it is important to carry out human factors research in the cockpit to acquire information about visual capability and limitations. Using this information in design, training and certification can increase the safety, comfort and efficiency of the aircraft [1].

We assuming that the visual capability of a subject varies according to the real task situation, and that the performance of the subject on the task is a very important measurement for estimation of the visual capability of the subject. To establish visual

© Springer International Publishing AG, part of Springer Nature 2019
I. L. Nunes (Ed.): AHFE 2018, AISC 781, pp. 26–32, 2019.
https://doi.org/10.1007/978-3-319-94334-3_4

capability, subjects are asked to wear devices, which will influence on the operation of the subjects to some degree. We want to use untouched methods to estimate visual capability. This research was to use the motor action pattern as the indicator to investigate the level of visual capability of the subjects in relation to the characteristics of the visual information, the nature of the task and state of the subject.

In order to establish the connection of visual capability and motor action pattern, firstly, we use a series of physiological parameters such as blink latency, fixation duration to qualify visual capability. These two physiological parameters can reflect the acquisition state of the subjects. [2] Then we use video image for capture the motion action patterns. We asked the subjects to perform flight tasks with different types. By analyzing the correlation of the change of bio-parameter and action, we build up connections of motor action patterns and visual capability.

The experiment is carried out on a simulator with high fidelity. The simulator consists of two parts, the outside view and the cockpit. The outside view is simulated and projected on a cylindrical screen which has a diameter of about 8 meters. In the cockpit, the arrangements are referred to ARJ21-700. There are control instruments and display instruments in the cockpit. The control instruments include the yoke, throttle, rudder pedal, flaps, landing gear, CDU and MCP. The display instruments include PFD, ND, EICAS, etc. The cockpit display and control interface was used as the visual information input, and the warning sign and period of appearance were used as the primary parameters of the indicator. The eye-tracking device and video camera-based motor action capture system were the main equipment used in the experiment. The subjects were asked to carry out tasks of different levels of difficulties. The performance of tasks was designed to highly related to how well the visual information was taken by the subjects. Correlation analysis were the main process to estimate the dynamic visual capability of the subject. Eleven pairs of experienced pilots carried out the test scenarios. All the experimental data was recorded and processed. The research found that the visual capability did varied during the task. There was a strong indication that variation of the visual capability was related to the nature of the task and the state of the subject. The motor action pattern was good indicator for visual capability.

2 Methods

2.1 Visual Capability Estimation

Pupil diameter, fixation duration and saccade frequency were recorded by Smart Eye Pro [3], the eye tracker with the sample rate of 30 Hz. Figure 1 shows eye movements measurements vary with the different phases of the flight.

Pupil diameter changes significantly after the stimulation occurs. Pupil diameter indicate the blink of the subjects. Thus, it can be informed that when pilots pay more effort to concentrate on visual information, their blink latency may become longer [4, 5]. Fixation duration can reflect the time for a participant to acquire information. At the beginning of the task, the flight task is simple, so the fixation duration is shorter. Right after the stimulation, the fixation is longer, this may because that to check the systems status is very important, the participants pay more attention. During the last part of the

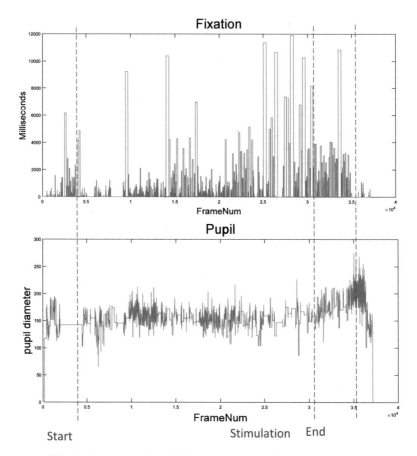

Fig. 1. Fixation and pupil diameter change during the experiments.

flight task, pilots have to continuously monitor several critical parameters, e.g. altitude, airspeed, heading, attitude, etc. Before the stimulation, pilots are relatively more relaxed as the task is simpler; there is small change of visual information. They didn't have to read a gauge accurately, and they glance at the whole cockpit and outside view. Thus, longer fixation duration can inform more cognitive activity.

To qualify visual capability, we use the following Eq. (1):

$$visualcapability = C_1 \frac{t_{cgn}}{T_{cgn}} + C_2 \frac{t_{fix}}{T_{fix}} \tag{1}$$

where t_{cgn} is the cognitive time for a participant to perceive, process the information and make the decision before making a control. t_{fix} is the duration of each fixation. t_{cgn} and t_{fix} are the baseline of cognitive time and fixation duration respectively, which are calculated from the training data. C_1 and C_2 are the weights to express the different contributions of cognitive time and fixation duration respectively. According to [1], they are empirically set by the reliability of the measurements. In this paper, C1 = 0.6, C2 = 0.4.

2.2 Motor Action Patterns

In our research, the main activities of the participants focused on hand activities. It's important to locate hand during the experiments. There are a lot of problems in detection. The light condition in the cockpit changes dramatically and blur caused by fast movement makes the detection failure. And the change of appearance of hand makes it difficult to achieve robust hand detection. Traditional image processing methods have limited feature representation ability to make accurate and efficient detection. In this research, we use the state-of-the-art deep convolutional features to represent hands and use the adaptive two-stage framework faster rcnn [6] to make robust detection. Figure 2 shows some detection results. There are some false positive detection, we will filter them out in the sequent processing. And for missing detection, we'll use Gaussian Process [8] to interpolate the missing detection.

Fig. 2. Pilot hand detection results.

In this research we simply use image coordinates (x_t, y_t) to indicate hand location at the frame of t. Then we define the following dimensions to represent motor action pattern {s, a, d}.

1. The magnitude of the motion is calculated by:

$$s = \sqrt{(x_t - x_{t-1})^2 + (y_t - y_{t-1})^2} \tag{2}$$

where (x_t, y_t) is the current location of hand and (x_{t-1}, y_{t-1}) is the location of the last frame. This is also the magnitude of velocity.

2. The orientation of movement:

$$a = \arctan(\frac{y_t - y_{t-1}}{x_t - x_{t-1}}) \qquad (3)$$

This is also the direction of velocity.

3. Hand duration. In our research, hand duration is defined as the duration of no operation. This parameter can indicate the free time of the subjects.

Then we use {s, a, d} to get the operate frequency. The detailed methods are, first we set up a threshold for s and a dimension to separate movement and stillness. The number of alternations between movement and stillness of hand is defined as operate frequency.

Figure 3 shows the frequency of operations.

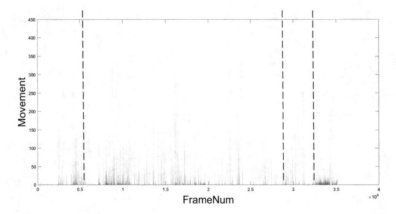

Fig. 3. Movement frequency. The value of the y axis indicates the movement magnitude. Darker colors mean more frequent operations.

3 Experiment

3.1 Participants and Apparatus

11 pairs of flight crew (pf and pnf) to take part in the flight simulation. The research will be focused on pf. They are aged from 30 to 60, with the average of 26. All of them are experienced pilots who have comprehensive knowledge about aviation and have been trained in simulated flight. The experiment is carried out on a simulator with high fidelity. There are two parts of the flight simulator: the outside view and the flight deck. Inside the flight deck, the arrangements of it consists of control instruments and display instruments. Thus, simulation environment will provide participants with a variety of visual information.

3.2 Procedure

In the experiment, each participant is asked to fly a complete flight task about 15 times. The flight task consists of normal and abnormal flight situations. At first, the participant was asked to take a normal flight. Then a simulated accident is input as the stimulation which will cause control interface instrument changes and alarm system issued a warning. These were used as visual information. The flight environment is simulated as in summer, at noon, sunny and no wind. During the experiment, several parameters are recorded. Video image recoded by a camera set up above the pilot. Blink, saccade, fixation and pupil diameter are recorded by SmartEye eye tracker.

3.3 Data Process

In our research, we used multiple sets of equipment to collect multi-channel data of different kinds. The sampling frequency of each device is different. So before analyzing the data, we need to synchronize all the data channel. For the noise and loss of data, we use Hanning window [7] to filter and smooth these physiological signal data.

4 Results and Discussion

The typical visual capability and motor action patterns are represented in Figs. 1 and 3. We can intuitively find out that motor action pattern has a correlation with visual capability. Changes in motor action patterns occur after changes in visual capability. This shows that when visual information is generated, subjects first notice the visual information, and then deal with this information, and then produce a behavioral response. At the beginning of the task, subjects were performing a relatively simpler task. Subjects do not need to respond to visual information at all times, he glanced at visual information but did not take the appropriate action. After the stimulation, Subjects need timely access to information and take action. These illustrate that t variation of the visual capability was related to the nature of the task. By comparing Figs. 1 and 3, we can also find that, visual information acquisition often follows the action pattern changes. In the experiment, subjects' motor action pattern can reflect cognitive state, and cognitive state can usually be characterized by visual capability. So, motor action pattern can be used as an indicator of visual capability.

Not all motor action features are related to visual capability, as can be seen from Fig. 3. The magnitude of hand movement is not significantly related to visual capability. This is because the magnitude of the motion is related to the cockpit layout rather than visual information. The frequency of action and free time operation is strongly related to visual capability.

We analyzed all experimental data. First, we time aligned all our experiments. Then we quantified the visual capability using Eq. (1), and calculated the motion frequency using methods introduced in Sect. 2.2. Finally, we averaged all the results showed in Fig. 4. We can notice that the motor action pattern was good indicator for visual capability.

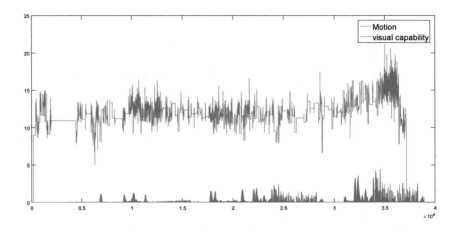

Fig. 4. The relationship between motor action pattern and visual capability

5   Conclusion and Future Work

In this study, several experiments are conduct in flight simulator. We used multiple data resource to estimate visual capability and capture motor action patterns. By qualify visual capability and extract correlated motion feature, we find that motor action patterns can be a good indicator for visual capability. The study provided a non-contact measurement method that did not affect the operation of the subjects. In this research, we only used a few physiological signals taken into consideration. Other parameters which are highly related to recognition state will be considered in our future work.

References

1. Wang, Z., Fu, S.: A layered multi-dimensional description of pilot's workload based on objective measures. In: International Conference on Engineering Psychology and Cognitive Ergonomics, pp. 203–211. Springer, Heidelberg (2013)
2. Farmer, E., Brownson, A.: QinetiQ: Review of Workload Measurement, Analysis and Interpretation Methods. European organization for the safety of air navigation (2003)
3. Smart Eye, A.B.: Smart-Eye Pro 5.6 User manual. Sweden-, Gothenburg, Smart Eye AB, Sweden Smart Eye AB (2009)
4. Wilson, G.F.: An analysis of mental workload in pilots during flight using multiple psychophysiological measures. Int. J. Aviat. Psychol. **12**(1), 3–18 (2002)
5. Van Orden, K.F., Limbert, W., Makeig, S., Jung, T.P.: Eye activity correlates of workload during a visuospatial memory task. Hum. Factors **43**(1), 111–121 (2001)
6. Ren, S., He, K., Girshick, R., Sun, J.: Faster R-CNN: towards real-time object detection with region proposal networks. In: Advances in Neural Information Processing Systems, pp. 91–99 (2015)
7. Tuovinen, J.E., Paas, F.: Exploring multidimensional approaches to the efficiency of instructional conditions. Instr. Sci. **32**, 133–152 (2004)
8. Rasmussen, C.E.: Gaussian processes in machine learning. In: Advanced Lectures on Machine Learning, pp. 63–71. Springer, Heidelberg (2004)

Pedestrian Perception of Autonomous Vehicles with External Interacting Features

Christopher R. Hudson[1], Shuchisnigdha Deb[1(✉)],
Daniel W. Carruth[1], John McGinley[1], and Darren Frey[2]

[1] Center for Advanced Vehicular Systems,
Mississippi State University, Starkville, USA
{chudson, deb, dwc2, jam}@cavs.msstate.edu
[2] Industrial and Systems Engineering,
Mississippi State University, Starkville, USA
dafl90@msstate.edu

Abstract. The increasing number of autonomous vehicles has raised questions regarding pedestrian interaction with autonomous vehicles. Researchers have studied external interfaces designed for vehicle operators and other road-users (e.g., pedestrians). Most past studies have considered the interaction between pedestrians and autonomous vehicles with no visible operator. However, pedestrian-autonomous vehicle interaction may be complicated when there is a human sitting in the conventional driver's seat of an autonomous vehicle. Such a scenario may cause some pedestrians to look to the passenger for cues when they should be looking for cues from the vehicle. The objective of the current study was to investigate pedestrians' perspective of autonomous vehicles based on the interaction effect between passenger status and external features on the vehicle. Sixteen pedestrians completed a VR experiment. The results provided important insight into the important question of pedestrian-autonomous vehicle interaction when passengers are present in the driver seat of the vehicle.

Keywords: Autonomous vehicles · Pedestrian-autonomous vehicle interaction
Virtual reality · Communication interface design

1 Introduction

An increasing number of autonomous vehicles are being integrated into every day traffic patterns. New advancements in self-driving vehicles are being rapidly developed and tested on roadways by major motor vehicle manufacturers and technology companies. TESLA has deployed semi-autonomous vehicle control systems to many of its vehicles with many more systems expected to follow suit. While many current systems are restricted to advanced driver assist features or basic autonomous behaviors such as lane keeping and collision avoidance, the long-term goal of vehicle manufacturers is to implement fully autonomous vehicles that minimize or eliminate direct human control. More than 20 states have authorized operation of autonomous vehicles since 2011 [1]. While these systems are experimental and have been involved in accidents [2, 3], as vehicles acquire more autonomy and operate in close proximity to other vulnerable

© Springer International Publishing AG, part of Springer Nature 2019
I. L. Nunes (Ed.): AHFE 2018, AISC 781, pp. 33–39, 2019.
https://doi.org/10.1007/978-3-319-94334-3_5

road users, the autonomous vehicles must successfully navigate the unspoken social etiquette that exists between passengers in vehicle and pedestrians.

The number of accidents at intersections involving pedestrians has increased over time. In 2015, there were 5376 pedestrian fatalities reported by the National Highway Traffic Safety Administration (NHTSA) [4]. This was a 9.5% increase over 2014 and resulted in the highest fatality rate since 1996. A total of 967 fatalities occurred at intersections with 76% of the fatalities occurring in urban areas. Autonomous vehicle technology promises to make roads safer for pedestrians and other vulnerable road users. However, pedestrians will need to interact safely with autonomous vehicles. Today, when a pedestrian interacts with a vehicle and its driver, the pedestrian and the driver may be able to make eye contact and exchange non-verbal cues or signals that help the pedestrian to know whether it is safe to cross. When a pedestrian meets a fully autonomous vehicle, pedestrians are unable to make "eye contact" with the operator as suggested by the NHSTA. One could argue that if no-one is in the vehicle, the pedestrian will know that a vehicle is autonomous and rely on other cues such as visibly slowing of the vehicle or displays on the autonomous vehicle. However, another scenario exists when a passenger in the autonomous vehicle is occupying the seat of the traditional vehicle operator. In this scenario, a pedestrian may be misled by the behavior of the apparent operator of the vehicle. For example, the passenger of the vehicle may be looking down and away from the road indicating that they are not aware of the pedestrian and the situation is not safe. However, the autonomous vehicle is aware of the pedestrian and will allow the pedestrian to safely cross the street. This scenario raises questions about how pedestrians will interpret conflicting signals from passengers and from the vehicle. We will explore pedestrians' responses to different features and autonomous vehicle behaviors to advance our understanding of autonomous vehicle and pedestrian interaction.

This study extends previous research in autonomous vehicle and pedestrian interaction at crosswalks in order to understand the interaction in the case where a passenger in the fully autonomous vehicle is seated in the traditional operator's seat and may be interpreted as being in control of the vehicle. We selected visual and audio features from previous survey and real-world studies on pedestrian crossing with an autonomous vehicle [5–7]. A vehicle passenger seated in the traditional operator's seat was added to the simulated vehicles in certain conditions. The passenger's status was one of three levels: no passenger, a passenger with head up and apparently paying attention, and a passenger with head down and distracted. Across all three passenger conditions, the vehicle had seven different external features: no feature, green *walk* in text, white walking silhouette, red upraised hand, stop sign, music, and verbal message saying *safe to cross*. For human safety, real world experimentation to investigate human-autonomous vehicle interaction cannot be approved by research ethics. Therefore, to create a controlled environment and expose participants to a fully autonomous vehicle without any deception, this study was performed in an ambulatory virtual reality environment [8].

2 Method

2.1 Participants

The study was approved by the Institutional Review Board at Mississippi State University. Sixteen participants (9 male and 7 female), ranging in age from 19 to 34 years old, were recruited from the Starkville, Mississippi area to participate in a virtual reality experiment. The participants were selected based on their language (English speakers), ability to walk at normal pace and gait, and normal or corrected-to-normal color vision.

2.2 Apparatus and Setting

Traffic environments were created using Unity 2017.3 and experienced with HTC Vive headset. Two lighthouse sensors of the HTC Vive tracked participants' head movements during their crossing across the virtual crosswalk. The lighthouse sensors were placed approximately 8 m apart, facing each other at opposite ends. The area tracked by the Vive was approximately 4 by 7 m that included a 5.5-meter-long crosswalk and sidewalks on both sides.

Artist models of the Google autonomous car, other cars and trucks, and a human were purchased from the Turbosquid website (www.turbosquid.com). Animations for the human models were created using Autodesk 3D Studio Max 2018. The attentive passenger's head was up and scanning left and right to give an impression that he was aware and observing the environment around the vehicle. The distracted passenger's head was looking down and his hands were moving giving the impression that he was working on a tablet.

2.3 Experimental Protocol

After the participants arrived at the lab, they read and signed an informed consent form. Then the participants completed a Simulation Sickness Questionnaire (SSQ, Kennedy et al., 1993) to establish a baseline score and to confirm their ability to participate in the study. After completing the SSQ survey, the participants put on the VR headset to experience and get used to moving and acting in virtual reality. The initial environment helps move participants past their initial excitement about VR and teach them to receive and respond to instruction within the virtual world. After their first exposure to virtual reality, the participants completed a second SSQ survey to assess any immediate discomfort from virtual reality. After this second SSQ, the participants were immersed in a virtual traffic environment with a two-lane four-way intersection and with two-way stop signs. They were positioned on a sidewalk facing a crosswalk, with the intersection and a two-way stop sign to their left. The vehicles were running only on the two lanes perpendicular to the participant's position and no stop sign was present in those lanes.

The first three trials provided practice crossing the street in the virtual scenario with three vehicles passing by from both directions of the two-way lane. The participants crossed the road when they felt it was safe to do so. Once the participants felt comfortable with the three practice crossings, they initiated the experimental trials by following instructions presented within the virtual world. During the experimental trials, the six vehicles (3 on each lane, as in the practice trial) and a fully autonomous car approached the crosswalk. The autonomous vehicle approached from the right and far lane. The autonomous car (trial car) was positioned in such a way within the scenario that even though all the other cars had passed by, the gap between the trial car and its lead vehicle was not sufficient for the participant to safely cross in front of the trial car if it did not stop. The car was equipped with one of seven levels of features: one no feature, two audible features, and four visual features. A display was installed on the front of the car to show no feature, green *walk* in text, white walking silhouette, red upraised hand, or stop sign. Sound cues provided audible features by playing music or a verbal message saying *safe to cross*. There were three different passenger statuses: no passenger, attentive passenger looking at the surrounding, and distracted passenger working on a tablet. All vehicles were moving at 30 mph at the start of the trial. The driverless car began to decelerate at around 175 feet away and to activate display features at around 210 feet away from the crosswalk. However, the car did not stop when the visual features were the upraised hand or the stop sign. Participants crossed the road for twenty-one experimental trials with 3×7 feature combinations. Figure 1 shows some of the features installed in cars with different passenger statuses. At the completion of the experiment, the participants responded to a five-item demographic survey and filled out rating scales for each of the features based on the different passenger statuses. Each of the participants were compensated with $10 for their participation (Fig. 1).

Fig. 1. Examples for experimental vehicles equipped with features

3   Results

3.1   Effect of Feature Type and Passenger Status

An ANOVA was conducted for the feature ratings considering both factors: feature type and passenger status. The analysis found no interaction effect between these factors with respect to the feature rating. However, significant main effects were observed for feature type [$F(6, 12) = 132.157$, $p = < 0.0001$] and passenger status [$F(2, 12) = 12.247$, $p < 0.01$].

Descriptive statistics for average feature ratings based on each of the driver statuses and each of the features types are presented in the following figures. The verbal message was rated as the most preferred audible feature and *walk in text* was rated the most preferred visual feature. For the features that implied "*do not cross*", the stop sign was more preferred than the upraised hand. In the case of different passenger statuses, features installed in no passenger and attentive passenger vehicles were preferred as compared to the distracted passenger vehicle (Fig. 2).

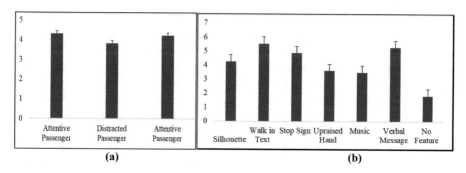

Fig. 2. Average feature ratings for different levels of (a) passenger statuses and (b) feature types.

3.2   Effect of Gender and Age

Another ANOVA was performed to observe demographic effect including data for gender (categorical variable) and age (continuous variable) along with the two factors. The results revealed that gender had a significant interaction effect with feature types to influence feature rating [$F(6, 12) = 3.571$, $p < 0.05$].

Average feature ratings for each of the feature types based on gender are shown in Fig. 3. The statistics shows that for no feature condition and for all of the display conditions where the car stopped at the intersection, males and females provided very similar ratings. However, for the display conditions where the car did not stop at the intersection, male participants rated both features with significant lower ratings as compared to the female participants. Both gender groups rated music with significantly lower scores as an option to feel comfortable crossing road as compared to the other audible feature.

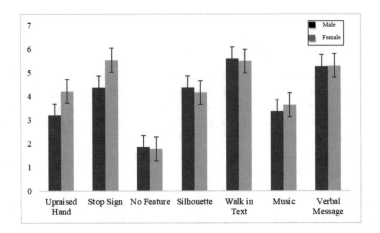

Fig. 3. Average feature ratings for different feature types based on gender

4   Discussion

Pedestrians' self-reported ratings were found to be influenced by both passenger status and feature type. In addition, there was no significant difference between feature ratings when the vehicle had no visible passenger and when the passenger of the vehicle appeared to be attentive. Pedestrians appear to feel comfortable communicating with an autonomous vehicle without any passenger as well as with a passenger who is paying attention. Therefore, it can be said that external features are useful replacements for cues that would normally be available from an attentive passenger. However, an autonomous vehicle with a distracted human passenger made the pedestrians' uncomfortable and cautious about the vehicle's intended action. This suggests that participants are confused when the autonomous vehicle is signaling intention to the pedestrian while cues from the apparent passenger indicate a lack of awareness of the pedestrian. Feature type also significantly influenced participants' rating for features. Participants were very comfortable with the walk, silhouette, and verbal message external features to confidently cross the road in front of an autonomous vehicle. Most of them also understood two negative signals, upraised hand and stop sign, as an indication for not to cross the road in front of that approaching autonomous vehicle. However, the upraised hand option was rated with significantly lower score than the stop sign option. In the case of the audible music, it might have interpreted by the participants as a negative and alarming signal to alert them not to cross the road and was thus rated lower as compared to the other audible feature. Lower scores on very familiar and standard image options like silhouette and upraised hand might indicate problems with visual presentation, which needs to be further investigated in the future research. While considering the effect of gender, it seemed like male participants were expecting the vehicles to stop all the time at the crosswalk and therefore, rated the features associated with failure to stop with lower ratings as compared to the female participants' ratings. The age range used in the study was not large enough (19–36 years) to show any significant effect of age.

5   Conclusion

The current study investigated participants' perspective toward an autonomous vehicle without any visible human passenger as well as with an attentive and a distracted passenger. The study allowed pedestrians to experience an autonomous vehicle in a simulated crosswalk and communicate with the vehicle. This research expanded Deb study [9] by including features that were recommended by most of the previous studies on this area. The consideration of different passenger statuses was a novel inclusion in this research.

The results came up with a short list of features for safe crossing conditions such as *Walk* in text, white walking silhouette, verbal message saying *safe to cross*, and stop sign for unsafe crossing condition. The authors plan to examine the effectiveness of these features on different type of autonomous vehicles (for example, olli, bus, etc.) with different positions and designs for display. The future research should also consider recruiting population with large age range to investigate age affect. Applied human factors researchers can use virtual reality experiments in user interface design, not limited to pedestrian-autonomous vehicle interaction.

References

1. National Highway Traffic Safety Administration: Traffic Safety Facts 2015 Data, February 2017. https://crashstats.nhtsa.dot.gov/Api/Public/ViewPublication/812375
2. Matthew, D.: Tesla's fatal Autopilot crash is a reminder that we are still a long way from truly autonomous vehicles (2017). http://www.businessinsider.com/tesla-fatal-autopilot-accident-autonomous-cars-years-away-2017-9
3. Samuel, G.: Self-driving bus involved in crash less than two hours after Las Vegas launch (2018). https://www.theguardian.com/technology/2017/nov/09/self-driving-bus-crashes-two-hours-after-las-vegas-launch-truck-autonomous-vehicle
4. National Conference of State Legislation. http://www.ncsl.org/research/transportation/autonomous-vehicles-self-driving-vehicles-enacted-legislation.aspx
5. Fridman, L., Mehler, B., Xia, L., Yang, Y., Facusse, L.Y., Reimer, B.: To Walk or Not to Walk: Crowdsourced Assessment of External Vehicle-to-Pedestrian Displays. arXiv preprint arXiv:1707.02698 (2017)
6. Deb, S., Warner, B., Poudel, S., Bhandari, S.: Identification of external design preferences in autonomous vehicles. In: Proceedings of the 2016 Industrial and Systems Engineering Research Conference, Anaheim, California (2016)
7. Clamann, M., Aubert, M., Cummings, M.L.: Evaluation of Vehicle-to-Pedestrian Communication Displays for Autonomous Vehicles (No. 17–02119) (2017)
8. Deb, S., Carruth, D.W., Sween, R., Strawderman, L., Garrison, T.M.: Efficacy of virtual reality in pedestrian safety research. Appl. Ergon. **65**, 449–460 (2017)
9. Deb, S.: Pedestrians' receptivity toward fully autonomous vehicles. Doctoral dissertation, Mississippi State University (2017)

Information and Communication Technology (ICT) Impact on Education and Achievement

Mahnoor Dar(⌧), Fatima Masood, Muhammad Ahmed,
Maryam Afzaal, Asad Ali, Zarina Bibi, Imran Kabir,
and Hafiz Usman Zia

University of Lahore, Lahore, Punjab, Pakistan
mahnoordar367@gmail.com, Faati67@gmail.com,
m.ahmed.3705@gmail.com, maryamafzaal610@gmail.com,
asadsheikh520@gmail.com, zarii768@gmail.com,
ims.manil993@gmail.com, usman.zia@cs.uol.edu.pk

Abstract. Information and Communication Technologies (ICTs) base research aim to force implementation in education and in teacher's training programs in Pakistan. It is important to know that how we can we take advantages from this course implementation in the well-known institutes to make the base line strong of under studies so they get flourish and improved their selves in further implementation. Ministry of Education focusing to explore the best practices for mainstreaming pilot projects containing interactive radio instruction (IRI, and how it can be manage and maintain time to time? How illiterate youth can get associate by using ICTs tools, to reach out and to teach? Potentials, Impact of ICTs policies and challenges are the critical factors investigated in pilot schools. Our focus and challenges we face regarding this is to strong ICT policy at education sector, which needs to answer a principal question. This examination contemplates research basic elements, possibilities, difficulties and effect of ICTs approach in schools.

Keywords: Interaction design · Rural development
Human computer interaction · User centered design
E-learning · ICT

1 Introduction

Thinking about the predominant worldwide patterns in training, computer study ought to be deal with as a necessary and understudies have access to computer and its application. ICT is progressively being utilizing to convey on guarantees of widespread instruction [2]. It discovered that if appropriate administration of ICT tools are set up, the understudies displays more propensities towards the comprehension and use of ICT apparatuses. Number of institutes see it extremely vital delicate aptitudes through ICT introduction to their understudies and fresh students. This needs prompt procurement of ICT aptitudes by students [3]. Learning is a steady procedure as learning process needs innovative artifacts and methods to secure the fresh data. It discovered that ICT is first step to get updated data over the world. ICT has fundamental importance towards the

© Springer International Publishing AG, part of Springer Nature 2019
I. L. Nunes (Ed.): AHFE 2018, AISC 781, pp. 40–45, 2019.
https://doi.org/10.1007/978-3-319-94334-3_6

establishment of society data. ICT has integrated with society for various purposes e.g. communication, amusement, Shopping and Education. ICT and instruction are growing after some time. In any case, the development rate is higher in ICT when contrasted with the training. In the field of information technology truth told Pakistan has grown up quickly. The principle difficulties and issues we cover in this examination think about are the ICT strategy, which need to answer foremost questions. The appropriate mix of ICT in training and its powerful use in learning and instructing, we have to screen utilizing solid and substantial pointers. Such sorts of policy makers can assist the arrangement creators with reviewing advancement of their nations' ICT improvement process and furthermore contrast and different nations ICT advance. There is absence of solid and nature of information accessible and in addition absence of research in the field of ICT for training in Pakistan. It has additionally absence of strategy and follow up techniques of ICT extends in schools [6], which put this part in more disappointment circumstance in imminent years [7]. ICTs providing number of projects to promote literacy its survey providing positive results. ICT and Education is an instrument for Pakistan to contend with rest of the world for the inclination of chances and flourishing. The diffusion of innovation into training is significant than innovation interests in the nation. Telecommunication has filled the need of ICT in the nation, which has participated in economic empowerment. It is reasonable setting to set up ICT in each division for improvement of the nation [4]. ICT usually utilized as a part of the training foundations in real urban areas of Pakistan. A few variables like absence of framework, absence of assets, poor strategy, oppose the entire use of ICT [5].

Previous knowledge of ICTs indicates that further study and employment would be very helpful, and it could upgrade the educational measure at optimize level, above the quality oriented results. Here because of analytical result there could be few classrooms where different activities/experiences are suggest in the proper unit, where mostly issues should be face that would be lack of power supply, scarcity of ICTs tools, lack of trained persons and ICTs application skills.

2 Literature Review

Today is the time of Information Communication Technology (ICT). Different ICT instruments utilized to teach and advise people. For ages, individuals have been living in total separation without any access to current media. Today's developing society relies upon the updated information. Despite the fact that we live in the innovative period, today, in the remote territories, people groups are experiencing different issues, for example, less availability to present day data sources. Concentrates apropos demonstrate that people groups are experiencing different sorts of issues because of absence of proper training and education. It shown that because of absence of learning about these inventions nobody taking advantage from these advancements in their working spots. ICT is demonstrating new methodologies for imparting and sharing the data. It enhances the knowledge and learning skills by utilizing such sort of advancements. Previously ICT tools used including the use for huge number of telephone, TV, Internet, cell phones, voice data frameworks, and fax [1].

3 Problem Statement

Education actually has been experiencing many cultural, social, economic and technical problems since the beginning of this century. With rapid inventions, new tools have been emerging continuously forcing educationist to include them in their teaching methodologies, adding to financial and management challenges of the institutes of today.

3.1 Financial Issue

Pakistani nation is in excess of 60% of its populace in provincial and remote regions. These provincial and remote regions have numerous issues and difficulties with respect to HR, socioeconomic, financial, correspondence learning and training. Nature of instruction in rustic Pakistan is in declining pattern due to previously mentioned reasons. Causative factors should be settled down for example, double medium of direction, blemished educational module, low quality of instructors, and packed classrooms shaping the educational modules and ICT integration to meet the needs [8].

3.2 Limited Internet Access

Teachers and students should go through computer labs gain access to a PC, tablet, or cell phone; broadband Internet availability; and to a school so that they practice well and maximize their learning ability. They should have some level of innovation capability keeping in mind the end goal to take an interest in online blogs, assignments, fill practice exercise and to build their profile. In any case, if there is no high accessibility of internet speed then it is crucial to practice for clients so that there are some minor issues that is proficient enough to deviate from the technology path.

3.3 Lack of Trained Staff

The instructors for the most centers focused to hands on the ICT skills though the ICT educational programs fixates on the incorporated utilization of ICT inside the learning and educating process. Mentors should have trained skills of trending technology. Even more in this way, the Trainer should be easy to handle that hand held tools, particularly when incorporated with student-focused guideline, frequently takes the learning and dialog an unexpected way in comparison to expected, in accomplishing the expressed learning goals in various courses.

3.4 Lack of Education Policy

ICT educational programs that 'interpret' the Pakistan national ICT-related activities into well-organized ICT design as a major aspect of the educational policy [9]. The viability of instructive process is enlarged, as a result of ICTs and policy makers should provide beneficial arrangements with respect to the utilization of ICT [10]. Pakistan, education is continuously being acknowledged as reflected by the latest National Education Policy by Ministry of Education in 2009.

4 Proposed Method

The world will become increasingly IT-driven in the years to come and the government focused on taking measures to equip the youth for this transformation. "ICT is the lever of economic progress. We are using this lever to empower the education system from the marginalized sections of society". To bridging the gap between the urban and remote areas of Pakistan, numbers of projects are working.

4.1 Smart Villages

Villages or regions that master challenges facilitating a digital transformation process. Use new methodologies of innovation and creativity and new sources of information. Often cloud-hosted solutions that connect devices/things and collect/combine/manage data of different rural domains and services within a smart ecosystem.

4.2 Implementing ICT Services

New Internet Technologies. Emerging remote conventions e.g. WiMAX are thought to hold much guarantee for giving network to remote territories and helps in developing Internet accessibility, but in pilot or other stages ventures using such advances generally, and facing hurdles to overcome these issues (Fig. 1).

4.3 Mobile Internet Centers Deployment at Rustic Areas

Little cost and effective information required to use such ventures in various activities using versatile Internet focuses has been guided in the past.

4.4 Hand Held Usage

On handheld devices (containing personal digital assistants and mobile phones), little research has been done that vitalize the importance of ICT tools.

Software's Availability. Free software usage is widely focused over the use of paid software's (particularly Microsoft items), yet explore around there is largely advocatory in nature.

5 Expected Result

Educational institutions failed to understand that inclusion of ICT tools into the education system do not make any significant difference until used efficiently and effectively. That is why those who say that traditional method is more effective than the modern, actually visualized ICT as a robot which will take their place, do their job and produce excellent results; ICT is not a robot at all. It can bring the fruits promised in the

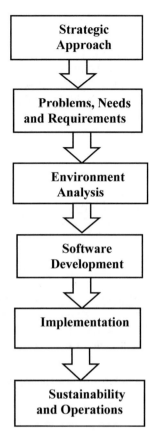

Fig. 1. ICT service model

ICT movement for education. Due to such preliminary requirements, there is a long list of hurdles that resist's in the integration of ICT in the developing countries like Pakistan [10].

6   Conclusion

The significance of ICT training cannot be ignored. Tracer investigations of the effect of ICTs on additionally study and work would be valuable, as this could be a helpful extra measure of instructive quality, past state administered testing comes about. The national ICT strategy design is likewise outdated and there is no subsequent technique in Pakistan but number of projects working for capacity building. It is essential component in the field of security, economy and education. ICT framework is helpful and effective if implemented in available resources that fluctuates with time but provide success in every occupation.

Acknowledgments. We authors acknowledge with thanks assistance rendered by Prof. Dr. Javed Anjum Sheikh, University of Lahore, Gujrat Campus for giving us deep insight during the course of the research work which really enhanced the manuscript.

References

1. Rooksby, E., Weckert, J., Lucas, R.: The rural digital divide. Rural Soc. **12**(3), 197–210 (2002)
2. Evoh, C.: Collaborative partnerships and the transformation of secondary education through ICTs in South Africa. Educ. Media Int. **44**(2), 81–98 (2007)
3. Adomi, E., Anie, S.: An assessment of computer literacy skills of professionals in Nigerian university libraries. Libr. Hi Tech News **23**(2), 10–14 (2006)
4. Shah, R.: Impact of higher education on earnings of women in the public sector educational institutions in Pakistan. Int. Bus. Econ. Res. J. (IBER) **6**(11), 117 (2011)
5. Fatima, H.Z., Shafique, F., Firdous, A.: ICT skills of LIS students: a survey of two library schools of the Punjab. Pakistan J. Libr. Inf. Sci. (2012)
6. Orike, S., Ahiakwo, C.O.: A sustainable model for information and communication technology in Nigeria's tertiary educational transformation. In: 2nd International Conference on Adaptive Science and Technology, 2009. ICAST 2009, pp. 204–210 (2009)
7. Hussain, S., Wang, Z., Rahim, S.: E-learning services for rural communities. Int. J. Comput. Appl. **68**(5), 15–20 (2013)
8. Sánchez, J., Salinas, Á., Harris, J.: Education with ICT in South Korea and Chile. Int. J. Educ. Dev. **31**(2), 126–148 (2011)
9. Tondeur, J., Valcke, M., Van Braak, J.: A multidimensional approach to determinants of computer use in primary education: teacher and school characteristics. J. Comput. Assist. Learn. **24**(6), 494–506 (2008)
10. King, J., Bond, T., Blandford, S.: An investigation of computer anxiety by gender and grade. Comput. Hum. Behav. **18**(1), 69–84 (2002)

The Use of Task-Flow Observation to Map Users' Experience and Interaction Touchpoints

Adriano B. Renzi[✉]

SENAC – Serviço Nacional de Aprendizagem Comercial, Rua Santa Luzia, 375,
Rio de Janeiro, Brazil
`adrianorenzi@gmail.com`

Abstract. As oppose to usability metrics, which tends to focus on specific tasks on specific digital apparatuses, the pervasive experience permeates different digital devices to build a journey with many interaction touchpoints and blended spaces. Sometimes the interactive experience occurs without any digital device at all, as a planned physical environment may fulfill the whole experience. This paper shows the use of Task-flow Observation in its full preparation and execution, while comparing three different approaches – all previously conducted in previous research projects – as a method to map users' behavior and their experience journey.

Keywords: User experience · Human factors · Pervasive interaction systems

1 Cross-Channel User Experience

Norman's precepts of system thinking [1], where a product is an integrated set of cohesive experiences, and Renzi's research [2] regarding pervasive user experience journey built by users in cross-channel scenarios, bring new thinking to user experience development. As interactions progress from the second wave into the third wave of computing, any attempt to understand user experience focused on isolated devices results in experiences with gaps, as noted by Mitchell's [3] on Human-computer interaction: "Today, things are increasingly smeared across multiple sites and moments in complex and often intermediate ways".

An experience journey is built from the blending of users' many short stories, permeating devices, physical and space environments, networks and people. As the stories blend into each other, they become an ecosystem to help fulfill the user experience [4]. The ecosystem can be one system, integrating many devices and many systems, to create a journey of sequential events: purchasing a flight ticket on a desktop computer, checking in from a tablet and presenting the boarding pass using a smartphone at the airport. But also, the journey can be a set of many different, not integrated, systems that build an experience, merged by one company, a service or a concept, such as the Disney experience [5].

Previous studies by Bødker and Kolomose [6] have presented the idea of an increasingly interaction with multiple interactive artifacts with overlapping capabilities during users daily activities. The authors first called this integrated interaction an artifact ecology, as the use of an interactive artifact cannot be understood in isolation

© Springer International Publishing AG, part of Springer Nature 2019
I. L. Nunes (Ed.): AHFE 2018, AISC 781, pp. 46–55, 2019.
https://doi.org/10.1007/978-3-319-94334-3_7

and each artifact influence the use of others. Each device used, affects users' perception [7], as presented in the Gibson approach to perception, whereas the world to be perceived is defined relative to the action repertoire of the perceiver. Changing that repertoire will change the affordances a user can encounter. With a device, the action repertoire is increased to include tool-enabled actions, so there ought to be new affordances to perceive.

The idea of building pervasive experience journeys from sets of human-computer-information-human interactions surfaces the acknowledgement that actions (short stories) may occur outside devices, in physical environments. Benyon [8] proposes that when the physical environment is carefully designed to commingle with a digital space, is a blended space "where a physical space is deliberately integrated in a close-knit way with a digital space". However, not always an environment is a blended space with a digital space, but nevertheless, is part of the user experience.

Throughout the whole experience, there may be short stories that don't need any digital device at all [4]. The cross-channel ecosystem includes users, devices, environments, information, software and actors. The concept of cross-channel experience builds journeys with less control and precision in favor of a more strategic view, understanding that experiences are user-produced more than user-centered [3]. The physical space, blended or not to a digital system, is part of a whole journey, and users' interactions with it can surface valuable information regarding possibilities of experiences.

On business related to developing digital products, methods to understand users' mental model, their expectations and motivations are often applied in the beginning of the project. The data collected helps guide the development of prototypes for further investigation on users' interaction and products' usability. Although Vermeeren *et al.* [9] had categorized 96 methods by their origin, type of collected data, type of execution, the information source, type of location, the experience period, developing phase etc., often, the methods focus on metrics related to single devices, without over seeing the whole experience, nor the environment of use. As oppose to usability metrics, the pervasive experience permeates different digital devices to build a journey with many interaction touchpoints. Each touchpoint representing a direct interaction of users with the system.

This paper presents the use of Task-flow Observation's procedures while comparing three different approaches, all previously conducted in previous research projects: (1) user flow mapping for comparative analysis of users search and purchase decisions in physical and online bookstores [10]; (2) physical environment experience in a shoes concept store and mapping of "hot" areas [11]; and (3) user behavior mapping while building a pervasive experience, with a food purchase app, at food courts.

2 Task-Flow Observation

The Task-flow observation proposes the observation of users' behavior in a specific context. The method has 4 phases for its full completion, but there are important precedents that directly influence the outcome of the observation:

1. Have delineated objectives for the research, as this will determine the tasks to be observed, the environment and users;
2. Determine if there are types of users to focus during the observation. The observation can be random or focused on gender, age, collectivity or physical attributes, as long as the characteristics can be identified visually. Any attempt to contact users prior to the observation, could distort results;
3. Select an assertive environment, proper to the proposal of the objectives and the selected users for observation. If there are many possible environments to achieve the proposed objectives, choose the places with particular characteristics that could bring factors of influence on human actions. This may surface interesting comparative data.

The observation at bookstores [10], for instance, had the objective of understanding the flow and actions of users, for further comparison with online action flow mapping. The choice of users to observe was random and started as soon as any person entered the bookstore. As it was important to map people's routes based on decision and behavior, no children were marked for observation.

The observation at the Melissa's concept store [11] focused on understanding users flow inside the shoe store. For this research, the comparison of the planned exhibit environment with users' hot spots of interaction was the major focus. The selection of users to observe was random, but the majority of people who entered the store were women.

For the food court observation, the objective was to map behavior and users' strategies on purchasing food and getting vacant tables for their food consumption. In overall, the choice of users was random, but it was important to have observations of people by themselves and people in groups.

The selection of environments can be influenced by different factors, mostly directly related to the objectives of the research. The Melissa observation was performed at the company's concept store in São Paulo, as there was no other option in Brazil. The food court observation could have been performed in any shopping mall, as the focus of the research would not analyze location factors. But the bookstore observation was performed in 3 different bookstores, as it was important to check the influence of the bookstore location and its external factors may have on users' behavior. The characteristics to choose the bookstores led to selecting a bookstore located near the beach of Ipanema, a bookstore located inside a shopping mall and a bookstore located in center city. The 3 different places brought up particularities of user behavior that were important for the research purpose.

After the definition of precedents, the researcher executes the method in 4 sequential phases: mapping, task-flow plan, observation and data analysis.

2.1   Mapping

Since the Task-flow observation prioritizes physical spaces, it is important that researches go to the proposed environment to map it. Sometimes, an authorization is required if the method is performed in a private space. As in the Melissa concept store, each visited bookstore needed an authorization from the main manager. Having the local manager aware and part of the process can help set the best days to execute the observation and gather information from him/her about users. In the bookstores, for example, it was important to observe users during peak days. And from the managers' experience in the business, all observations were set on weekends in order to gather information on busy periods. The same way, a previous interview with the manager of the bookstore in center city, showed that the peak periods were throughout working days.

A previous visit to the environment can also help pinpoint the spots for better observation and movement during users' actions. Drawing the schematics of the place is helpful to mark the flow and actions of users. Figure 1 is the map used in one of the bookstores observation (Travessa bookstore in Ipanema). In this bookstore, the best starting point is marked in Fig. 1 by the letter "A". As the observation goes on, the researcher moves around as necessity urges. To write the spot A is not mandatory, but to have a map upfront helps the researcher think about possibilities and plan accordingly.

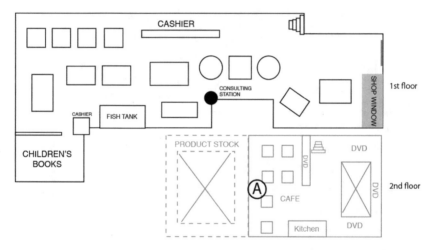

Fig. 1. Top view map of Travessa bookstore, in Ipanema, Rio de Janeiro. The 2nd floor was the best starting point of observation.

2.2   Task-Flow Plan

If based on Malinowski's method of anthropology observation, whose work brought great impact to anthropology research, the observation would have a broader range of awareness. It would need to cover not only behavior, but also a study of local customs, cultural tradition, values, society relations and rituals. But the Task-flow observation needs to be more centered to its purpose. As much as other usability research methods, it must have established a set o points of interest. This will allow the researcher have a script to help fasten his/her notifications during the observation. Specifications regarding types of users to be observed, actions to have special attention or any interaction in parts of the whole environment, should also be part of the task-flow set up plan.

The bookstore observations were centered on users' trajectory, browsing actions, pick up actions, reading actions, seek help actions, users' length of time in the store, the use of the cafe and the final purchase of books. A script was included in the observation sheet to help fasten the notifications (Fig. 2).

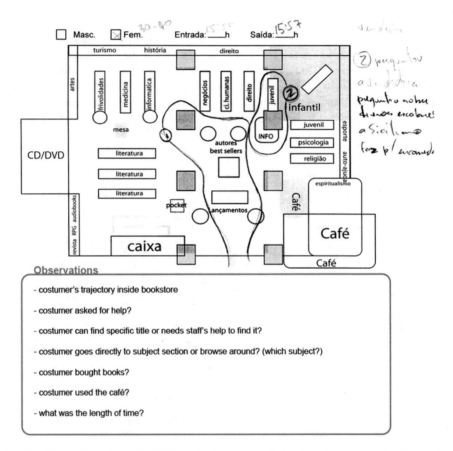

Fig. 2. Observation sheet for Sciliano bookstore, at Praia shopping mall, Rio de Janeiro. Notifications added around the sheet and trajectory marked on map.

A simpler focus point was set for Melissa concept store, as the possible actions seemed simpler: users' trajectory, points of interest, interaction with the exhibit, trying on shoes actions and final purchase.

In a sense, although with a smaller script to follow, the observation on food courts had a broader reach. It centered on mapping the sequential actions of users and understanding their social behavior and strategies to get available seats on a crowded environment.

2.3   Observation

Once the plan and rules of observation are established, the researcher must follow the initial setup. If it is set in the script that people would be observed randomly and the length of period is important, as soon as the researcher is ready to start a new observation proceeding, he/she should look for the first person to enters the environment, mark the entrance time and follow the person throughout the whole trajectory. It is important to not let users notice they are being observed or it will distort the results. Observations and notifications must be done from medium or far viewpoints, and disguised as best as possible. If a person notices the obvious observations from the researcher, the notes for the particular user should be discarded and a new user should be chosen at the starting point. If a subject of observation is lost for a period of time, the researcher must consider if the length of time has any implications to the observation. When the disappearance affects the proceedings, the notes on the missed user should be discarded.

Sometimes, as the Task-flow observation brings unexpected information, the rules can be adapted to better gather data. For instance, although ruled that users would be picked randomly at bookstores, it was noticed in the first day of observation that adults accompanied by their siblings could not be part of the study. As the children run around the store, pausing on colorful books and run again between shelves, the parents would run along to check on them. So, their trajectories were not from search decisions, nor were their actions. The actions were fully commanded by their parents' protective urge of looking after their children. Finding the children's influence factor resulted in leaving out of the observation, any adult accompanied with child.

2.4   Results Analysis

Having the map written with real-time observations, the relevant information stands out visually. In all three cases, the most relevant user flows and how it related to users' intentions and actions became clear. The Melissa concept store flow stands out the hot spots and the core flow (Fig. 3), which helped expose the less important spaces and identify the relation between the exhibit and purchase.

Fig. 3. Observation sheet for Sciliano bookstore, in a shopping mall, Rio de Janeiro. Notifications added around the sheet and trajectory marked on map.

52   A. B. Renzi

Analyzing the observed flows from bookstores, the maps and the data from interviews, surfaced a relation between the flow drawn on the sheets and users' two types of book search: exploratory and objective. Every user who bought a book was interviewed outside the bookstore in order to understand users' objectives when entering the store, their type of search and their perspective on staff's help. The flow chart (Fig. 4) explicit the 2 types of search, diverged visually: exploratory search took users throughout a variety of subjects and shelves, while users in objective search went directly to the subject of their interest. If the title was not easily found, they went to the closest staff for help.

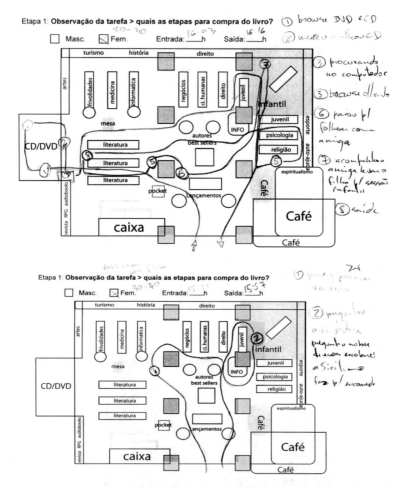

Fig. 4. Flow chart on the left is a representation of exploratory search, from users who entered the bookstore with no intention of purchase and to browse around. Flow chart on the right is a representation of objective search, from users who entered the store with a book of category in mind. They go directly to a specific section, check if there is a copy of the desired title, confirm information with staff and leave if there is no copy.

Although the observation at the food court did not have emphasis on users routes, the notes on users' trajectory helped understand their behavior and strategies (Fig. 5). The results brought insights about user behavior and strategies towards food purchase and accommodation search. Nevertheless, the marked map evinced the obvious different entrances and users' trajectories to browse food options and save a seat.

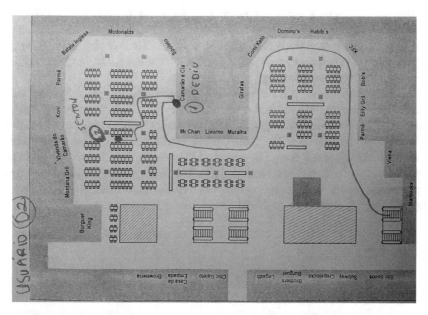

Fig. 5. Users may arrive from 5 possible entrances. In this example, a single user arrives from a stairway, browses the food options, reaches Shrimp & Co, looks for a seat and find one not in front of his food choice. If it was a more crowded day and in a group of friends, the user's route and actions would be different.

3   Conclusion and Discussion

The Flow-task observation brings a broad range of possibilities as a research tool. As interactions have already expanded out of devices in the last few years, and technology experts' predictions [12] say it will be vastly integrated to our physical surroundings in the near future, the user experience is hardly disassociated from the environment and pervasive experience theories.

Depending on the objectives of the research, it may be important to add other methods to better understand the users' behavior. The food court observation helped to gather important data about users' journey, which served as basis to develop a prototype. In order to understand the full experience adding a device to the experience, the researcher conducted a Think-aloud Protocol. The results helped put together the whole experience. Citing the research at bookstores [10] as a second example, interviewing book buyers after they left the environment was important to compare their declared motives and expectations to what was observed inside the bookstore. If an interview is intended to be part of the sequential observation, it should be included in the plan sheet, in order to fasten the notes and easily relate the answers to the respective observation when analyzing results. Sometimes, altering part of observation rules may be necessary, as spontaneous situations may occur or a further investigation may help elucidate a context, such as a quick consulting with staff to check what kind of help was asked by consumers.

The three examples show the method obvious use to map users' behavior in physical environments and compare factors that may influence users' actions. But apart from just observing how users perform actions in a selected environment, it can help pinpoint situations that may urge users to connect to the system (touchpoints) and map the blurred interactions throughout blended spaces.

If considering Benyon and Resmini's description of touchpoints in blended spaces, the interaction can go beyond a series of devices, and also include actors and materials. The idea of system ecologies expanding from digital devices to physical materials and actors, who are set as part of the journey experience, evince the sense of things at the Disney experience [5] and bring UX design and Service design closer, a proposed point of view, presented on previous research [13]. The future merge of these two subjects may certainly set the Flow-task observation as a fundamental method to understand the integration of the experience and help create interaction strategies.

References

1. Norman, D.: Systems thinking: a product is more than the product. Interactions **16**(5), 52–54 (2009). http://interactions.acm.org/content/?p=1286
2. Renzi, A.B.: Experiência do usuário: construção da jornada pervasiva em um ecossitema. In: Proceedings SPGD 2017, vol. 1, Rio de Janeiro (2017). https://www.even3.com.br/anais/spgd_2017/59572-experiencia-do-usuario--construcao-da-jornada-pervasiva-em-um-ecossistema
3. Benyon, D., Resmini, A.: User experience in cross-channel ecosystems. That Paper was Presented at the 31st British Human Computer Interaction Conference, Sunderland, UK, 3–6 July 2017
4. Renzi, A.B.: UX Heuristics for cross-channel interactive scenarios. In: Design, User Experience, and Usability: Theory, Methodology, and Management, Vancouver, BC, pp. 481–491 (2017)
5. Renzi, A.B., Sande, A., Schnaider, S.: Experience, usability and the sense of things. In: Design, User Experience, and Usability: Designing Pleasurable Experiences. HCI International 2017 Proceedings, part II, Vancouver, BC, pp. 77–86, 9–14 July 2017
6. Bødker, S., Klokmose, C.N.: Dynamics in artifact ecologies. In: Proceedings of NordiCHI 2012, pp. 448–457 (2012)
7. Kirsh, D.: Embodied cognition and the magical future of interaction design. ACM Trans. Comput.-Hum. Interact. **20**, 3:1–3:30 (2013)
8. Benyon, D.R.: Spaces of Interaction, Places for Experience. Morgan and Claypool, San Rafael (2014)
9. Vermeeren, A.P.O.S., Law, E.L.C., Roto, V., Obrist, M., Hoonhout, J., Vananen-Vainio-Mattila, K.: User experience evaluation methods: current state and development needs. In: Proceedings: NordiCHI 2010, 16–20 October 2010

10. Renzi, A.B., Freitas, S.: Comparativo de interação e fluxo de leitores em livrarias online e livrarias físicas durante etapas de aquisição de informação e compra. In: Interaction South America Proceedings, Belo Horizonte, MG (2011)
11. Hermida, S., Renzi, A.B.: Melissa's concept store: physical environment for experience. In: Design, User Experience, and Usability: Wearables and Fashion Technology. HCI International 2017 Proceedings, part II, Vancouver, BC, pp. 983–703, 9–14 July 2017
12. Renzi, A.B., Freitas, S.: Delphi method to explore future scenario possibilities on technology and HCI. In: Design, User Experience, and Usability: Wearables and Fashion Technology. HCI International 2015 Proceedings, Los Angeles, CA, pp. 644–653 (2015)
13. Renzi, A.B.: Experiência do ususário: a jornada de Designers nos processos de gestão de suas empresas de pequeno porte utilizando sistema fantasiado em ecossistema de interação cross-channel. DSc. thesis. 239 p., Escola Superior de Desenho Industrial. Rio de Janeiro, RJ (2016)

Serbian and Libyan Female Drivers' Anthropometric Measurements in the Light of the Third Autonomy Level Vehicles

Ahmed Essdai, Vesna Spasojevic Brkic$^{(\boxtimes)}$, and Zorica Veljkovic

Faculty of Mechanical Engineering, University of Belgrade,
Kraljice Marije 16, 11120 Belgrade 35, Serbia
essdai64@hotmail.com,
{vspasojevic, zveljkovic}@mas.bg.ac.rs

Abstract. At level 3 the driver still remains "in-the-loop" due to the inability of the automated system to manage a particular driving situation/environment and anthropometric data are still essential in design. In that aim anthropometric measurements of 193 Serbian and 50 Libyan female drivers have been collected using standard anthropometric instruments and compared using statistical methods - descriptive statistics with Kolmogorov test for normality, linear regression and correlation analysis and comparison of measured anthropometrics using z test. Results obtained in this study are useful to passenger car designers aimed to compete on autonomy level 3 and show that there exist significant differences between Serbian and Libyan female drivers. Future research on other rarely available anthropometric data for other nationalities is recommended, due to globalization trends and constant migrations.

Keywords: Serbian and Libyan anthropometric data · Descriptive statistics
Linear regression · Correlation analysis · Z-test for difference of means

1 Introduction

Automotive sector is one of the most fascinating fields for applications of novel technologies within autonomous robotics [1–4]. Passenger cars fall into four levels of autonomy - starting from driving assistance systems to independent decision-making vehicles [4–6]. Although today we are approaching Level 2 of autonomy, there is only a very limited understanding of drivers' behavior and performance during this level of automation [1, 5, 7]. At level 3 the driver still remains "in-the-loop" due to the inability of the automated system to manage a particular driving situation/environment [7, 8]. Anthropometric data are essential in order to design safe and efficient driving/working place, equipment and tools [9–19]. Also, it is still important to have fresh anthropometric data [19].

Serbian and Libyan anthropometric data are very rarely available [9, 13, 17, 20], so this study focuses on those data collection and comparison. Also, it is evident in recent years that there is also a rapid increase in number of female drivers together with different behavior and attitudes compared to male drivers [21, 22].

© Springer International Publishing AG, part of Springer Nature 2019
I. L. Nunes (Ed.): AHFE 2018, AISC 781, pp. 56–68, 2019.
https://doi.org/10.1007/978-3-319-94334-3_8

Anthropometric measurements of 193 Serbian and 50 Libyan female drivers have been collected. The taken anthropometric measures were foot length, standing height, sitting height, leg length, with lower leg length and upper leg lengths, hip breadth, arm length, eye sitting height and shoulder with. The static anthropometry method, which implies measuring in the erect position during standing and sitting [13, 23] has been used. The standard anthropometric instruments used in this study were anthropometer, a beam caliper, sliding calipers and steel tape. Other instruments included a weight scale and a stool for seated measurement. The participants remained in their clothes and shoes during the measurement. Collected data have been analyzed using statistical methods explained in the next section and conclusions have been derived at the end. Proposals for further research are also given.

2   Statistical Data Analysis

Statistical analysis includes:

- Descriptive statistics for all measured anthropometric measurements for Serbian and Libyan female drivers with Kolmogorov test for normality;
- Linear regression and correlation analysis; and
- Comparison of measured anthropometrics using z test.

Anthropometric measurements were given for examined 193 Serbian and 50 Libyan female drivers. They included weight (*WEI*), standing height (*STH*), sitting height (*SIH*), lower leg length (*LLL*), upper leg length (*ULL*), shoulder width (*SHW*), hip breadth (*HIB*), arm length (*ARL*), and foot length (*FOL*).

All measures are given in millimeters, except weight which is expressed in kilograms.

2.1   Descriptive Statistics

Descriptive statistics for both drivers group includes sample size (*N*), mean, standard deviation (*SD*), median (*Me*), minimal (*Min.*) and maximal (*Max.*) values with their range (*R*) and coefficient of variation (c_v). In both cases - Serbian (Table 1) and Libyan (Table 2) female drivers all values of coefficient of variation are significantly less than 30%, leading to conclusion that data are homogenous [24].

Further analysis includes additional testing of data normality using Kolmogorov test which led to non-significant p values, with conclusion that all examined parameters are normally distributed, i.e. for further examination of data parameter method were used [24].

Descriptive statistics data for Serbian female drivers are shown in Table 1, while for Libyan female drivers are given in Table 2.

Table 1. Descriptive statistics of Serbian female drivers

Meas.	N	Mean	SD	Me	Min.	Max.	R	c_v(%)	D	p	SIGn.	VT
WEI	193	65.539	11.565	64.0	45	115	70	17.65	0.238	1	n.s.	Parameter
STH	193	1694.378	61.465	1700.0	1520	1880	360	3.63	0.138	1	n.s.	Parameter
SIH	193	866.088	44.943	870.0	560	950	390	5.19	0.197	1	n.s.	Parameter
LLL	193	557.409	36.297	560.0	370	710	340	6.51	0.299	1	n.s.	Parameter
ULL	193	592.627	50.368	590.0	384	780	396	8.50	0.270	1	n.s.	Parameter
SHW	193	412.596	34.391	400.0	358	580	222	8.34	0.271	1	n.s.	Parameter
HIB	193	370.036	42.700	360.0	290	520	230	11.54	0.154	1	n.s.	Parameter
ARL	193	652.202	47.296	650.0	410	795	385	7.25	0.265	1	n.s.	Parameter
FOL	193	249.793	13.108	255.0	225	285	60	5.25	0.185	1	n.s.	Parameter

Table 2. Descriptive statistics of Libyan female drivers

Meas.	N	Mean	SD	Me	Min.	Max.	R	c_v(%)	D	p	SIGn.	VT
WEI	50	73.140	73.5	9.394	54	90	36	12.84	0.173	1	n.s.	Parameter
STH	50	1663.780	1660	53.796	1510	1780	270	3.23	0.168	1	n.s.	Parameter
SIH	50	824.4	845	73.656	670	960	290	8.93	0.144	1	n.s.	Parameter
LLL	50	512.8	500	41.652	450	630	180	8.12	0.226	1	n.s.	Parameter
ULL	50	565.8	570	41.654	490	670	180	7.36	0.172	1	n.s.	Parameter
SHW	50	402.8	400	30.973	340	500	160	7.69	0.194	1	n.s.	Parameter
HIB	50	386.6	390	31.727	320	460	140	8.21	0.196	1	n.s.	Parameter
ARL	50	617.4	620	36.579	530	680	150	5.92	0.178	1	n.s.	Parameter
FOL	50	252.4	255	13.141	230	275	45	5.21	0.188	1	n.s.	Parameter

2.2 Regression Analysis

Since all measurements are normally distributed and data are randomly obtained, linear regression and correlation analysis were conducted, including coefficient of determination calculation between all anthropometric measurements and for the both examined groups of drivers. Both groups include as per 38 regression comparisons. Criteria applied for existing correlations are:

- $|r| \in [0.0, 0.5)$ there is no correlation;
- $|r| \in [0.5, 0.7)$ there is weak correlation;
- $|r| \in [0.7, 0.9)$ there is strong correlation; and
- $|r| \in [0.9, 1.0)$ there is absolute correlation.

For Serbian female drivers' correlation analysis data are shown in Table 3, while for Libyan female drivers data are presented in Table 4.

For Serbian female drivers' correlation results indicate that all existing correlations, all seven of them, are weak, i.e. that correlation exist. Correlation coefficients are between [0.5–0.7), i.e. coefficient of determination is in range 25.20%–47.89%. Correlations exist between weight and shoulder width, weight and hip breadth, as well as with weight and

Table 3. The correlation between anthropometric measurements for Serbian female drivers

Regression	r	$r^2(\%)$	Corr. signif.	Regression	r	$r^2(\%)$	Corr. signif.
WEI v.s. STH	0.315	9.92	n.s.	SIH v.s. HIB	0.124	1.54	n.s.
WEI v.s. SIH	0.140	1.96	n.s.	SIH v.s. ARL	0.221	4.88	n.s.
WEI v.s. LLL	0.282	7.95	n.s.	SIH v.s. FOL	0.110	1.21	n.s.
WEI v.s. ULL	0.351	12.32	n.s.	LLL v.s. ULL	0.692	47.89	Exists
WEI v.s. SHW	0.548	30.03	Exists	LLL v.s. SHW	0.122	1.49	n.s.
WEI v.s. HIB	0.658	43.30	Exists	LLL v.s. HIB	0.080	0.64	n.s.
WEI v.s. ARL	0.316	9.99	n.s.	LLL v.s. ARL	0.361	13.03	n.s.
WEI v.s. FOL	0.516	26.63	Exists	LLL v.s. FOL	0.333	11.09	n.s.
STH v.s. SIH	0.422	17.81	n.s.	ULL v.s. SHW	0.314	9.86	n.s.
STH v.s. LLL	0.469	22.00	n.s.	ULL v.s. HIB	0.185	3.42	n.s.
STH v.s. ULL	0.435	18.92	n.s.	ULL v.s. ARL	0.462	21.34	n.s.
STH v.s. SHW	0.270	7.29	n.s.	ULL v.s. FOL	0.359	12.89	n.s.
STH v.s. HIB	0.192	3.69	n.s.	SHW v.s. HIB	0.626	39.19	Exists
STH v.s. ARL	0.450	20.25	n.s.	SHW v.s. ARL	0.502	25.20	Exists
STH v.s. FOL	0.594	35.28	Exists	SHW v.s. FOL	0.417	17.39	n.s.
SIH v.s. LLL	0.399	15.92	n.s.	HIB v.s. ARL	0.382	14.59	n.s.
SIH v.s. ULL	0.290	8.41	n.s.	HIB v.s. FOL	0.403	16.24	n.s.
SIH v.s. SHW	0.097	0.94	n.s.	ARL v.s. FOL	0.366	13.40	n.s.

Table 4. The correlations between anthropometric measurements for Libyan female drivers

Regression	r	$r^2(\%)$	Corr. signif.	Regression	r	$r^2(\%)$	Corr. signif.
WEI v.s. STH	0.094	0.88	n.s.	SIH v.s. HIB	0.068	0.46	n.s.
WEI v.s. SIH	0.072	0.52	n.s.	SIH v.s. ARL	0.219	4.80	n.s.
WEI v.s. LLL	0.079	0.62	n.s.	SIH v.s. FOL	0.117	1.37	n.s.
WEI v.s. ULL	0.048	0.23	n.s.	LLL v.s. ULL	0.815	66.42	Strong
WEI v.s. SHW	0.638	40.70	Weak	LLL v.s. SHW	0.108	1.17	n.s.
WEI v.s. HIB	0.810	65.61	Strong	LLL v.s. HIB	0.003	0.00	n.s.
WEI v.s. ARL	0.228	5.20	n.s.	LLL v.s. ARL	0.289	8.35	n.s.
WEI v.s. FOL	0.368	13.54	n.s.	LLL v.s. FOL	0.118	1.39	n.s.
STH v.s. SIH	0.704	49.56	Strong	ULL v.s. SHW	0.010	0.01	n.s.
STH v.s. LLL	0.541	29.27	Weak	ULL v.s. HIB	0.026	0.07	n.s.
STH v.s. ULL	0.521	27.14	Weak	ULL v.s. ARL	0.289	8.35	n.s.
STH v.s. SHW	0.053	0.28	n.s.	ULL v.s. FOL	0.024	0.06	n.s.
STH v.s. HIB	0.162	2.62	n.s.	SHW v.s. HIB	0.695	48.30	Weak
STH v.s. ARL	0.216	4.67	n.s.	SHW v.s. ARL	0.193	3.72	n.s.
STH v.s. FOL	0.117	1.37	n.s.	SHW v.s. FOL	0.442	19.54	n.s.
SIH v.s. LLL	0.468	21.90	n.s.	HIB v.s. ARL	0.240	5.76	n.s.
SIH v.s. ULL	0.486	23.62	n.s.	HIB v.s. FOL	0.380	14.44	n.s.
SIH v.s. SHW	0.089	0.79	n.s.	ARL v.s. FOL	0.190	3.61	n.s.

foot length. Results indicate that correlation also exists between standing height and foot length, lower and upper leg length, shoulder width and hip breadth and shoulder width and arm length. Other correlations are non-significant - 81.58% (Table 3).

For Libyan female drivers' correlation indicate weak links in four cases with coefficient of determination in range from 27.14% to 48.30%, while strong correlation exists in three cases ranking coefficient of determination between 49.56% and 66.42%. Weak correlations exist between weight and shoulder width, standing high and lower leg length, standing height and upper leg length, shoulder width and hip breadth. Anthropometric measurements for Libyan females' sample are not correlated in 80.56% cases.

Comparing of existing regression and correlation results for Serbian and Libyan female drivers shows that four comparisons are the same for measured anthropometric parameters, i.e. between weight and shoulder width, weight and hip breadth, lower and upper leg lengths, and shoulder width and hip breadth.

Existing regression and correlation only for Serbian female drivers include relationships between weight and foot length, standing height and foot length, and shoulder width with arm length.

For Libyan female drivers existing correlations that differ from results of Serbian female drivers are standing height and sitting height, standing height and lower and upper leg length respectively.

Regression graphs (including both examined groups - Serbian and Libyan) where at least one correlation exists are presented at Figs. 1, 2, 3, 4, 5, 6, 7, 8, 9 and 10, with regression lines drawn only when correlations exist, while otherwise only scatter plots are drawn for observed sample.

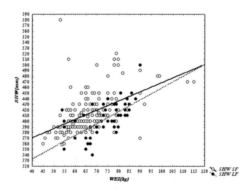

Fig. 1. Scatter plots between *WEI* and *SHW*

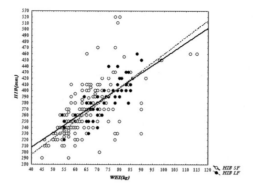

Fig. 2. Scatter plots between *WEI* and *HIB*

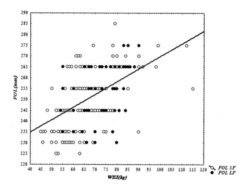

Fig. 3. Scatter plots between *WEI* and *FOL*

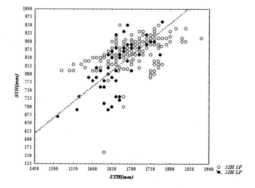

Fig. 4. Scatter plots between *STH* and *SIH*

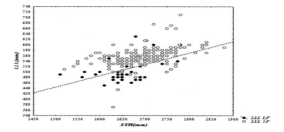

Fig. 5. Scatter plots between *STH* and *LLL*

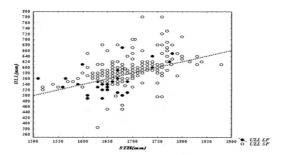

Fig. 6. Scatter plots between *STH* and *ULL*

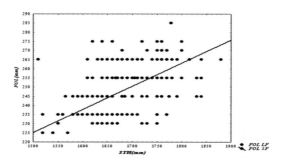

Fig. 7. Scatter plots between *STH* and *FOL*

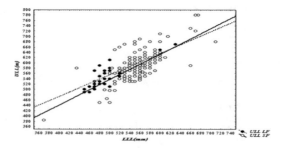

Fig. 8. Scatter plots between *ULL* and *LLL*

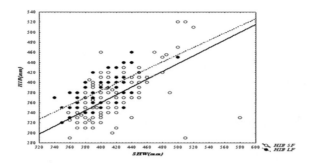

Fig. 9. Scatter plots between *SHW* and *HIB*

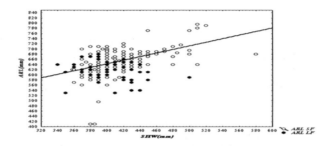

Fig. 10. Scatter plots between *SHW* and *ARL*

2.3 Hypothesis Testing

Since all measured anthropometric parameters are normally distributed, for their comparisons z test for difference of means is used for their, as it is shown at Table 5.

Conclusions of z test for comparison of the anthropometric measurements are based on relations of test p values regarding the following criteria:

- $P > 0.05$ non-significance, marked as n.s. - non-significant and with =
- $P < 0.05$ some significant difference exists, marked as significant and with >
- $P < 0.01$ strong significance exists, strong and with >>
- $P < 0.001$ absolute significance exists, marked with absolute and with >>>.

Comparisons obtained by hypothesis testing indicate that Libyan female drivers have statistically absolutely higher weight values than Serbian female drivers, in average for 10.39% (Table 5, Fig. 10).

Also, Libyan female drivers have statistically significantly wider hip breadth than Serbian in average for 4.28% (Table 5, Fig. 11).

Table 5. Hypothesis testing of differences between anthropometric measurements for Serbian and Libyan female drivers

Serbian female		Libyan female	p-value	Criteria	Significance
WEI SF	<<<	WEI LF	0.000	p < 0.001	Absolute
STH SF	>>>	STH LF	0.0005	p < 0.001	Absolute
SIH SF	>>>	SIH LF	0.0001	p < 0.001	Absolute
LLL SF	>>>	LLL LF	0.000	p < 0.001	Absolute
ULL SF	>>>	ULL LF	0.0001	p < 0.001	Absolute
SHW SF	=	SHW LF	0.0517		n.s.
HIB SF	<<	HIB LF	0.0023	p < 0.01	Significant
ARL SF	>>>	ARL LF	0.000	p < 0.001	Absolute
FOL SF	=	FOL LF	0.2105		n.s.

Serbian females are statistically absolutely higher, have higher values of sitting high, longer lower and upper legs, as well as length of arms - in averages for 1.84%, 5.06%, 8.70%, 4.74% and 5.64% respectively (Table 5, Fig. 12).

Average values of shoulder width and foot length are the same for both groups of female drivers (Table 5, Figs. 13 and 14).

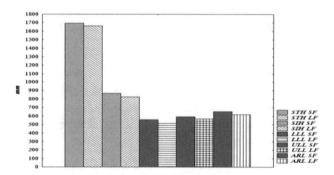

Fig. 11. Mean anthropometric measurements between for *SF* and *LF* drivers

Fig. 12. Difference of mean weights *SF* and *LF* drivers

Fig. 13. Difference of mean in *HIB* measurements for *SF* then the *LF* drivers

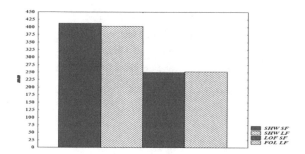

Fig. 14. Anthropometric measurements with same values for *SF* and *LF* drivers

Table 6. Nomenclature

Abbreviations	
SF	Serbian female drivers
LF	Libyan female drivers
Anthropometric measurements abbreviations	
WEI	weight
STH	standing height
SIH	sitting height
LLL	lower leg length
ULL	upper leg length
SHW	shoulder width
HIB	hip breadth
ARL	arm length
FOL	foot length

(*continued*)

Table 6. (*continued*)

Abbreviations	
Statistical abbreviations	
p	p value of test
r^2	coefficient of determination
corr.signif.	significance of correlation
SD	standard deviation
Me	median
Min.	minimum
Max.	maximum
R	range
c_v	coefficient of variation
D	Kolmogorov statistics
p	p value of test
SIGn	significance of test
r	coefficient of correlation
VT	type of variable

3 Conclusions

Results obtained in this study are useful for passenger car designers aimed to compete on autonomy level 3 and show that there exist significant differences between Serbian and Libyan female drivers.

Conclusions about results from regression analysis show that:

- Small percentage of correlation between anthropometric measurements exists for both examined groups;
- In case of Serbian female drivers, all existing correlations (18.42%) are weak;
- In case of Libyan female drivers, for existing correlations (19.44%) there is a more correlations that are strong then a weak correlation;
- Less than 20% of measured anthropometric measurements are mutually correlated, for both groups; and
- There exists difference in correlation of some anthropometric measurements between Serbian and Libyan female drivers, which indicate possibility of different relationships of anthropometry for different populations.

Conclusions in the field of hypothesis testing about difference in anthropometric measurements show the following:

- Most anthropometric measurements are statistically absolutely larger for Serbian than for Libyan female drivers;
- Libyan female drivers have larger weight and hip breadth than Serbian; and
- Foot lengths and shoulder width are the same for both samples.

However, more research on other rarely available anthropometric data on other nationalities are needed, due to globalization trends and constant migrations and that is proposal as avenue for future surveys. Also, more research is needed on the understanding the human factors of how drivers are involved in the "occasional control" of the level 3 vehicles.

References

1. Goodrich, M.A., Olsen, D.R., Crandall, J.W., Palmer, T.J.: Experiments in adjustable autonomy. In: IJCAI Workshop on Autonomy, Delegation and Control: Interacting with Intelligent Agents. American Association for Artificial Intelligence Press, Seattle, WA (2001)
2. Heide, A., Henning, K.: The "cognitive car": a roadmap for research issues in the automotive sector. Annu. Rev. Control **30**, 197–203 (2006)
3. Jullien, B., Pardi, T.: Structuring new automotive industries, restructuring old automotive industries and the new geopolitics of the global automotive sector. Int. J. Automot. Technol. Manag. **13**, 96–113 (2013)
4. Lu, Z., Happee, R., Cabrall, C.D., Kyriakidis, M., de Winter, J.C.: Human factors of transitions in automated driving: a general framework and literature survey. Transp. Res. Part F: Traffic Psychol. Behav. **43**, 183–198 (2016)
5. Greenblatt, J.B., Shaheen, S.: Automated vehicles, on-demand mobility, and environmental impacts. Curr. Sustain./Renew. Energy Rep. **2**, 74–81 (2015)
6. Reimer, B.: Driver assistance systems and the transition to automated vehicles: a path to increase older adult safety and mobility? Public Policy Aging Rep. **24**, 27–31 (2014)
7. Louw, T., Merat, N.: Are you in the loop? Using gaze dispersion to understand driver visual attention during vehicle automation. Transp. Res. Part C: Emerg. Technol. **76**, 35–50 (2017)
8. Purucker, C., Rüger, F., Schneider, N., Neukum, A., Fdrber, B.: Comparing the perception of critical longitudinal distances between dynamic driving simulation, test track and Vehicle in the Loop. In: 5th AHFE Conference-Advances in Human Aspects of Transportation, pp. 421–430 (2014)
9. Klarin, M.M., Cvijanovic, J.M., Brkić, V.S.: Additional adjustment of the driver seat in accordance with the latest anthropometric measurements of drivers in Belgrade. Proc. Inst. Mech. Eng. Part D: J. Automob. Eng. **215**, 709–712 (2001)
10. Klarin, M.M., Spasojević-Brkić, V.K., Stanojević, P.D., Sajfert, Z.D.: Anthropometrical limitations in the construction of passenger vehicles: case study. Proc. Inst. Mech. Eng. Part D: J. Automob. Eng. **222**, 1409–1419 (2008)
11. Klarin, M.M., Spasojević-Brkić, V.K., Sajfert, Z.D., Djordjević, D.B., Nikolić, M.S., Ćoćkalo, D.Z.: Determining the width of the optimal space needed to accommodate the drivers of passenger vehicles using the analogy of anthropometric measurement dynamics and mechanical mechanisms. Proc. Inst. Mech. Eng. Part D: J. Automob. Eng. **225**, 425–440 (2011)
12. Klarin, M.M., Spasojević-Brkić, V.K., Sajfert, Z.D., Žunjić, A.G., Nikolić, M.S.: Determination of passenger car interior space for foot controls accommodation. Proc. Inst. Mech. Eng. Part D: J. Automob. Eng. **223**, 1529–1547 (2009)
13. Spasojevic Brkić, V., Klarin, M.M., Brkić, A.D.: Ergonomic design of crane cabin interior: the path to improved safety. Saf. Sci. **73**, 43–51 (2015)
14. Spasojević-Brkić, V.K., Klarin, M.M., Brkić, A.D., Sajfert, Z.D.: Designing interior space for drivers of passenger vehicle. Tehnika **69**, 317–325 (2014)

15. Hamza, K., Hossoy, I., Reyes-Luna, J.F., Papalambros, P.Y.: Combined maximisation of interior comfort and frontal crashworthiness in preliminary vehicle design. Int. J. Veh. Des. **35**, 167–185 (2004)
16. Spasojević-Brkić, V., Milazzo, F.M., Brkić, A., Maneski, T.: Emerging risks in smart process industry cranes survey: SAFERA research project SPRINCE. Serbian J. Manag. **10**, 247–254 (2015)
17. Omić, S., Brkić, V.K., Golubović, T.A., Brkić, A.D., Klarin, M.M.: An anthropometric study of Serbian metal industry workers. Work **56**, 257–265 (2017)
18. Spasojević Brkić, V.K., Veljković, Z.A., Golubović, T., Brkić, A.D., Kosić Šotić, I.: Workspace design for crane cabins applying a combined traditional approach and the Taguchi method for design of experiments. JOSE **22**, 228–240 (2016)
19. Simões-Marques, M., Nunes, I.L.: Application of a user-centered design approach to the definition of a knowledge base development tool. Adv. Ergon. Des. Usability Spec. Popul.: Part II **17**, 443 (2014)
20. Dondur, N., Spasojević-Brkić, V., Brkić, A.: Crane cabins with integrated visual systems for the detection and interpretation of environment-economic appraisal. J. Appl. Eng. Sci. **10**, 191–196 (2012)
21. Laapotti, S., Keskinen, E.: Has the difference in accident patterns between male and female drivers changed between 1984 and 2000? Accid. Anal. Prev. **36**, 577–584 (2004)
22. Simon, F., Corbett, C.: Road traffic offending, stress, age, and accident history among male and female drivers. Ergonomics **39**, 757–780 (1996)
23. Barroso, M.P., Arezes, P.M., da Costa, L.G., Miguel, A.S.: Anthropometric study of portuguese workers. Int. J. Ind. Ergon. **35**, 401–410 (2005)
24. Montgomery, D.C., Runger, G.C.: Applied Statistics and Probability for Engineers. Wiley, Hoboken (2010)

#LookWhatIDidNotBuy: Mitigating Excessive Consumption Through Mobile Social Media

Pedro Campos[1], Luisa Soares[1], Sara Moniz[1],
and Arminda Guerra Lopes[1,2](✉)

[1] Madeira-ITI, Larsys, Universidade da Madeira, Funchal, Portugal
aguerralopes@gmail.com
[2] Instituto Politecnico de Castelo Branco, Castelo Branco, Portugal

Abstract. Excessively consuming essentially futile or unnecessary assets has turned into a true epidemic, at least in the richest and more developed countries in the world. These consuming habits, when excessive, can really lead to serious problems, of psychological nature, therefore converting excessive consumption into an effective public (mental) health problem. We studied the factors that drive people to consume more or less, and the psychologists in our team conducted an empirical study in order to inform the design of novel interactive technologies for mitigating this problem. #LookWhatIDidNotBuy is a new psychological counseling app that promotes the social media sharing of the photos of goods that the user managed not to buy, thus resisting the temptation. The app also provides advice using positive reinforcement, daily challenges, and tips. Our design goes against the dominant narrative of goal-setting apps and goal-setting theory, advocating that sharing the media of goods not bought can induce positive behavior change.

Keywords: Mobile media · Social media · Excessive consumption
Behavioral psychology · Ubiquitous computing · User interfaces

1 Introduction

Over the past 100 years, consumption has been accepted as a cultural mean of seeking success, happiness, and the populist notion of "good life" [1]. As reported by Brewer and Porter "Our lives today are dominated by the material objects that proliferate all around us, including the prospects and problems they afford" [2]. In fact, the Industrial Revolution enabled the development of new production techniques that promoted mass production and technological advances that revolutionized the transport sector, which was essential for better movement of goods and raw materials and disposal of products. Advertising, marketing sales techniques and payment facilities incite consumption and make individuals consume more than they really need, emerging the society of consumption. Hence consumption starts to occupy a central place in the lives of individuals [3]. Several studies show that excessive consumption has a negative impact, not only in terms of individual well-being but also environmentally (ex.: the overuse of natural resources can lead to lack of resources). Credit and payment facilities amplify the existence of indiscriminate consumption and situations of over-indebtedness. Only a

© Springer International Publishing AG, part of Springer Nature 2019
I. L. Nunes (Ed.): AHFE 2018, AISC 781, pp. 69–75, 2019.
https://doi.org/10.1007/978-3-319-94334-3_9

few number of studies have been conducted on how to motivate people to change their lifestyles and reduce consumption in general. In fact, the current consumer society contributes to the development of buying behaviors aimed at the possession of material goods, so as to give apparently, social status, success and well-being [4]. The belief that well-being can be enhanced through one's relationship with objects is one central characteristic of highly materialistic individuals. Research suggests that these individuals are less happy and more unsatisfied, facing a greater risk of psychological disorders such as depression and neuroticism compared with less materialistic individuals [3].

On the other hand, the influence of social media in peoples' everyday lives has now reached unprecedented levels [5]. Researchers in mobile multimedia have investigated interesting and timely questions, such as Twitter's influence in the health domain [6], how the design of social media influences the way people interact with each other online and ultimately how it shapes our society, and there are even studies about the ecological impact of social media and fashion consumption [7]. Nevertheless, to the present date and to our best knowledge, there hasn't been a study about mobile media as a way to mitigate the problem of excessive consumption. This work addresses specifically this topic in a narrow but different way. By studying the factors that drive people to consume more, we gained insights that were then used to inform the design of a new mobile app we call #LookWhatIDidNotBuy. In terms of mobile and ubiquitous media, this goes against the dominant narrative of goal-setting apps and goal-setting theory, advocating that sharing the media of goods that the user did not buy can actually induce positive behavior change.

2   Clinical Aspects of Excessive Consumption

When reading about the factors that influence people's well-being, there are two recommended habits that can help in a positive way. The first is to take notice, to be aware and to process what happens around you. Being aware and knowing how to deal with situations, helps build good practices that will help people overcome more sensitive situations. The other is to keep learning, to absorb and integrate new things, and to keep being active [3].

Well-being consists of three components: cognitive evaluations of the conditions of one's life (e.g., overall life satisfaction); positive affective states (e.g., happiness) and negative affective states (e.g., depression) [1]. Twell-being on a micro level includes happiness, life satisfaction, and subjective and objective well-being oppositely to well-being on a macro level which includes societal, environmental and political issues [8]. Lee and Ahn [9], Iyer and Muncy [10] provided conceptual framework that confirms that by reducing one's level of consumption, subjective well-being increases on the micro level [8]. As stated by Lee and Ahn [9] in their research, "excessive consumption negatively affects consumer well-being", although ironically, consumers in developed countries still consider that the primary source of happiness consists in the possession of material goods. In fact, materialistic consumers experience lack of control and autonomy in consumption. Studies demonstrate that a greater level of control promote self-determination and self-actualization leading to a higher level of consumer well-being [9]. The least materialistic people report the most life satisfaction.

Shopping experiences provide pleasure and relaxation, but when excessive is a costly way of life [7]. Likewise, compulsive buyers organize their lives around shopping experiences, promoting concerns that can lead to clinical disorders. Compulsive buying disorders or "*oniomania*" are not yet included in contemporary diagnostic systems, such as the Diagnostic and Statistical Manual of Mental Disorders (DSM-IV) (American Psychiatric Association 2000) or the International Classification of Diseases, 10th edition (World Health Organization 1992). Oniomania leads to irrational contraction of debts, occupational, interpersonal, marital, social, and spiritual distress. In fact, Rook [11] postulates that impulsive buying is composed by five elements, namely: *spontaneity*, which is characterizes by the sudden need to act triggered by a visual stimulus (e.g.: a promotional activity); *the feeling of psychological imbalance* caused by the intensity of the desire to suddenly acquire a product; *the psychological conflict* experienced by consumers when considering receiving a direct reward from their impulsiveness and on the other hand, the negative consequences that may arise from the purchase impulsively, *decreased cognitive ability* to make assessments, which begins in the intensity of consumer emotional stages resulting in lack of cognitive control over their final decision and the *disregard of the consequences* resulting anxiety caused by impulsiveness, since the consumer does not perform a careful evaluation of the purchase alternatives, disregarding the possible consequences (mainly negative) of the purchase.

3   A Mobile Social Media Solution

Taking into consideration the work described, we set out to designing and implementing a mobile tool that could help mitigate this problem. #LookWhatIDidNotBuy is a novel psychological counseling app, designed in close cooperation with clinical psychologists who investigated this matter as part of this project. The app is different from existing interventions, in the sense that it that promotes the social media sharing of the photos of goods that the user managed not to buy, thus resisting the temptation. The app also provides advice using positive reinforcement, daily challenges, and tips. However, our design goes against the dominant narrative of goal-setting apps and goal-setting theory, advocating that sharing the media of goods not bought is more effective to induce positive behavior change. Figure 1 shows a screenshot that illustrates how the app works.

The design of this mobile app was inspired by Instagram and Prisma. The goal is to foster quick and easy access to the psychological recommendations, but also to "spread the word": the first screen is essentially an Instagram-like photo shoot mode, where the user simply points the phone's camera to some good (e.g. a needlessly expensive fur coat) and is then shown the screen depicted in Fig. 1. There is an overlay automatically imposed in the photo that is then shared to the user's social media channels, with the hashtag #lookwhatididnotbuy and an accompanying "did you know" fact, a long list of which is embedded into the app.

• Did you know that 20% of the world's
people living in rich countries account for
86% of total global consumer spending!
#lookwhatididnotbuy

Share

Fig. 1. An example of a photo taken with the app and ready to be shared using our proposed hashtag.

The key factor here is that the spreading of mobile media in the form of photos of needless goods is a way to mitigate the problem of excessive consumption. The interaction designers, together with the psychologists in this project decided to also add a second option of shooting, stamping and sharing. A "I fell into temptation" button and logo. Both options are depicted in Fig. 2 and illustrate the psychology of the design idea.

3.1   Mobile Media Sharing

The most important factor in #LookWhatIDidNotBuy is the exploitation of the power in mobile social media, which is very significant these days. We exploit that power as a psychological way of mitigating this problem that very few people seem to be aware of.

Upon sharing the photo of some item the user resisted the temptation to buy, the app includes in the post a "did you know" fact (see Fig. 1). Examples of these curious facts include: *Did you know the average woman will consume over 2.7 kg (6 lb) of lipstick in their lifetime? Did you know consumers spend between $1,200 and $1,300 on online shopping per year? Did you know that the typical American now first takes on debt(s) – usually a credit card and/or car loan – while still in high school?*

This is aimed at increasing the attention of social media viewers in order to raise awareness of the problem. However, it has a double effect of also improving the behavior of the user: the act of sharing per se can induce a positive behavior change and help the user to buy less.

Fig. 2. The two different logos superimposed in the photos taken with the app.

3.2   Mobile Psychological Advice and Tips

The second most important factor in #LookWhatIDidNotBuy is the psychological advice and tips that the app has embedded. Every day, a new notification is presented to the user, giving an incentive for completing a challenge, as opposed to engaging in compulsive or excessive buying behaviors, in what we call "daily challenges". Some examples of these daily challenges include: *Find the most beautiful place within 100 miles and go there; find a bench and take in the view; Ride your bike. Just make sure to wear a helmet; Volunteer. There are hundreds of opportunities daily to make a difference; Call your grandparents. They miss you*; and many others. Currently we include 31 daily challenges that sum up to a month of intervention.

4   Evaluation

As illustrated in Fig. 3, we have been evaluating the impact of the app on the users' consumption habits. Our sample is composed of 24 voluntary participants to whom we administered a questionnaire based on the work of psychologists about consumer habits. In the future, we will conduct a post test to compare differences in the behavior of the participants. They are now in the phase of exploration with using the app

#LookWhatIDidNotBuy and in 2 months from now we will administer the question-naire again and compare results. Our hypothesis is that participants will be more aware of consumer habits and diminish their shopping habits after using the app.

Fig. 3. A user with the #LookWhatIDidNotBuy app.

So far, the feedback has been quite positive and the majority of users intends to continue using the app and reflecting upon his or her own buying behaviors. There are, essentially, three main aspects we analyzed from a qualitative perspective: the effec-tiveness of the daily challenges; the effectiveness of the "did you know" facts; and the effectiveness of the shooting/sharing media of goods. Our results clearly suggest that this latter factor is extremely positive, as most users have reported feelings of engagement towards using the photo-shoot and share with both logos: the "Look What I Did Not Buy" but also the "I Fell Into Temptation". This might be explained from the fact that people are already so used to shooting photos and sharing them that the mere action of sharing something with the hashtag and a funny logo can improve self-reflection upon one's buying behaviors. Not to mention it also helps disseminating the problem of excessive consumption through social media networks.

5 Conclusion

Work around mobile and ubiquitous multimedia shows a lack of concrete applications targeted at solving mental health problems, as well as social problems. There is an enormous amount of knowledge about excessive consumption in the Psychology field, but current interventions are limited to using human skills in the form of clinical psychologists. In this paper we show a focused approach that illustrates the power of mobile social media in terms of increasing awareness for this problem but also for mitigating it. The act of sharing media of the goods, coupled with sound psychological advice can be incorporated in a mobile app in order to induce a positive behavior change. Future work will naturally include analyzing results from post-tests and comparing them with the pre-tests in order to better assess the effectiveness of this approach in mitigating the problem of excessive consumption.

Acknowledgements. This research was partially funded by Madeira-ITI, through LARSYS, Projeto Estratégico LA9-UID/EEA/50009/2013.

References

1. Burroughs, J.E., Rindfleisch, A.: Materialism and well-being: a conflicting values perspective. J. Consum. Res. **29**(3), 348–370 (2002)
2. Brewer, J., Porter, R.: Introduction. In: Brewer, J., Porter, R. (eds.) Consumption and the World of Goods. Routledge, London (1993)
3. Allen, R.: The British industrial revolution in global perspective: how commerce created the industrial revolution and modern economic growth a meta-analysis. J. Pers. Soc. Psychol. **107**, 879–974 (2006)
4. Dittmar, H.: Consumer Culture, Identity and Well-Being: The Search for the 'Good Life' and the 'Body Perfect'. Psychology Press, New York (2011)
5. Pan, Y., Thomas, J.: Hot or not: a qualitative study on ecological impact of social media & fashion consumption. In: Proceedings of the ACM 2012 Conference on Computer Supported Cooperative Work Companion, pp. 293–300. ACM, New York (2012)
6. McNeill, A.R., Briggs, P.: Understanding Twitter influence in the health domain: a social-psychological contribution. In: Proceedings of the 23rd International Conference on World Wide Web, pp. 673–678. ACM, New York (2014)
7. Elliott, R.: Addictive consumption: function and fragmentation in postmodernity. J. Consum. Policy **17**(2), 159–179 (1994)
8. Hoffmann, S., Lee, M.S.W.: Consume less and be happy? Consume less to be happy! An introduction to the special issue on anti-consumption and consumer well-being. J. Consum. Aff. **50**(1), 3–17 (2016)
9. Lee, M.S.W., Ahn, C.: Anti-consumption, materialism, and consumer well-being. J. Consum. Aff. **50**(1), 18–47 (2016). Special Issue on Anti-Consumption and Consumer Well-Being
10. Iyer, R., Muncy, J.A.: Purpose and object of anti consumption. J. Bus. Res. **62**(2), 160–168 (2009)
11. Rook, D.: The buying impulse. J. Consum. Res. **14**, 189–199 (1987)

The Antecedents of Intelligent Personal Assistants Adoption

Tihomir Orehovački[1]([✉]), Darko Etinger[1], and Snježana Babić[2]

[1] Faculty of Informatics, Juraj Dobrila University of Pula,
Zagrebačka 30, 52100 Pula, Croatia
{tihomir.orehovacki,darko.etinger}@unipu.hr
[2] Polytechnic of Rijeka, Trpimirova 2/V, 51000 Rijeka, Croatia
snjezana.babic@veleri.hr

Abstract. Intelligent personal assistants (IPAs) are software agents designed to provide the aid to the users in conducting daily routines such as answering phone calls, taking notes, shopping, making appointments, findings places and answers, web browsing, and alike. They are therefore commonly applied in all aspects of human endeavor. End-user acceptance is an important determinant of the success of each piece of software, and IPAs are no exception. Considering that literature lacks studies on IPAs acceptance, this paper introduces a nomological network of factors that determine users' satisfaction and behavioral intention related to IPAs. After completing the scenario based activities by means of Google Assistant as representative sample of IPAs, students from two higher education institutions were asked to fill out a post-use questionnaire. The aim of this paper is to examine the psychometric characteristics of the research framework which reflects the interplay among eight relevant aspects of several theories and models related to the adoption and user acceptance of new technologies and innovations. The psychometric features of the model were explored by means of the partial least squares (PLS) approach to structural equation modeling (SEM). Reported findings have important implications for practitioners engaged in the development of IPAs as well as for researchers dealing with similar studies in the field.

Keywords: Intelligent personal assistants (IPAs) · Educational ecosystem
Adoption · Theory and models · Post-use questionnaire · Empirical findings

1 Introduction

An intelligent personal assistant (IPA) is "an implementation of an intelligent social agent that assists a user in operating a computing device and using application programs on a computing device". In this context, it is very important that social intelligence of the agent has the ability to be "appealing, affective, adaptive, and appropriate when interacting with the user" [1]. According to Venkatesh and Davis [2] "technology adoption and use in the workplace remains a central concern of information systems research and practice". The fundamental problem for an interactive system is, according to Li et al. [3], understanding and modeling user behavior. Based on the set forth, explaining the users' adoption and use of IPAs is very important for their

© Springer International Publishing AG, part of Springer Nature 2019
I. L. Nunes (Ed.): AHFE 2018, AISC 781, pp. 76–87, 2019.
https://doi.org/10.1007/978-3-319-94334-3_10

implementation in different contexts of people's activities (e.g. education, marketing, trade, traffic, health, etc.) [4, 5]. IPAs provide natural communication between the user and the agent which places the understanding of the natural language at the heart of the human computer interaction in that respect [6]. In addition, IPAs support numerous scenarios of interaction. Drawing on comparison of several IPAs (Google Assistant, Amazon Alexa, Apple Siri, and Microsoft Cortana), Reis et al. [7] concluded that these services have many features in common (such as playing music, online search, and playing games) but the most important is to determine the foundation of data acquisition, i.e. user's state of mind or context of information relevant to the user. Ponciano et al. [8] focused their study on IPAs for Internet of Things (IoT) environments and argue in this context that a good IPA has "the capability of surveying its user behaviour and suggest tasks or make decisions with the intention of simplifying the user interaction with his/her surroundings". Li et al. [3] found following three aspects of the disadvantages of using only voice user interfaces (VUIs) in IPAs: (1) Conversational content: "IPAs are incapable of conveying emotion"; (2) Performance: "responses from IPAs demonstrate poor variety of social behavior, preventing users to engage in a long-time interaction with them"; (3) Function: "IPAs are not good at comprehending users intent and can only carry out limited conversation spanning one or a few turns". Han and Yang [9] developed a comprehensive research model that can explain customers' continuance intentions to adopt and use IPAs. Results of their research have shown that interpersonal attraction (task attraction, social attraction, and physical attraction) and security/privacy risk are important factors affecting the adoption of IPAs. Siddike et al. [10] found that reliability, attractiveness, and emotional attachments are influential factors for generating the trustworthiness of people toward using cognitive assistants (CAs) and that the innovativeness positively moderates the intention of people to use CAs.

The remainder of the paper is structured as follows. Theoretical foundation of our work is briefly described in the following section. Employed research framework is introduced in the third section. Study findings are outlined in the fourth section. Concluding remarks and future work directions are provided in the last section.

2   Background to the Research

2.1   Adoption of Technologies

When software adoption is tackled, several theories and models have to be taken into account. Theory of Reasoned Action (TRA) [11] is one of the fundamental models in social psychology designed for prediction and clarification of voluntary behaviors. According to the underlying assumptions of TRA, the behavioral intention (BI) of the individual is affected by his/her attitude and expectations of significant other persons (subjective norm, SN). The extension of TRA with perceived behavioral control as third determinant of BI is referred to as Theory of Planned Behavior (TPB) [12]. By adapting TRA to the context of information systems, Technology Acceptance Model (TAM) [13] has emerged. The main postulates of TAM [13] indicate that both perceived ease of use (PEOU) and perceived usefulness (PU) are significant antecedents of

BI which in turn contributes to the actual software use. TAM2 [2], as the first extension of TAM, incorporated dimensions of social influence (SN, voluntariness, and image) and cognitive instrumental facets (job relevance, output quality, result demonstrability, and PEOU) as determinants of PU. As a follow up, TAM2 was enhanced with the anchoring (self-efficacy, anxiety, playfulness, and perceptions of external control) and adjustment (perceived enjoyment and objective usability) related predecessors of PEOU, which resulted in TAM3 [14] as the most recent version of TAM. Unified Theory of Acceptance and Use of Technology (UTAUT) [15] originated as an outcome of longitudinal study on empirical validation of eight prominent models and their extensions dealing with user acceptance. This model pointed out that performance expectancy, effort expectancy, social influence, and facilitating conditions affect BI whereas facilitating conditions and BI contribute to the use behavior. In UTAUT2 [16], hedonic motivation, price value, and habit are added as predictors of BI, and habit as determinant of use behavior. Expectation-Confirmation Theory (ECT) [17] posits that users' satisfaction is affected by interplay of prior expectation and perception of performance. Positive disconfirmation (performance > expectation) leads to satisfaction while negative disconfirmation (performance < expectation) results in dissatisfaction. On the other hand, Expectation-Confirmation Model (ECM) stipulates that users' continuance intension is predicted by the level of his/her satisfaction, extent to which his/her expectations are confirmed, and post-usage perceived usefulness.

2.2   Research Model and Hypotheses

The aim of the research presented in this paper was to identify the antecedents of IPAs adoption. For that purpose, an empirical study was carried out in which students employed Google Assistant in educational settings and examined pragmatic and hedonic facets of its quality by means of the measuring instrument in the form of post-use questionnaire. Based on the set of quality attributes proposed in [18] and questionnaire items introduced in [19], an evaluation framework composed of 8 constructs (effectiveness, controllability, reliability, accuracy, ease of use, usefulness, satisfaction, and loyalty) was created. Ten hypotheses that reflect an interplay among the aforementioned constructs were defined as an outcome of comprehensive literature review.

Effectiveness denotes an extent to which an IPA enables users to execute tasks accurately and completely on a mobile device. IPAs interface should contain all the functionalities needed for task execution to enhance user's effectiveness [18]. The results of study carried out by Kiseleva et al. [20] indicate that many participants were impressed by the effectiveness of IPAs, although they were beginners in using those applications. According to Venkatesh et al. [16], perceived effectiveness contributes to performance expectancy (usefulness). In that respect we propose following hypothesis:

H1. Effectiveness positively influences Usefulness.

Controllability measures the degree to which users have full freedom in executing tasks by means of the IPA on mobile device. Based on TRA model [11], the control is divided into perception of internal control (computer self-efficacy) and perception of external control (facilitating conditions) as proposed in the TAM model [13]. According to Venkatesh and Bala [14], perception of external control is an important determinant in predicting perceived usefulness and perceived ease of use. Results of empirical study conducted by Faqih and Jaradat [21] indicate that perception of external control has a positive influence on the perceived ease of use. Taking into account the aforementioned, we propose the following hypotheses:

H2a. Controllability positively influences Usefulness.
H2b. Controllability positively influences Ease of Use.

Reliability denotes the extent to which IPA is dependable, stable, and bug-free when employed on a mobile device. Siddike et al. [10] found that perceived reliability positively affects the intention of users to use cognitive assistants (CAs). Reliability is one of the main dimensions of system quality in DeLone and McLean's IS success model [22]. Drawing on the integration of TAM [13] and IS success model [22], Pai and Huang [23] found that system quality significantly affects perceived ease of use. In that respect, we hypothesize:

H3. Reliability positively influences Ease of Use.

Accuracy means that information provided by IPA is accurate, precise, and free from errors. According to DeLone and McLean [22], accuracy is one of the essential aspects of information quality. If the IPA returns accurate outcomes, users have to invest less physical and mental activities in order to interact with it. Therefore, we hypothesize that:

H4. Accuracy positively influences Ease of Use.

Ease of Use represents an extent to which interaction with the IPA is free of effort. Many studies confirmed that perceived ease of use has positive effect on perceived usefulness (e.g. [2, 13, 24, 25]). Orehovački and Babić [19] found that perceived ease of use significantly affects students' satisfaction with the employment of Web 2.0 application meant for collaborative writing. Hence, we propose the following hypotheses:

H5a. Ease of Use positively influences Usefulness.
H5b. Ease of Use positively influences Satisfaction.

Usefulness refers to the degree to which using the IPA improves the user performance in the context of tasks execution on mobile device [18]. Amoroso and Lim [26] found that perceived usefulness positively affects users' satisfaction with mobile-based applications. On the other hand, Orehovački and Babić [27] discovered that perceived usefulness of Google Docs plays an important role in predicting students' continuance intention (loyalty) in the context of this Web 2.0 application. Therefore, we propose the following hypotheses:

H6a. Usefulness positively influences Satisfaction.
H6b. Usefulness positively influences Loyalty.

Satisfaction represents the degree to which interaction with the IPA has met users' expectations.

Loyalty refers to the extent to which the users have the intention to continue to use the IPA and recommend it to others [18]. Current studies (e.g. [18, 28, 29]) confirmed that satisfaction has a significant impact on continuance intention. In that respect, we propose the following hypothesis:

H7. Satisfaction positively influences Loyalty.

The research model that illustrates interrelations among described constructs in terms of the proposed hypotheses is presented in Fig. 1.

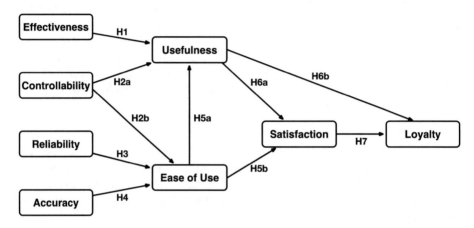

Fig. 1. Research model with proposed hypotheses

3   Research Method and Findings

3.1   Data Collection Procedures and Data Analysis

In order to examine the proposed hypotheses, data was collected using an online questionnaire which was composed of multiple choice questions (to collect demographic information) and Likert scales (to measure users' perceptions and use behavior of the Google Assistant as a representative sample of IPAs). In the process of development of the questionnaire items, the constructs' indicators were adapted from previous instruments based on the literature review. Each question was tailored to fit the context of this study. The questionnaire items related to each of the constructs included in the model were measured using a five-point Likert scale (items ranged from 1 - strongly disagree to 5 - strongly agree).

3.2   Study Participants

A total of 309 students (63.4% male, 36.6% female), aged 19.94 years (SD = 2.770) on average, took part in the study. At the time study took place, most of the sample (57.9%) was enrolled to one of the undergraduate study programs at Polytechnic of Rijeka while remaining 42.1% were undergraduate students at the Faculty of Informatics, Juraj Dobrila University of Pula. The majority (89%) were full-time students.

Most of respondents (83.5%) completed the scenario of interaction with IPAs on smartphones that run the Android operating system.

3.3   Model Assessment

The research model represents the relationships among the eight proposed constructs. Since the constructs are not directly measured, we specified a measurement model for each construct. All eight constructs have reflective measurement models. Each construct is measured with reflective items which relate to corresponding questionnaire items: Effectiveness (eff1–eff7), Controllability (control1–control3), Reliability (reliabil1–reliabil5), Accuracy (accur1–accur6), Usefulness (usef1–usef6), Ease of Use (eou1–eou8), Satisfaction (satisf1–satisf8), and Loyalty (loy1–loy7).

The psychometric features of the model were explored using the Partial Least Squares Structural Equation Modeling (PLS-SEM). The software tool SmartPLS 3.2.7 [30] was used to assess the measurement and the structural model. PLS-SEM maximizes the explained variance of the endogenous latent variables by estimating partial model relationships in an iterative sequence of ordinary least squares (OLS) regressions [31]. An important characteristic of PLS-SEM is that it estimates latent variable scores as exact linear combinations of their associated manifest variables [31] and treats them as perfect substitutes for the manifest variables.

The assessment of reflective outer models involves determining indicator reliability (squared standardized outer loadings), internal consistency reliability (composite reliability), convergent validity (average variance extracted - AVE), and discriminant validity (Fornell-Larcker criterion, cross-loadings, heterotrait-monotrait - HTMT ratio of correlations) [32]. A confirmatory factor analysis (CFA) was employed to establish the reliability of the items and the convergent and discriminant validity of the constructs. The factor structure matrix of item loadings and cross-loadings (presented in Table 1) confirms that the convergent validity of each construct is achieved as the item loadings for each construct are above the threshold of 0.708 [32]. The items eff1, eff7, eou1, eou3, satisf2 have not met the minimum requirements and were thus removed from the model. Since Effectiveness, Ease of Use and Satisfaction are explained by more than 3 items, the removal of the items not meeting the required criteria for inclusion did not represent a severe problem. On the contrary, it helped boosting the AVE above the threshold (>0.5) and helped achieving a better representation of the constructs.

The verification of the reliability of indicators was obtained using Cronbach's alpha coefficient, testing the contribution made by each indicator to be similar, as well as the composite reliability coefficient which takes respective indicators into account. Convergent validity, measured by Average Variance Extracted, represents the common variance between the indicators and their construct and should be higher than 0.5 [32].

In order to confirm the discriminant validity among constructs (Fornell-Larcker criterion) the AVE square root must be superior to the correlation between constructs. Findings on the assessment of measurement model in terms of the Cronbach's Alpha (CA) coefficient, Composite reliability (CR) coefficient, Average Variance Extracted (AVE) along with the square roots of the AVE (highlighted numbers in the diagonal) and the correlation between constructs are presented in Table 2.

Table 1. Factor structure matrix of loadings and cross-loadings

	(1)	(2)	(3)	(4)	(5)	(6)	(7)	(8)
accur1	**0.893**	0.540	0.436	0.449	0.299	0.430	0.524	0.419
accur2	**0.883**	0.481	0.412	0.455	0.251	0.481	0.474	0.385
accur3	**0.874**	0.540	0.431	0.546	0.311	0.457	0.504	0.384
accur4	**0.898**	0.570	0.461	0.548	0.307	0.498	0.542	0.460
accur5	**0.897**	0.534	0.440	0.417	0.285	0.459	0.501	0.403
accur6	**0.913**	0.513	0.441	0.452	0.316	0.431	0.524	0.443
control1	0.580	**0.836**	0.510	0.548	0.431	0.379	0.527	0.488
control2	0.422	**0.790**	0.362	0.403	0.206	0.379	0.368	0.334
control3	0.429	**0.821**	0.309	0.426	0.222	0.311	0.369	0.403
eff1	0.262	0.313	**0.790**	0.261	0.449	0.176	0.444	0.549
eff2	0.285	0.350	**0.800**	0.336	0.404	0.216	0.415	0.583
eff3	0.491	0.407	**0.787**	0.371	0.351	0.351	0.511	0.501
eff4	0.458	0.461	**0.762**	0.386	0.336	0.306	0.422	0.491
eff5	0.408	0.389	**0.775**	0.348	0.395	0.264	0.466	0.428
eff6	0.418	0.400	**0.744**	0.379	0.333	0.260	0.432	0.463
eou2	0.425	0.424	0.368	**0.828**	0.273	0.307	0.412	0.405
eou4	0.417	0.410	0.360	**0.812**	0.271	0.331	0.399	0.362
eou5	0.361	0.436	0.308	**0.795**	0.191	0.415	0.339	0.299
eou6	0.374	0.415	0.224	**0.775**	0.133	0.443	0.264	0.267
eou7	0.545	0.542	0.382	**0.828**	0.249	0.449	0.433	0.378
eou8	0.453	0.503	0.449	**0.782**	0.306	0.344	0.402	0.561
loy1	0.250	0.268	0.411	0.216	**0.874**	0.134	0.466	0.420
loy2	0.239	0.307	0.414	0.216	**0.883**	0.135	0.484	0.379
loy3	0.334	0.363	0.435	0.302	**0.873**	0.238	0.593	0.403
loy4	0.361	0.318	0.441	0.301	**0.750**	0.170	0.557	0.298
loy5	0.224	0.293	0.413	0.235	**0.909**	0.122	0.480	0.349
loy6	0.322	0.371	0.444	0.286	**0.877**	0.199	0.550	0.397
loy7	0.249	0.280	0.382	0.250	**0.871**	0.156	0.479	0.315
reliabil1	0.294	0.240	0.123	0.307	0.060	**0.749**	0.194	0.148
reliabil2	0.425	0.358	0.281	0.399	0.172	**0.839**	0.334	0.211
reliabil3	0.488	0.443	0.376	0.405	0.235	**0.852**	0.397	0.324
reliabil4	0.477	0.391	0.348	0.452	0.148	**0.908**	0.382	0.294
reliabil5	0.448	0.371	0.227	0.394	0.180	**0.826**	0.319	0.283
satisf1	0.549	0.478	0.517	0.479	0.521	0.382	**0.847**	0.440
satisf3	0.589	0.542	0.501	0.475	0.528	0.357	**0.847**	0.435
satisf4	0.477	0.433	0.522	0.386	0.558	0.312	**0.874**	0.439
satisf5	0.423	0.446	0.505	0.391	0.546	0.303	**0.868**	0.426
satisf6	0.447	0.385	0.474	0.339	0.515	0.327	**0.872**	0.409
satisf7	0.491	0.442	0.495	0.374	0.514	0.355	**0.887**	0.422
satisf8	0.455	0.429	0.423	0.383	0.412	0.340	**0.799**	0.363

(*continued*)

Table 1. (*continued*)

	(1)	(2)	(3)	(4)	(5)	(6)	(7)	(8)
usef1	0.356	0.376	0.551	0.360	0.364	0.210	0.389	**0.852**
usef2	0.355	0.375	0.527	0.322	0.313	0.209	0.368	**0.847**
usef3	0.449	0.519	0.621	0.502	0.370	0.349	0.442	**0.873**
usef4	0.515	0.503	0.559	0.515	0.342	0.338	0.472	**0.804**
usef5	0.274	0.359	0.441	0.285	0.400	0.150	0.369	**0.780**
usef6	0.354	0.393	0.552	0.396	0.349	0.249	0.404	**0.853**

Table 2. Measurement model assessment and discriminant validity of the constructs

	(1)	(2)	(3)	(4)	(5)	(6)	(7)	(8)
CA	0.949	0.753	0.869	0.891	0.943	0.892	0.939	0.913
CR	0.959	0.856	0.901	0.916	0.953	0.921	0.951	0.933
AVE	0.798	0.665	0.603	0.646	0.746	0.700	0.734	0.698
Accuracy (1)	**0.893**							
Controllability (2)	0.595	**0.816**						
Effectiveness (3)	0.490	0.494	**0.776**					
Ease of Use (4)	0.541	0.572	0.443	**0.804**				
Loyalty (5)	0.332	0.368	0.489	0.302	**0.864**			
Reliability (6)	0.516	0.437	0.333	0.473	0.194	**0.836**		
Satisfaction (7)	0.574	0.528	0.575	0.473	0.602	0.396	**0.857**	
Usefulness (8)	0.466	0.510	0.653	0.483	0.426	0.307	0.491	**0.835**

*square root of AVE on diagonal

Recent advances in PLS-SEM [32] propose the heterotrait-monotrait - HTMT ratio of correlations for testing the discriminant validity of the constructs. Results outlined in Table 3 indicate that the conservative threshold of 0.85 is not met thus confirming the discriminant validity of the proposed eight constructs.

After establishing the reliability for the indicators and the convergent and discriminant validity of the constructs, we examined the structural model. The results of the PLS analysis for the proposed hypotheses are shown in Fig. 2. The structural model shows a significant positive relationship between all constructs. Therefore, hypotheses H1, H2a, H2b, H3, H4, H5a, H5b, H6a, H6b and H7 are all supported.

By analyzing the direct, indirect and total effects, it is evident that the strongest relationship is the one between Satisfaction and Loyalty ($\beta = 0.518$, $p < 0.001$). Usefulness is strongly affected by Effectiveness ($\beta = 0.495$, $p < 0.001$). Along with moderate relationships between constructs, weak significant relationships are present: the one between Usefulness and Loyalty ($\beta = 0.172$, $p < 0.01$) and between Controllability and Usefulness ($\beta = 0.17$, $p < 0.01$). The obtained R-squared coefficients of determination reflect the amount of variance explained by the model, thus indicating its

Table 3. Heterotrait-monotrait ratio (HTMT)

	(1)	(2)	(3)	(4)	(5)	(6)	(7)	(8)
Accuracy (1)	1							
Controllability (2)	0.687	1						
Effectiveness (3)	0.547	0.599	1					
Ease of Use (4)	0.573	0.678	0.496	1				
Loyalty (5)	0.345	0.411	0.535	0.319	1			
Reliability (6)	0.552	0.526	0.372	0.528	0.204	1		
Satisfaction (7)	0.605	0.611	0.637	0.508	0.632	0.425	1	
Usefulness (8)	0.492	0.595	0.721	0.513	0.458	0.327	0.525	1

* below the threshold of 0.85

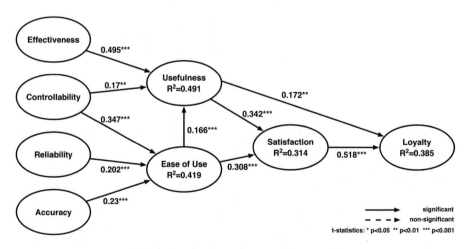

Fig. 2. Structural model evaluation results

predictive power. The model explains 38.5% of the variance in Loyalty, 31.4% of the variance in Satisfaction, 49.1% of the variance in Usefulness, and 41.9% of variance in Ease of Use.

The effect size f^2 coefficients were calculated for the relationships between constructs. This measure takes into consideration whether the exogenous construct has a substantive impact on the endogenous construct [32], if omitted from the model. The results show weak effect size coefficients except for the following relationships: Usefulness is moderately influenced by Effectiveness ($f^2 = 0.346$), and Loyalty is moderately influenced by Satisfaction ($f^2 = 0.331$).

4 Conclusion

The study presented in this paper investigated the factors that affect the adoption of the intelligent personal assistants (IPAs). Participants in the study were students from two higher education institutions while Google Assistant served as a representative sample of IPAs. Drawing on the literature review, an evaluation framework composed of 8 constructs (Effectiveness, Controllability, Reliability, Accuracy, Ease of Use, Usefulness, Satisfaction, and Loyalty) was designed. Psychometric features of the model were examined by means of the partial least squares (PLS) approach to structural equation modeling (SEM). Findings of the study indicate that if the IPA enhances students' performance in executing tasks on a mobile device, enables them to complete tasks at hand in the manner they want, and students find interaction with it simple, then the IPA will be perceived as beneficial mobile applications. It was also discovered that if students can decide in what manner they will perform tasks on their mobile devices by means of the IPA, interaction with the IPA is free of errors, and the information the IPA provides is accurate, then students will perceive the IPA as mobile application that can be employed without significant effort. The results of the model analysis are also implying that the IPA meets students expectations if they perceive it as useful and easy to use. Finally, it was uncovered that students are willing to continue to use the IPA if they find it advantageous and if they are content with it. All the above-presented findings are in line with results reported by other studies in the literature. However, we must emphasize that this is the first study to show that Accuracy has a direct and significant impact on Ease of Use with respect to interaction with the IPA.

This study provides useful contributions and implications to both researchers and practitioners. From a theoretical point of view, an interplay of constructs adopted from TAM [13] and Expectation-Confirmation Theory [17] was extended with four constructs (Effectiveness, Controllability, Reliability, and Accuracy), which turned out to have relevant power in the explanation of the introduced model. Since all the hypothesized relationships were supported, the validated model can be used for predicting students' adoption of IPAs as well as diagnosing possible reasons for the lack of their adoption. In order to increase the level to which users are satisfied with IPAs and are willing te become their loyal consumers, practitioners must not focus only on fulfilling users' expectations related to perceived usefulness and perceived ease of use but develop reliable and accurate IPAs that would improve the extent of users' effectiveness and controllability in conducting task on mobile devices.

The findings of this research must be considered in light of its limitations. First, generalizability is an issue that plagues most empirical studies. Given that participants in our study were undergraduate information science students, sample composed of more heterogenous users could provide different answers to questionnaire items. Keeping that in mind, the empirical results should be interpreted carefully. Similarly, only one IPA was evaluated in this study, whereas an assessment of several different

IPAs may reveal different results. Secondly, only one research method (post-use questionnaire) was employed, potentially leading to a bias due to common method variance.

Taking the above into consideration, further studies should be carried out in order to draw generalizable sound conclusions and to examine the robustness of findings. In that respect, our future work will be devoted to employing several research methods in order to determine to what extent findings presented in this paper can be applied to other samples, contexts of use, and IPA types. Finally, we are planning to extend the proposed model with other constructs that may have impact on the adoption of IPAs.

References

1. Gong, L.: Intelligent personal assistants. U.S. Patent Application No. 10/158, 213 (2003)
2. Venkatesh, V., Davis, F.D.: A theoretical extension of the technology acceptance model: four longitudinal field studies. Manag. Sci. **46**(2), 186–204 (2000)
3. Li, J., District, C., Lee, Y.Y.: Multimodal interaction and believability. In: The 31st British Human Computer Interaction Conference, pp. 1–4 (2017)
4. Oskouei, R.J., Varzeghani, H.N., Samadyar, Z.: Intelligent agents: a comprehensive survey. Int. J. Electron. Commun. Comput. Eng. **5**(4), 790–798 (2014)
5. Siddike, M., Kalam, A., Spohrer, J., Demirkan, H., Kohda, Y.: People's interactions with cognitive assistants for enhanced performances. In: Proceedings of the 51st Hawaii International Conference on System Sciences, pp. 1642–1648, January 2018
6. Kim, Y.B., Rochette, A., Sarikaya, R.: Natural language model re-usability for scaling to different domains. In: Proceedings of the 2016 Conference on Empirical Methods in Natural Language Processing, pp. 2071–2076 (2016)
7. Reis, A., Paulino, D., Paredes, H., Barroso, J.: Using intelligent personal assistants to strengthen the elderlies' social bonds. In: International Conference on Universal Access in Human-Computer Interaction, pp. 593–602. Springer, Cham (2017)
8. Ponciano, R., Pais, S., Casal, J.: Using accuracy analysis to find the best classifier for intelligent personal assistants. Procedia Comput. Sci. **52**, 310–317 (2015)
9. Han, S., Yang, H.: Understanding adoption of intelligent personal assistants: a parasocial relationship perspective. Ind. Manag. Data Syst. **118**(3), 618–636 (2018)
10. Siddike, M., Kalam, A., Kohda, Y.: Towards a framework of trust determinants in people and cognitive assistants interactions. In: Proceedings of the 51st Hawaii International Conference on System Sciences, pp. 5394–5401 (2018)
11. Fishbein, M., Ajzen, I.: Belief, Attitude, Intention and Behavior: An Introduction to Theory and Research. Addison-Wesley, Reading (1975)
12. Ajzen, I.: The theory of planned behavior. Organ. Behav. Hum. Decis. Process. **50**(2), 179–211 (1991)
13. Davis, F.D.: Perceived usefulness, perceived ease of use, and user acceptance of information technology. MIS Q. **13**(3), 319–340 (1989)
14. Venkatesh, V., Bala, H.: Technology acceptance model 3 and a research agenda on interventions. Decis. Sci. **39**(2), 273–315 (2008)
15. Venkatesh, V., Morris, M.G., Davis, G.B., Davis, F.D.: User acceptance of information technology: toward a unified view. MIS Q. **27**(3), 425–478 (2003)

16. Venkatesh, V., Thong, J.Y., Xu, X.: Consumer acceptance and use of information technology: extending the unified theory of acceptance and use of technology. MIS Q. **36**, 157–178 (2012)
17. Oliver, R.L.: A cognitive model for the antecedents and consequences of satisfaction. J. Mark. Res. **17**, 460–469 (1980)
18. Orehovački, T., Granić, A., Kermek, D.: Evaluating the perceived and estimated quality in use of Web 2.0 applications. J. Syst. Softw. **86**(12), 3039–3059 (2013)
19. Orehovački, T., Babić, S.: Identifying the relevance of quality dimensions contributing to universal access of social web applications for collaborative writing on mobile devices: an empirical study. In: Universal Access in the Information Society (2017). https://doi.org/10.1007/s10209-017-0555-7
20. Kiseleva, J., Williams, K., Jiang, J., Hassan Awadallah, A., Crook, A.C., Zitouni, I., Anastasakos, T.: Understanding user satisfaction with intelligent assistants. In: Proceedings of the 2016 ACM on Conference on Human Information Interaction and Retrieval, pp. 121–130. ACM (2016)
21. Faqih, K.M., Jaradat, M.I.R.M.: Assessing the moderating effect of gender differences and individualism-collectivism at individual-level on the adoption of mobile commerce technology: TAM3 perspective. J. Retail. Consum. Serv. **22**, 37–52 (2015)
22. DeLone, W.H., McLean, E.R.: The DeLone and McLean model of information systems success: a ten-year update. J. Manag. Inf. Syst. **19**(4), 9–30 (2003)
23. Pai, F.Y., Huang, K.I.: Applying the technology acceptance model to the introduction of healthcare information systems. Technol. Forecast. Soc. Chang. **78**(4), 650–660 (2011)
24. Davis, F.D., Bagozzi, R.P., Warshaw, P.R.: User acceptance of computer technology: a comparison of two theoretical models. Manag. Sci. **35**, 982–1003 (1989)
25. Taylor, S., Todd, P.A.: Understanding information technology usage: a test of competing models. Inf. Syst. Res. **6**(4), 144–176 (1995)
26. Amoroso, D., Lim, R.: Exploring the personal innovativeness construct: the roles of ease of use, satisfaction and attitudes. Asian Pacific J. Inf. Syst. **25**(4), 662–685 (2016)
27. Orehovački, T., Babić, S.: Predicting students' continuance intention related to the use of collaborative Web 2.0 applications. In: 23rd International Conference on Information Systems Development, pp. 112–123 (2014)
28. Casaló, L., Flavián, C., Guinalíu, M.: The role of perceived usability, reputation, satisfaction and consumer familiarity on the website loyalty formation process. Comput. Hum. Behav. **24**(2), 325–345 (2008)
29. Chou, S.-W., Min, H.-T., Chang, Y.-C., Lin, C.-T.: Understanding continuance intention of knowledge creation using extended expectation–confirmation theory: an empirical study of Taiwan and China online communities. Behav. Inf. Technol. **29**(6), 557–570 (2010)
30. Ringle, C.M., Wende, S., Becker, J.-M.: SmartPLS 3. Boenningstedt: SmartPLS GmbH (2015). http://www.smartpls.com
31. Hair, J.F., Sarstedt, M., Ringle, C.M., Mena, J.A.: An assessment of the use of partial least squares structural equation modeling in marketing research. J. Acad. Mark. Sci. **40**(3), 414–433 (2012)
32. Hair, J.F., Hult, G.T.M., Ringle, C.M., Sarstedt, M.: A Primer on Partial Least Squares Structural Equation Modeling (PLS-SEM). Sage Publications Inc., Los Angeles (2017)

Research on Comparison Experiment of Humanized Interface Design of Smart TV Based on User Experience

Na Lin[1,2], Haimei Wu[1,2], Huimin Hu[1,2(✉)], and Wei Li[1,2]

[1] AQSIQ Key Laboratory of Human Factors and Ergonomics (CNIS),
No. 36 Yongan Road, Changping District, Beijing, China
{linna,wuhm,huhm,liwei}@cnis.gov.cn
[2] Human Factor and Ergonomics Laboratory, China National Institute of
Standardization, Beijing, China

Abstract. 5 Smart TV products (represented by the letter A–E) were selected in this study to compare the interface design. Twelve volunteers (6 males and 6 females) from 18 to 40 years old (M = 27.92, SD = 5.85) were recruited to complete the task of "playing an online movie" in a simulated living room environment. The evaluation results between 5 Smart TV products that based on user experience were significant differences, which might be the result of the distinction of humanized design. The layout of homepages, terminologies of icons to enter the subordinate pages, menu formats and the entrance position of the subordinate pages all affected the results.

Keywords: Humanized interface design · User experience · Smart TV

1 Introduction

By integrating the internet into television sets, Smart TVs allow consumers to use on-demand streaming media services, listen to radio, access interactive media, use social networks, and download applications. Nowadays, Smart TVs not only offer access to the internet and legacy web services, but also provide content services that are immediately coupled to broadcast content that is rendered on the terminal device. To provide more and better services, a Smart TV must have a menu system and user interface that can be navigated to complete a task. Therefore, the design of Smart TV user interfaces should focus on usability [1], which refers to "the extent to which a product can be used by specified users to achieve specified goals with effectiveness, efficiency and satisfaction in a specified context of use." [2] As several researched have noted, an intuitive and easily navigated human-computer interaction and user interface are critical to a good user experience of a Smart TV [3]. However, usability issues

This work is supported by 2017 National Quality Infrastructure (2017NQI) project (2017YFF0206603 and 2017YFF0206506), AQSIQ science and technology planning project (2016QK177), and China National Institute of Standardization through the "special funds for the basic R&D undertakings by welfare research institutions" (522016Y-4488).

© Springer International Publishing AG, part of Springer Nature 2019
I. L. Nunes (Ed.): AHFE 2018, AISC 781, pp. 88–97, 2019.
https://doi.org/10.1007/978-3-319-94334-3_11

commonly arise concerning interaction and user interface navigation that has been poorly designed, typically because of the organization, placement, visual de-sign, or terminology involved.

At present, the interface design of smart TV draw lessons from the design of web pages and human–computer interfaces. Because of the development of the Internet industry, there is great effort in the academic field of human–computer interfaces design.

In 2003, The Eyetrack III [4, 5] research released by The Poynter Institute, the Estlow Center for Journalism & New Media tested participants' eye movements across several news homepage designs. The researchers noticed a common pattern: The eyes most often fixated first in the upper left of the page, then hovered in that area before going left to right. Only after perusing the top portion of the page for some time did their eyes explore further down the page (Fig. 1).

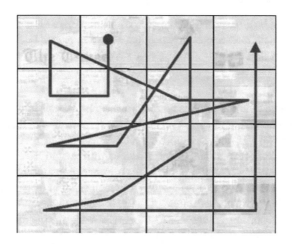

Fig. 1. The most common eye-movement pattern across multiple homepage designs.

In the view of Fig. 2, the experimenters summarized the importance of the zones of the news homepage designs.

Analogously, Eyetracking research showed that people scan webpages and phone screens in the F-shaped scanning pattern [6]. In the F-shaped scanning pattern is characterized by many fixations concentrated at the top and the left side of the page. Users first read in a horizontal movement, across the upper part of the content area. This initial element formed the F's top bar. Next, users moved down the page a bit and then read across in a second horizontal movement that typically covered a shorter area than the previous movement. This additional element formed the F's lower bar. Finally, users scanned the content's left side in a vertical movement. Sometimes this was a slow and systematic scan that appeared as a solid stripe on an eyetracking heatmap. Other times users moved faster, creating a spottier heatmap. This last element formed the F's stem.

Fig. 2. The zones of importance formulated from the Eyetrack data. The dark grey rectangle of Priority 1 represented the most important region; the white rectangle of Priority 2 represented the sub important region; the light grey rectangle of Priority 3 represented the most unimportant region.

In addition, time of completion is crucial in research; the speed with which a person identify the user interface contents is related to the ease with which they perceived the corresponding visual aid, and therefore to affordance. When a choice was required during a task in addition to a movement, a choice time would be affected by both the number of icons and their configuration. Design implications were that the number of icons should be minimized in menus and that the icons should be arranged in a manner reflective of the shape of the useful field of view. Backs et al. [7] reported that finding a target object and reporting an associated numerical value in menus was significantly faster in vertical than in horizontal configurations. In the work of Deininger [8], the task performance differences were not statistically meaningful among the best configurations including horizontal, vertical, and square arrangements. Molina-Rueda et al. [9] conducted a study that compared the use of horizontal and vertical surfaces for which a group of people was selected to interact with the same application and the same surface but with differing orientations. But Naäsänen et al. [10] found that far more time was needed to select elements from sets oriented vertically,mean operation times were decidedly shortest for square shape panels. Another study [11] revealed that the dialogue windows, especially with a larger number of icons, should be built in compact, perhaps square, configurations. If it was not possible to use the square icon configuration, it was desirable to use the horizontal configuration, which could be utilized more efficiently than the vertical configuration. In addition, for design of interfaces, the implication is that frequently used menus or icon arrays should be permanently visible to minimize performance time.

We can argue about the validity of these results when applied to smart TV interfaces, but it is an important step towards a more objective assessment of this issue.

The purpose of this study was to investigate the effects of graphical interface characteristics such as design blocks and elements on interaction task efficiency.

Efficiency can be described as the speed and accuracy with which users can complete tasks for which they use the user interfaces. Therefore, we analyzed task completion times which represented interaction task efficiency and calculated task accomplishment ratios. In addition, the subjective scores were conducted to compare the satisfaction of the subjects in the process of completing specified goals.

2 Method

2.1 Experiment Design

5 Smart TV products were selected in this study to compare the interface design. Volunteers were recruited to complete the task of "playing an online movie" which different from the functions of traditional TV. Experiments were conducted in a quiet and bright environment. 3 indices were recorded to compare the advantages and dis-advantages of 5 Smart TV products.

2.2 Apparatus

Put Smart TV products (represented by the letter A–E, see Table 1) in the room which simulated living room environment. The environmental illuminance was strictly controlled between 195lx to 214lx [12, 13]. The distance between the eyes of the observer and the screen of TV products was fixed (about 2.5 m).

Table 1. Parameters of 5 smart TV products

Symbol of product	Resolution	Screen size
A	3840 * 2160·pix	50·in.
B	3840 * 2160·pix	49·in.
C	3840 * 2160·pix	50·in.
D	3840 * 2160·pix	50·in.
E	3840 * 2160·pix	49·in.

2.3 Participants

Twelve ordinary adults from 18 to 40 years old (6 males and 6 females, mean age = 27.92, standard deviation of age = 5.85) were recruited and paid to participate in the experiment. All had normal or corrected-to-normal visual acuities and healthy physical conditions, without ophthalmic diseases. They did not have any history of neurological and mental diseases.

2.4 Procedure

Participants were first presented with a set of instructions to understand the procedures and the purpose of this experiment before formal tests. The instructions also explained that participants were allowed to ask help form experimenter or refer to the product manual when they are not able to complete the tasks independently. Then participants had to be familiar with the buttons on the telecontroller until the volunteers stated that they were ready to do the real tasks.

In the formal test, volunteers were invited to complete the task of "playing an online movie" through the interfaces of 5 smart TV respectively. They need to enter the subordinate category of movie from home page, then choose and play the third Chinese films in the subcategory named Chinese films. 3 indices including interaction task efficiency (task completion time), task accomplishment ratio and the subjective satisfaction scores (from 1 to 5, where 1 means very dissatisfied and 5 means very satisfied) of interfaces design were recorded. We only calculated the number in one situation that the subject accomplished the task independently when we count the completion rate. Some subject asked help from experimenter or refer to the product manual in the test, this kind of situation was estimated as "unable to accomplish the task independently". In addition, we also asked the subjects to give an oral statement of dissatisfactions and suggestions of humanized interface design.

2.5 Data Analysis

According to the experimental design, the differences among the task completion times and subjective satisfaction scores of 5 Smart TV were analyzed by IBM SPSS 20 Statistics software (IBM-SPSS Inc. Chicago, IL). The repeated-measures analysis of variance (MANOVA) was applied to analysis these two indicators, with a post hoc test of LSD. Greenhouse–Geisser correction was applied to p values associated with multiple df repeated measures comparisons where appropriate on the index of task completion times.

3   Results

3.1   Repeated-Measure ANOVA of Task Completion Times and Subjective Satisfaction Scores

Repeated-measure ANOVA was applied to task completion times and subjective satisfaction scores of 5 Smart TV products.

With regard to the task completion times, a repeated-measure ANOVA was applied to the task completion times of 5 Smart TV products, a significant main effect $(F(2.02,22.26) = 8.41, p < 0.001)$ was found. The post hoc test showed that the task completion time of product B was obviously shorter than others $(p < .05)$, that of product D was obviously longer than others $(p < .05)$, that of products A C and E were obviously longer than B but shorter than D $(ps < .05)$.

With regard to the subjective satisfaction scores, a repeated-measure ANOVA was also applied to the subjective satisfaction scores of the different products, a significant main effect $(F(4,44) = 7.79, p < 0.001)$ was found. The post hoc test revealed that the subjective satisfaction score of product B was obviously higher than the rest of the products $(p < .05)$, that of product D was obviously lower than others $(p < .05)$, that of products A C and E were obviously lower than B but higher than D $(ps < .05)$.

Basic descriptive statistical parameters including the means and Standard errors which displayed in parentheses regarding obtained task completion times and subjective satisfaction scores of 5 Smart TV products are presented in Table 2.

Table 2. The mean task completion time and mean satisfaction score of 5 Smart TV products (Standard errors are given in parentheses)

Symbol of product	Mean task completion time	Mean subjective satisfaction score
A	124.58 (20.94)	4.20 (0.24)
B	52.08 (5.92)	4.70 (0.12)
C	90.42 (14.73)	4.03 (0.19)
D	210.25 (38.93)	3.44 (0.17)
E	131.50 (35.08)	4.06 (0.17)

3.2 Task Accomplishment Ratio Statistics

We calculated the number in the situation that the subjects accomplished the task independently when we count the completion rate of each product. The results are summarized in Table 3.

Table 3. Task accomplishment ratio of the five products

Symbol of product	A	B	C	D	E
E task accomplishment ratio	75.00%	100.00%	75.00%	50.00%	66.67%

4 Discussion

Effectiveness is the driving force behind successful task completion and helps users to complete their goals. The effectiveness of these Smart TV user interfaces is determined by whether users can locate and use the navigation option to take them to the expected location. The evaluation results between 5 Smart TV products that based on user experience were significant differences, which might be the result of the distinction of humanized design [14, 15].

The statistical analysis of task completion times (interaction task efficiency) and subjective satisfaction scores revealed consistent patterns. The results demonstrate that the participants could use interface of product B more efficiently than others, and all subjects completed the task successfully. At the same time, the participants were more satisfied with the interface design of product B than others.

This result might be related to that the icon to enter the subordinate page occupied a large area and located in sub important region [5]. Naäsänen et al. [10] observed the decrease in search time with the increasing character sizes. Michalski's [16] findings are generally in agreement with results obtained by Naäsänen et al. Therefore, although the entrance position (grey rectangle) of the subordinate page was not in the most important area, the size of the icon offset the disadvantage of the location.

The terminologie of icon to enter the subordinate page accorded with the cognitive habits of the subjects was another advantage. In addition, product B used vertical segmentation homepage design [7]. The contents of homepage were divided into three parts in the lateral direction and were arranged in the form of mosaics (Fig. 3).

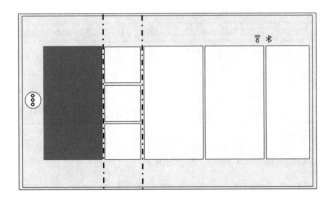

Fig. 3. The homepage wireframe of product B.

The efficiency and satisfaction indices of products A, C and E were higher than D and lower than B, but no significant difference between the three. About 67% to 75% subjects completed the task successfully on the three products. The contents were functionally grouped on the homepages of products A, C and E, and the structure of menu navigation was conducive to visual search with its conciseness. The orderliness of menu navigation balanced the disadvantage of the entrance position(grey rectangle or icons) of the subordinate pages which located in sub important region. It should be noted that some participants complained about the terminologies of icons to enter the subordinate pages. They claimed that the terminologies were not intuitive enough, which reduced efficiency and satisfaction. Some researchers put forward an assumption that users of any new interface need to quickly gain an understanding of which elements on the screen can be used. Users frequently take only a few seconds to familiarize themselves with all of the elements on the page and then establish a mental plan of the interface in a very short time. Therefore, design blocks or elements that are the most visually prominent attract the most attention and will help to shape a user's perception of the interface [17, 18]. The results reflected that the terminologies of icons needed to be improved. In addition, products A, C and E all used horizontal segmentation homepage design [12] (Fig. 4), and the contents were divided into two parts in the longitudinal direction. On the homepages of products A and E, the contents of the upper part were arranged in the form of matrix, and the icons were located at the lower part in the form of horizontal arrangement. On the homepage of product C, the options that adopted rectangles of color were functionally grouped and located at upper part and the left of the lower part. The remaining options that located at the right of the lower part were displayed by vivid photos.

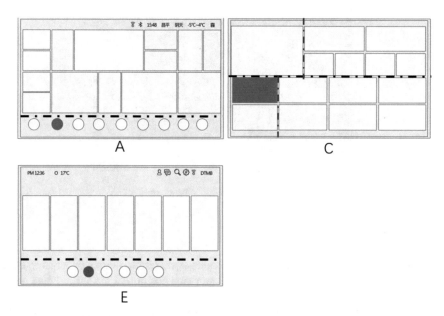

Fig. 4. The homepages wireframes of products A, C and E.

The efficiency and satisfaction indices of product D were lower than other products. Only 50% subjects completed the task successfully when they use interface of product D. In the Eyetrack III research [5], the result displayed that navigation placed at the top of a homepage performed best. Goldberg and Kotval [4, 19, 20] compared two sets of 11 computer buttons used commonly in graphical programs. The first collection was functionally grouped, while in the second one icons were placed randomly. Researchers employed eyeball movement measures to analyze data gathered during experiments made on 12 subjects and proved that the functionally grouped set of icons was operated more efficiently. By contrast, the menu on the homepage of product D was tiled layout at the bottom of the screen (Fig. 5) and the icons of the menu were placed randomly, which caused a lot of complaints, such as the sequence of icons was lack of logic and the location was not consistent with browsing habits. Therefore, the participants needed longer time and made more mistakes when they made their choices even in a small number of options. In addition, due to the terminology did not conform to the cognitive habits of the subjects, participants could hardly build up connection between the purpose of the task and the terminology of icon, which caused the result that only 50% subjects completed the task successfully.

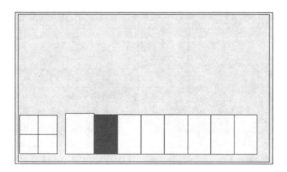

Fig. 5. The homepage wireframe of product D.

5 Conclusion

In summary, we believe that the layout of homepages, terminologies of icons to enter the subordinate pages, menu formats and the entrance position of the subordinate pages all affected the results.

In consideration of configuration of homepages design, there is great effort in the academic field. Considering the contradiction and limitation of these researches, the results should be verified strictly before been translated into new design.

In the field of design, people are encouraged to create their own solutions to problems, but appropriate methods are highly recommended to be considered before any design is conducted. A design process is implemented, should base on the output of design thinking, and the results be evaluated and analyzed also, interdisciplinary collaboration among people from various fields and backgrounds should be engaged to ensure that the proper design approaches were taken. Therefore, a scheme that can be widely used should be developed in the future,which comprehends incorporates users' experiences and allows problems with the user interface to be identified and the user interface to be assessed.

Acknowledgements. We would like to thank the participants who took part in the experiment. We gratefully acknowledge the financial support from 2017 National Quality Infrastructure (2017NQI) project (2017YFF0206603 and 2017YFF0206506), AQSIQ science and technology planning project (2016QK177), and China National Institute of Standardization through the "special funds for the basic R&D undertakings by welfare research institutions" (522016Y-4488).

References

1. Zhang, X.: The Analysis of Usability in Intereaction Design of Smart TV. Shandong Polytechnic University (2015)
2. Schall, A.: Eye tracking insights into effective navigation design. In: Design, User Experience, and Usability. Theories, Methods, and Tools for Designing the User Experience, pp. 363–370. Springer (2014)

3. Blackler, A.L., Popovic, V., Mahar, D.P.: Applying and testing design for intuitive interaction. Int. J. Des. Sci. Technol. **20**(1), 7–26 (2014)
4. Zhang, R.: The studies about interface design of smart TV based on user experience. Hefei University of Technology (2015)
5. Outing, S., Ruel, L.: The best of eyetrack III: what we saw when we looked through their eyes. Poynter Institute (2004). Accessed 20 June 2006
6. Nielsen, J.: F-Shaped Pattern For Reading Web Content. http://www.useit.com/alertbox/reading_pattern.html
7. Backs, R.W., Walrath, L.C., Hancock, G.A.: Comparison of horizontal and vertical menu formats. In: Proceedings of the Human Factors & Ergonomics Society Annual Meeting, vol. 31, pp. 715–717 (1987)
8. Deininger, R.L.: Human factors engineering studies of the design and use of pushbutton telephone sets. Bell Labs Tech. J. **39**(4), 995–1012 (1960)
9. Molina-Rueda, A., Magallanes, Y., Sánchez, J.A., Enriquez, D.F.: Using heat maps for studying user preferences in vertical and horizontal multi-touch surfaces. In: International Conference on Electronics, Communications and Computing (2013)
10. Näsänen, R., Karlsson, J., Ojanpää, H.: Display quality and the speed of visual letter search. Displays **22**(4), 107–113 (2001)
11. Grobelny, J., Karwowski, W., Drury, C.: Usability of graphical icons in the design of human-computer interfaces. Int. J. Hum. Comput. Interact. **18**(2), 167–182 (2005)
12. Ye, C., Liu, Z., Zhang, J.Q., et al.: Effect of environmental illumination on optimal value of mobile phone brightness. Chin. J. Liq. Cryst. Displays **29**(6), 1042–1049 (2014)
13. Li, H.T., Zhang, Y.X., Xu, W.D., et al.: Study on the optimal parameters of mobile phone screen brightness of difference environment illumination. Psychol. Sci. **36**(5), 1110–1116 (2013)
14. Nielsen, J.: Usability Engineering. Morgan Kaufmann, San Francisco (1993)
15. Shneiderman, B.: Designing the User Interface: Strategies for Effective Human-Computer Interaction, 5th edn. Person, London (2009)
16. Michalski, R., Grobelny, J., Karwowski, W.: The effects of graphical interface design characteristics on human–computer interaction task efficiency. Int. J. Ind. Ergon. **36**(11), 959–977 (2006)
17. Zhang, Q., Cui, L.X.: Reach of concreteness effects in semantic processing. J. Beijing Normal Univ. (Soc. Sci.) **4**, 28–34 (2002)
18. Wang, S.M.: Integrating service design and eye tracking insight for designing smart tv user interfaces. Int. J. Adv. Comput. Sci. Appl. **6**, 7 (2015)
19. Gong, Y.: Research on Graphic Symbols Design Recognition. Zhe Jang University (2012)
20. Goldberg, J.H., Kotval, X.P.: Computer interface evaluation using eye movements: methods and constructs. Int. J. Ind. Ergon. **24**(6), 631–645 (1999)

Human Factors and System Interactions in Complex Systems

Moving Forward with Autonomous Systems: Ethical Dilemmas

Aysen K. Taylor$^{(\boxtimes)}$ and Sarah Bouazzaoui

Engineering Management and Systems Engineering,
Old Dominion University, Norfolk, VA 23529, USA
aysenafk@cox.net

Abstract. Automation has improved transportation systems in various domains over the last several decades. Increasing autonomy in these systems has gradually reduced the role of the human operator to that of system monitor with the ultimate goal of eliminating the human from the control system entirely. Commercial aviation has benefited from automation, but it operates with the support of a broad infrastructure of safety when compared to vehicular road traffic. While not designed to operate in a fully autonomous mode, the computer, sensor, and software technology developed for aircraft are being applied to self-driving cars with the expectation that driving will also see significant improvements in accident rates and efficiency through the elimination of human error and negligence [1]. A sophisticated combination of hardware sensors and computer software analyzes the environment and controls the speed and direction of the car without input from its human occupants and their opaque interactions increase the complexity of the system. This approach has potential benefits but also potential problems. Autonomous vehicles will present ethical challenges while being developed and after deployment. The purpose of this paper is to consider the many ethical implications involved with the implementation and oversight of autonomous vehicle (AV) technology. This paper examines primary ethical dilemmas present in the use of autonomous cars including liability and moral agency.

Keywords: Autonomous · Cars · Ethics

1 Introduction

Cars with some degree of autonomous driving ability have been in development since the 1920's [2]. None of these were adopted for general use and reflect various attempts to solve the control problem with the technology of the day. Since none became operational, the complex regulatory, ethical, and philosophical issues were not sorted out. The laudable benefits of reducing traffic deaths and injuries by removing human error and negligence were extolled with little mention of the issues of liability and how a computer controlled car should make decisions which imply inevitable harm to one or more human beings in or near an AV.

Current AV development relies on an array of sensors and mapping data to guide the vehicle through a mix of AV and human controlled vehicles. Typical equipment

© Springer International Publishing AG, part of Springer Nature 2019
I. L. Nunes (Ed.): AHFE 2018, AISC 781, pp. 101–108, 2019.
https://doi.org/10.1007/978-3-319-94334-3_12

used on an AV includes a GPS receiver, a Light Detection and Ranging (LIDAR) sensor, video cameras, and radars. GPS information is compared to highly detailed custom maps to determine the best route to take. Radars on the front and back of the car keep track of other vehicles. LIDAR rotates 360° to track surrounding objects. The video camera is designed to read road signs and traffic lights. This data is fed into a computer that processes the information using algorithms to make decisions about steering, braking, and acceleration. Unlike the varied human responses seen when they drive a car, ethical choices in AVs are made in advance of the accident. The ethical dilemma of AVs most often cited involves an AV encountering a single or multiple pedestrians in the road and a pedestrian on the side of the road. It can continue its path and hit multiple pedestrians, swerve left and hit the single pedestrian or swerve right and hit a brick wall. Each choice implies harm to one or more human beings. This is a modern variation of the classic trolley car dilemma where the trolley driver sees 5 people on his track and 1 person on an adjoining track. He can continue his path and kill 5 people or switch to the alternate track and kill 1. Some have suggested this scenario is too unlikely to allow it guide the development of AVs. They claim the trolley car scenario is rare and AVs will make them rarer still. Many are concerned that being preoccupied with this ethical dilemma could delay the introduction of AVs and result in many avoidable human traffic deaths as a result [3]. Whatever the opinion on the validity of this ethical dilemma, the makers of AVs will have to program algorithms to make such ethical choices. These ethical decisions can be rooted in philosophical doctrine. According to Jeremy Bentham's philosophy, the AV should follow utilitarian ethics, i.e. should take the action that will minimize total harm even if that decision kills a bystander or a passenger. When Immanuel Kant's philosophy is applied the action taken by an AV should follow duty bound principals. This doctrine means the AV should not take an action that explicitly harms a human and the car should follow its course, regardless of what harm results [4]. How does human instinct compare to programed forethought regarding who will be harmed in the event of an accident? Drivers who must make split second decisions are not held to the same standard as programmed decision-making that may have been done years before its logic results in a fatality. How will the public receive a new paradigm that takes choice away from the human being and replaces that with an experience similar to airline passengers in that they have no direct control of the transport they are riding in? Surveys show people want others to ride in AVs using utilitarian algorithms but they would prefer to buy AVs that protect their passengers before considering the welfare of other people on the road [5].

Cars in use today that have driver assist features such as Tesla's "autopilot" mode provide a preview of some of the ethical questions raised by the technology and how the industry responds to these issues. When a Tesla owner died in Florida while using autopilot, the company reiterated that their system requires the human driver to remain vigilant while in the autopilot mode. This reaction indicates that Tesla has a poor understanding of the human response when tasks they once performed are replaced in part by an automated machine. Humans tend to over rely on highly automated systems [6, 7]. German authorities considered this when they banned Tesla from referring to their driver assist software by the name "autopilot" [8].

The ultimate goal of self-driving car projects is to remove the human from the control loop entirely. This eliminates issues of human-machine interaction but also denies industry its typical response to blame the human component of the system when a partially automated machine is involved in an accident [9].

Figure 1 illustrates the trend in US traffic fatality rates since 1972 and the path to fully autonomous vehicles.

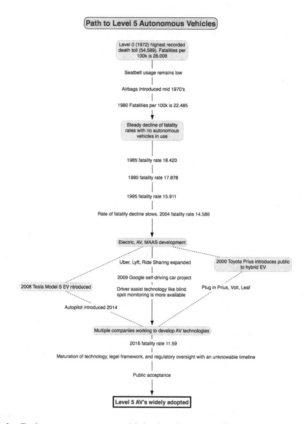

Fig. 1. Path to autonomous vehicle development (Source: Aysen Taylor)

Regulatory fragmentation due to individual state insurance requirements and separate state inspection standards will complicate the implementation of fully autonomous cars. Researchers have proposed that a federal authority be established to have a national regulatory framework to overcome the problems with state specific rules and standards.

Various definitions have been published over the last 20 years regarding different levels of automated controls in vehicles. The National Highway Transportation Safety Administration has adopted the Society of Automotive Engineers (SAE) International definitions for levels of automation as defined in Table 1 below.

Table 1. Various automation levels for automobiles (Source: NHTSA 2016)

SAE Level 0	The human driver does everything
SAE Level 1	An automated system on the vehicle can sometimes assist the human driver conduct some parts of the driving task
SAE Level 2	An automated system on the vehicle can actually conduct some parts of the driving task, while the human continues to monitor the driving environment and performs the rest of the driving task
SAE Level 3	An automated system can both actually conduct some parts of the driving task and monitor the driving environment in some instances, but the human driver must be ready to take back control when the automated system requests
SAE Level 4	An automated system can conduct the driving task and monitor the driving environment, and the human need not take back control, but the automated system can operate only in certain environments and under certain conditions
SAE Level 5	The automated system can perform all driving tasks, under all conditions that a human driver could perform them

Level 5 AVs are the end goal of all AV development projects but in the interim some makers such as Tesla are deploying intermediate level vehicles that provide assistance to a human driver (Level 2) while others such as Google's Waymo project are focusing solely on Level 5 technologies. In either case, the deployed technology will co-exist with Level 0 and 1 vehicles for decades.

Developers of this technology promote vehicle to vehicle and vehicle to environment technologies that could provide many benefits but also expand concerns relating to privacy and cyber security.

The discussion of a transition from human driven vehicles to AVs should include the needs of the many people who cannot afford or want to use a car. Some prefer bicycles or live in economic zones that are underserved by public transit or ride-sharing services like Zip Car. Baltimore, Maryland has developed a plan called Baltimore Complete Streets that seeks to consider the needs of this often overlooked part of the transiting public [10].

The potential benefits of AVs include a large reduction in accident rates, reduced fuel consumption, and more intelligent use of parking facilities. It could also extend mobility to the elderly and provide unprecedented mobility to the disabled. These gains come with significant initial economic cost and complexity. The need to support both traditional vehicles and AV traffic for decades and the historical precedent for a reactive approach from industry and regulations means the full scope of the task of advancing AV technology cannot be known. A systems approach is needed to avoid creating deeper problems later in the process that could be difficult to resolve. Field testing is being conducted and more graduated rollouts of AVs will be needed but this comes with its own hazards and ethical challenges. When AVs are tested or fully deployed, the other traffic is participating in a grand experiment without having provided their explicit consent. An Institutional Review Board (IRB) must approve the must mundane experiments involving human subjects but with AVs the general public is automatically opted in for the developmental testing of the greatest change in transportation since the invention of the car itself.

2 Liability and Regulatory Frameworks

When an AV is involved in an accident that results in property damage, injury, or death, who will be held accountable? Each state establishes laws that guide the process of redress to anyone who alleges harm stemming from a vehicular accident. New Hampshire does not require owners to have insurance but requires them to have sufficient resources to pay claims against them. It would be difficult for AVs to be deployed in the current framework where each state regulates car insurance within its boundaries [11].

It is unlikely that AV manufactures would accept the full burden of liability for any harm that resulted from the use of their products. Some historical precedent may provide insight into how this liability issue might be resolved. Congress passed the 1986 Vaccine Injury Compensation program (VICP) in 1986 in response to vaccine manufacturers concerns about lawsuits filed against them when their products injured or killed someone. Pharmaceutical companies threatened to stop making vaccines if they were held liable for injuries vaccines caused. This new law barred anyone who alleges harm after receiving vaccines from suing in a state or federal court and directs all legal actions to be heard in the U.S. Court of Federal Claims, sitting without a jury. Vaccine manufacturers cannot be held liable for any harm their products cause. A tax on each vaccine is used to fund compensation for those successful in wining their cases in this special court. Will an AV court be proposed using the vaccine court as a model?

3 Reducing Traffic Fatalities

All developers of AVs emphasize the reduction of traffic fatalities first and foremost when promoting their projects. The potential benefit of drastic reductions in vehicle deaths is laudable but it is frequently forgotten that simpler and cheaper changes that can be implemented immediately with available technology are not being carried out. The nonprofit organization founded by Congress called the National Safety Council (NSC) has recommended the following changes to reduce highway fatalities [12].

- Mandate ignition interlocks for convicted drunk drivers
- Install automated enforcement techniques to reduce speeders
- Extend laws banning all cell phone use and upgrade enforcement from secondary to primary
- Upgrade seat belt laws from secondary to primary and extend laws to every passenger in every seating position is all types of vehicles
- Adopt a three tiered licensing system for all new drivers under 21
- Standardize and accelerate adoption safety technologies such as blind spot monitoring, automatic emergency braking, land departure warning, and adaptive headlights
- Pass or reinstate motorcycle helmet laws
- Adopt comprehensive programs for pedestrian safety.

The cost of the NSC proposals is much less than the cost of implementing AVs and could save many lives while AV technology is developed. What are the ethical implications of not making these meaningful safety changes to the current transportation system while AVs are being vigorously promoted for their potential to save lives? If there is a lack of political will to make many of these changes, what possibility exists that the paradigm shift needed for wide spread adoption of AVs can be successful?

A study by Stimpson et al. [13] recommended incentivizing use of mass transit where available since their data analysis showed that increasing the share of mass transit miles traveled compared to vehicle miles traveled could reduce traffic fatality rates. While AVs show great potential to make transportation safer, more public discussion of how to better utilize the existing transportation system to reduce deaths now should be encouraged. Government authorities should require more robust validations of the safety claims AVs developers are promising.

4 Concerns about Data Privacy, Cybersecurity, and Technological Limitations

The 9/11 attacks have been used as a justification to surveil all manner of digital communications and connected AVs will generate more data about our daily lives that will likely be gathered for commercial, law enforcement, or national security purposes. How will individual privacy rights be balanced against these objectives?

The German government has published ethics reports about AVs that state drivers must always make the decision as to how or if their data is forwarded to others and how it is used generally [14]. AVs are expected to utilize vehicle to vehicle and vehicle to infrastructure communication systems and these could be vulnerable to malicious attacks from individuals, terrorists, or state actors. This may pose the biggest challenge to the implementation of AVs as cybersecurity threats have only grown as more infrastructure components are taken online. Modern electrical grids and water systems are subject to hacking attacks that would have been impossible 50 years ago. The era when you needed physical access to infrastructure to break them is over. Machine learning is expected to help AVs improve their performance but at the cost of possibly not knowing how it is making decisions. The automobile business is highly competitive and the sharing of technical solutions amongst manufactures is uncommon. The pressure to win sales has already resulted in some companies such as Volkswagen hacking their own software to fool regulators when they checked their diesel cars' pollution emissions. This ethical lapse of great proportions does not inspire confidence in the industry that will bring us safe and secure AV solutions.

Elon Musk stated that Tesla cars without LIDAR would someday safely operate in a fully autonomous mode. He believes cameras and computer vision will suffice [15]. Researchers have made small changes to stop signs that look like graffiti or abstract art that have fooled deep neural networks that try to interpret the meaning of the stop sign. Instead of reading it as a STOP sign, it reads this modified sign as SPEED LIMIT 45. The figure below shows a sign that was successful in this type of attack. The deep neural network misread this sign 73% of the time [16]. A human driver would have no trouble understanding the sign but the current technology used in AVs is unreliable in this essential task (Fig. 2).

Fig. 2. Stop sign used to confuse the autonomous system (Source: Evtimov et al. [16])

5 Conclusions

A single occupancy vehicle, whether it has a human driver or is autonomous, is an inefficient means to move people. Pollution, congestions of roadways, and parking challenges will not be helped with AVs alone. Weiland et al. [17] have suggested that AVs will only be successful if they are also electric powered and used in the broader mobility as a service (MAAS) model. This model sees people using a combination of ride sharing, public transport, and AVs to get where they want to go. This requires multiple paradigm shifts to occur simultaneously. Individual car ownership is still the preferred way to operate cars for most people. Public transit options vary a lot from one municipality to another but are not currently available for the majority of people. Where available, they are very expensive to build and have a long return on investment. The ethical decisions AVs must make are hard to gather a consensus for and may require government agency mandates and congressional guidance to establish liability when AVs are involved in accidents. States rights could be tested if automakers claim that conforming to 50 varying rules for licensure and certification is too burdensome. Privacy advocates will have a vast new area of concern as the technology enabling AVs could be used to track our movements in a way that makes many people uncomfortable. The introduction of AVs will cost many professional drivers their jobs, especially truck drivers. Goldman Sachs estimates that 3.1 million truck drivers would lose work if AV trucks replaced trucks driven by humans [18]. This represents 2 percent of current total employment. While the promised benefits of AVs are great, the task to bring them to a reality in meaningful numbers is daunting and some of the toughest questions to solve are not only technical in nature but involve social change and a complete reworking of liability when something bad happens on the road. A national discussion of these social, privacy, legal, and ethical challenges should be encouraged to help society shape the development of this technology in such a way that they can both accept AVs and derive the maximum benefit from them.

References

1. Negroni, C.: Lessons from aviation for Tesla and self-driving cars (2016). https://www.forbes.com/sites/christinenegroni/2016/07/07/danger-lurks-at-intersection-of-human-and-self-driving-car/#1d8f66294fd7
2. Phantom auto will tour city. The Milwaukee Sentinel, p. 14, 8 December 1926. https://news.google.com/newspapers?id=unBQAAAAIBAJ&sjid=QQ8EAAAAIBAJ&pg=7304,3766749
3. Iagnemma, K.: Why we have the ethics of self-driving cars all wrong (2018). https://www.weforum.org/agenda/2018/01/why-we-have-the-ethics-of-self-driving-cars-all-wrong/
4. Rahwan, I.: The Social Dilemma Of Driverless Cars [Video file], 16 November 2016. https://www.youtube.com/watch?v=nhCh1pBsS80
5. Bonnefon, J., Shariff, A., Rahwan, I.: The social dilemma of autonomous vehicles. Science **352**, 1573–1576 (2016)
6. Parasuraman, R., Riley, V.: Humans and automation: use, misuse, disuse, abuse. Hum. Factors **39**(2), 230–253 (1997)
7. Endsley, M.R.: Level of automation: integrating humans and automated systems. In: Proceedings of the human factors and ergonomics society. Ergonomics Society, pp. 200–204 (1997). http://dx.doi.org/10.1177/107118139704100146
8. Bigelow, P.: Don't call it autopilot, German authorities tell Tesla. Car and Driver, 17 October 2016. https://blog.caranddriver.com/dont-call-it-autopilot-german-authorities-tell-tesla/
9. National Tesla crash report (2017). https://static.nhtsa.gov/odi/inv/2016/INCLA-PE16007-7876.PDF
10. Dorsey, R.: A complete streets law for Baltimore: a design solution to a transportation crisis (2017). https://www.baltimorecompletestreets.com/policy-brief/
11. Fagnant, D.J., Kockelman, K.: Preparing a nation for autonomous vehicles: opportunities, barriers and policy recommendations. Transp. Res. Part A: Policy Pract. **77**, 167–181 (2015)
12. Motor vehicle deaths in 2016 estimated to be highest in nine years (2017). http://www.nsc.org/Connect/NSCNewsReleases/Lists/Posts/Post.aspx?ID=180
13. Stimpson, J.P., Wilson, F.A., Araz, O.M., Pagan, J.A.: Share of mass transit miles traveled and reduced motor vehicle fatalities in major cities of the United States. J. Urban Health **91**, 1136–1143 (2014)
14. Ethics commission automated and connected driving (2017). https://www.bmvi.de/SharedDocs/EN/publications/report-ethics-commission.pdf?__blob=publicationFile
15. iGadgetPro. Elon Musk on disadvantages of self driving cars TED 2017 [Video file], 1 May 2017. https://www.youtube.com/watch?v=h_cnISo_Qxk
16. Evtimov, I., Eykholt, K., Fernandes, E., Kohno, T., Li, B., Prakash, A., Rahmati, A., Song, D.: Robust physical-world attacks on deep learning models. Cryptogr. Secur. (2017)
17. Weiland, J., Rucks, G., Walker, J.: How the U.S. transportation system can save $1 trillion, 2 billion barrels of oil, and 1 gigaton of carbon emissions (2015). https://www.rmi.org/news/u-s-transportation-system-can-save-1-trillion-2-billion-barrels-oil-1-gigaton-carbon-emissions-annually/
18. Balakrishnan, A.: Self-driving cars could cost America's professional drivers up to 25,000 jobs a month, Goldman Sachs says (2017). https://www.cnbc.com/2017/05/22/goldman-sachs-analysis-of-autonomous-vehicle-job-loss.html

Signal-Processing Transformation from Smartwatch to Arm Movement Gestures

Franca Rupprecht[1](✉), Bo Heck[1], Bernd Hamann[2], and Achim Ebert[1]

[1] Computer Graphics and HCI, Technische Universität Kaiserslautern,
67663 Kaiserslautern, Germany
{rupprecht,ebert}@cs.uni-kl.de, bheck@rhrk.uni-kl.de
[2] Department of Computer Science,
University of California, Davis, CA 95616, USA
hamann@cs.ucdavis.edu

Abstract. This paper concerns virtual reality (VR) environments and innovative, natural interaction techniques for them. The presented research was driven by the goal to enable users to invoke actions with their body physically, causing the correct action of the VR environment. The paper introduces a system that tracks a user's movements that are recognized as specific gestures. Smartwatches are promising new devices enabling new modes of interaction. They can support natural, hands-free interaction. The presented effort is concerned with the replacement of common touch input gestures with body movement gestures. Missing or insufficiently precise sensor data are a challenge, e.g., gyroscope and magnetometer data. This data is needed, together with acceleration data, to compute orientation and motion of the device. A transformation of recorded smartwatch data to arm movement gestures is introduced, involving data smoothing and gesture state machines.

Keywords: Intuitive and natural interaction · Low budget interaction devices
Mobile devices · Virtual reality · Body movement gestures
Gesture recognition

1 Introduction

Mobile devices are almost ambiguous today and feature a wide range of input and output capabilities like touch screens, cameras, accelerometer, microphones, speakers, near-field communication, Wi-Fi, etc. The usage of smart-devices is easy and intuitive, and they offer a wide range of interaction metaphors, which can lead to a more natural and intuitive interaction as well as a broad array of control elements [1]. Especially the smartwatch as latest technology in that field gives new possibilities of interaction techniques. As the watch is fix on the wrist, the hands are free what leads to a more natural interaction in the meaning of body gestures, also other technology like finger tracking can be combined and new interaction techniques will be enabled. Next to the common touch gestures which are performed very frequently on the smart device's display we developed additional movement gestures which enriches the input capabilities of the smartwatch significant. Virtual Reality (VR) visual interaction

© Springer International Publishing AG, part of Springer Nature 2019
I. L. Nunes (Ed.): AHFE 2018, AISC 781, pp. 109–121, 2019.
https://doi.org/10.1007/978-3-319-94334-3_13

environments make possible the sensation of being physically present in a non-physical world [2]. The value of this experience is a better perception and comprehension of complex data based on simulation and visualization from a near-real-world perspective [3]. A user's sense of immersion, the perception of being physically present in a non-physical world, increases when the used devices are efficient, intuitive, and as "natural" as possible. The most natural and intuitive way to interact with data in a VR environment is to perform the actual real-world interaction [4]. For example, gamers are typically clicking the same mouse button to swing a sword in different directions. However, the natural interaction to swing a sword in a VR application is to actually swing the arm in the physically correct direction as the sword is an extension of the user's arm. Therefore, intuitive and natural interaction techniques for VR applications can be achieved by using the human body itself as an input device [5]. VR devices are usually specialized to support one interaction modality used only in VR environments. Substantial research has been done in this field, yet VR input devices still lack highly desirable intuitive, natural, and multi-modal interaction capabilities, offered at reasonable, low cost. In a preliminary study [6] we figured out that our low budget setup and the implemented in air gestures are comparable to common VR input technology, specifically body tracking enabled by a 3D camera. Thereby, we could state that a combination of smartphone and smartwatch capabilities, outperforming a comparable common VR input device. We have demonstrated the effective use for a simple application. The main advantages of our framework for highly effective and intuitive gesture-based interaction are:

- Location independence
- Simplicity-of-use
- Intuitive usability
- Eyes-free interaction capability

- Support for several different inputs
- High degree of flexibility
- Potential to reduce motion sickness
- Elegant combination with existing input technology

In this paper, we present the improvements of our gesture recognition algorithms and the adaption of enhanced gestures accordingly. For the investigation, the Apple Watch (watchOS2) is used to develop different arm movement gestures to enable new natural interaction mechanisms. Unfortunately, the only sensor data which can be proceed from the Apple Watch are the acceleration data as gyroscope and magnetometer was not accessible to the time of our approach. Those three sensors are used to calculate the orientation and motion dynamics of the device. The challenge is described by missing sensor data, precisely gyroscope and magnetometer data, which need to be compensated in order to calculate orientation and motion dynamics of the device. The aim of this paper is to create a system that is able to recognize arm gestures only using the accelerometer data of the device. This system has to allow a person, skilled in programming, to define their own gestures. Based on six key values and a statement sequence we are able to define precise arm movement gestures, exemplary demonstrated on seven different gestures. More gestures are conceivable and easily adoptable with our approach. Subsequent, we are able to transform the giving signal-processing from the smartwatch into arm movement gestures with the use of smoothing algorithms and gesture state machines which lead to the actual gesture recognition. After giving a short description of the signal-processing model, and the associated existing shortfalls, we will demonstrate the resulting adaption of gestures in detail. Furthermore, an evaluation of the gesture recognizer is conducted and the results are presented.

2   Related Work

Current research covers many aspects of interaction in VRs, being of great interest to our work. Bergé et al. [7] stated that mid-air hand and mid-air phone gestures perform better than touchscreen input implying that users were able to per- form the tasks without training. Tregillus and Folmer [8] affirmed that walking-in-place as a natural and immersive way to navigate in VR potentially reduce VRISE (Virtual reality induced symptoms and effects [9]) but they also address difficulties that come along with the implementation of this interaction technique. Freeman et al. [10] addressed the issue of missing awareness of the physical space when performing in-air gestures with a multi-modal feedback system. In order to overcome the lack of current display touch sensors to equip a user with further input manipulators, Wilkinson et al. utilized wrist-worn motion sensors as additional input devices [11]. Driven by the limited input space of common smart watches, the design of non-touchscreen gestures are examined [12]. Houben et al. prototyped cross-device applications with a focus on smartwatches. In their work, they provided a toolkit to accelerate application development process by using hardware emulations and a UI framework [13]. Similar to this work, several investigations concerning interaction techniques with wrist-worn devices such as smartwatches and fitness trackers have been made. In general, two types of recognizing techniques can be differentiated: (1) machine learning techniques base on a (high) number of training samples from which features are extracted and gestures with the use of probability classifiers identified and (2) simple pattern recognition with predefined features. Mace et al. [14] compared naive Bayesian classification with feature sepa-rability weighting against dynamic time warping. The extremely differing gesture types (circle, figure eight, square, and star) could be recognized with an average accuracy of 97% for the feature separability weighted Bayesian Classifier, and 95% for the dynamic time warping with only five gesture samples. Mänyjärvi et al. [15] presented a hidden markov model in order to define continuous gesture recognition based on user defined gestures for primitive gestures used to remotely control a DVD player. Their model could reach an accuracy value of 90% to 95%. The investigation performed in [16] employs a hidden markov model for user dependent gesture recognition. The models are used for training and recognition of user-chosen gestures performed with a Wii remote controller. Only few gestures are tested, which differ extremely. Shortcomings of high computational power are mentioned. Compared to the Wii remote controller, the smartwatch data does not show comparable high peaks. Thus, the model is not suitable for the underlying kind of data. Methods from machine learning have a high flexibility and find application especially at end user side, where user do not know the data nor are able to identify features. Those techniques can lead to excellent accuracy rates with an exceeding number of training samples. However, classifying the training data and identifying the correct gesture with machine learning techniques are resource-intensive. Considering, the less computation power of a smartwatch and that the end user is not the point of interest in this work, machine learning techniques are not applicable. Work done in [19] presents a frame-based feature extraction stage to accelerometer-based gesture recognition performed with the Wii remote controller. A gesture descriptor combining spectral features and temporal features is presented. However, the recognition starting point is activated with a mouse click and not only due to the recognition. Chu et al. [17] used a fuzzy control technique to classify different gestures based on acceleration data from smartphones. The gestures defined in the study are totally different, it would be interesting if there is still a precision rate of 91% with gestures more similar to each other.

3   Setup

Our approach uses common technologies, at relatively low cost, supporting intuitive, basic interaction techniques already known. A smartphone fixed in an HD viewer serves as fully operational HMD and allows one to experience a virtual environment in 3D space. The smartphone holds the VR application and communicates directly with a smartwatch. Wearing a smartwatch with in-built sensors "moves" the user into the interaction device and leads to a more natural interaction experience. In order to support control capabilities to a great extent, we consider all input capabilities supported by the smartphone and the smartwatch. In addition to touch display and crown, we considered accelerometer, gyroscope and magnetometer, as they are built-in sensors. Our watch setup consists of two components: (1) A smartwatch, the Apple Watch Sport 38 mm Generation 1 and (2) an HMD. The watch's dimensions are 38.6 mm x 33.3 mm x 10.5 mm. Neither watch nor HMD are tethered, and there is no technical limitation to the tracking area. Also, the battery is no limiting factor in our investigation. A user's range of movement is defined by the actual physical space. One considerable limitation is the fact that body movement gestures are limited to one arm. This limitation implies that all other body parts cannot be utilized for gesturing. Body movements and gestures involving more body parts, like legs, both arms, or torso, would enable a more natural user interface experience. Smart watch and smartphone are connected in our framework via Bluetooth, making possible a continuous communication. Accelerometer data collected by the watch are communicated to the phone that computes and detects defined gestures, making use of the smartphone's computation power. It is challenging to devise an algorithm to transform the raw stream of accelerometer data into explicit gestures. Gestures should not interfere with each other, and the system must compute and detect gestures in real time. The resulting data stream to be transmitted and the resulting computation time required for data processing can lead to potential bottlenecks. In previous work, we designed seven distinct gestures dedicated to VR modes of orientation, navigation, and manipulation. evaluated them comparatively to common VR input technology, specifically body tracking enabled by a 3D camera. During the experiment the user was located in a VE constituting a factory building. Latter is an accurate 3D model of a machine hall existing in real world. *Orientation* is implemented through head-tracking. A user can look around and orientate oneself. The smartphone uses built-in sensors, like accelerometer and gyroscope, to deter- mine orientation and motion (of the devices), permitting translation, done by the game engine, into the user's viewpoint in a virtual scene. *Navigation* is implemented by two interaction techniques: (1) In the watch setup, a user looks in walking direction, and single-touch taps the watch to indicate begin or end of movement. (2) In the 3D Camera setup, a user "walks on the spot". *Manipulation* refers to the interaction with objects in a scene.

4   System Model

In order to describe the model, we define a gesture as following: A gesture is a pattern of wrist movements. Patterns of interest are characterized as intentionally performed, easily memorable, and easily performed by a wide range of users. Furthermore, a gesture pattern is a sequence of states based on key information of the sensor data. Our approach based on one central class *GestureRecognizer*, that collects, refines and translates the sensor data. Every gesture is an object of the type *Gesture*, that implements a state machine (SM), that takes the refined sensor data and per- forms state changes accordingly. To enable the recognition of a specific gesture, the corresponding class gets registered in the `GestureRecognizer`, so that its update gets called periodically. The recognized gesture is send to the `UnityEngine`, where corresponding functionality is executed. In terms of computation time and performance, it is recommended to register only the gestures that should be performed and as long as needed. Afterwards they should be released again.

State Machine. A state machine (SM) is a model that describes the output of a control system based on the incoming stimuli from the past. States represent all possible situations in which the state machine may ever be. The incoming inputs and the sequence of inputs in the past determine the state in the system and lead to the corresponding output of reaching that state [18]. If the number of distinguishable situations for a given state machine is finite, the number of states is finite and the model is called a finite state machine.

Key Values. Every time a state machine gets updated, the following key values are used to determine the state and corresponding transition:

• Direction of acceleration	• Value of acceleration
• Direction of velocity	• Value of velocity
• Direction of gravity	• Time

In each frame the extracted sensor data are sent to all state machines (SMs) as new incoming inputs. In general, an update of the SMs occurs every time new input arrives; some SMs only get updated if a key value changes. An update of the SM does not imply a state change. Two different types of SMs are used in the system model. The first type of SM defines states based on segment positions. The second type of SM defines states based on the number of reached segments. For the purpose of the definition, they are both true SMs, but we can use a lot less states this way, as some gestures do not need an exact position but a specific number of direction changes. If one gesture is recognized, all SMs get reset and a message is sent to execute the desired interaction in the VR environment. As condition of our approach, we merely use the patterns generated by a single axis accelerometer and try to extract the information needed to define gestures. The key values are calculated based on the following sensor data: (1) gravitation, (2) acceleration, and (3) velocity. The gravitation value is extracted in order to be used as reference of the watch's posture with which we can align the sensor data and ultimately because the gravitation is polluting the sensor-data. Acceleration data is used to compute the path of the wrist in 3D space monitored over time. The velocity is derived from the acceleration data and used to define additional state transition conditions. An overview of all forces and their dependencies are depicted in Fig. 1.

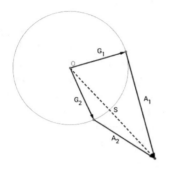

$\vec{S}$ = Raw sensor data
$\vec{G}$ = Gravitation
$\vec{A}$ = Acceleration
$\vec{V}$ = Velocity

We know:

$$\vec{S} = \vec{G} + \vec{A}$$
$$\|G\| = 1$$

Fig. 1. Key values of gesture recognizer. Different $\vec{G}$ and $\vec{A}$ can add up to $\vec{S}$.

If we measure $\vec{S}$ then we know that $\vec{G}$ also points in the same direction, meaning $\vec{G} = \frac{\vec{S}}{\|S\|}$. However, adopting the direction of $\vec{G}$ from $\vec{S}$ is only correct if the watch stands still. Therefore, it is challenging to find the right moment to calculate the direction. Our approach is to implement an adoption-rate describing the degree of how much we trust in $\frac{\vec{S}}{\|S\|}$ to be the same as the actual/real gravitation force with the following steps:

$$\vec{G}_{new} = \vec{G}_{old}(1 - weight) + weight \frac{\vec{S}}{\|\vec{S}\|} \tag{1}$$

$$weight = 0.3 \cdot \exp\left(-\left(\|\vec{S}\| - 1\right)^2 \cdot 14\right) \tag{2}$$

The *weight* distribution corresponds to a Gaussian bell curve, which has its peak in $\|\vec{S}\| = 1$. This function guarantees that $\vec{G}$ is rapidly corrected once the user stays still and that the changing rate from $\vec{G}$ is lowered while gestures are performed. Computations according to Eq. 1 are performed every frame with a rate of frames per second; after merely frames *Gold* is only covered by $(1 - 0.3)^8 = 5.7\%$ if $\|\vec{S}\|$ is close to 1. After computing $\vec{G}$ we know:

$$\vec{A} = \vec{S} - \vec{G}_{new} \tag{3}$$

This approximation still does not consider the position of the smartwatch on the wrist and therefore $\vec{G}_{new}$ could be pointing anywhere. Taking this into account, we apply a rotation R to $\vec{A}$ with the property $(0, 0, -1) = \frac{\vec{G}_{new}}{\|\vec{G}_{new}\|} \cdot R$. Thus, the robustness of the gesture recognition is enhanced.

States. In order to define states for the state machine in an easy computable form, areas on a sphere are defined into sectors. Every vector is transformed into an identifier representing if this vector lies in that sector. In total, we defined 28 sectors: 5 sectors in longitude axis, whereby the 3 middle slices are divided into 8 sectors in latitude axis. Tested in a preliminary study, we figured out that those sectors are the optimal number of segmentations, which are comfortable to reach and that provide an adequate number of possible permutations and therefore gesture states. Based on the defined sectors, the following refined denotations of the key values are used: $Acceleration =$ Sector to which $\vec{A}$ points to; $Velocity =$ Sector to which $\vec{V}$ points to; $Gravity =$ Sector to which $\vec{G}$ points to; $\|Acceleration\| =$ Value of Acceleration $= \|\vec{A}\|$; $\|Velocity\| =$ Value of Velocity $= \|\vec{V}\|$. Next, due to the keen accelerometer sensors and repetitive assimilation of data in- accuracies over time, so-called drift of the computed acceleration occurs. This well-known problem appears by using sensors without the ability to re-calibrate. Multiple factors lead to this inaccuracy. The system is often slightly lagging behind a movement and hence computes faulty values. If the hand is rotated by degree, the sign of the number changes. As the systems sensor is slightly lagging behind, the sign shift is not recognized for a short time. But we cannot avoid the sensor lag, as we cannot trust that the direction of $\vec{G}$ and the direction of $\vec{S}$ is the same. As defined, $\vec{S} = \vec{A}_{real} + \vec{G}$ and $\vec{A}_{polluted} = \vec{S} - \vec{G}$, for that short moment, one has $\vec{A}_{polluted} = \vec{S} + \vec{G} = \vec{A}_{real} + 2\vec{G}$, and $\vec{G}$ gets adjusted. Furthermore, over time the $\sum \vec{A} \neq 0$, therefore the velocity of the object, also is $\|Vel\| \neq 0$. In long movements, the velocity drifts extremely as the sensor cannot be calibrated $\vec{G}$ is not adjusted. It cannot be stressed enough that rotating the wrist causes anomalies due to above mentioned problem, which implies that $\vec{A}$ and $\vec{V}$ cannot be trusted for/of a second after a full rotation is performed. In order to effect higher accuracy, the following adjustment to the velocity vector is performed every frame:

$$\vec{V}_{new} = \vec{V}_{old} \cdot reductionFactor_1 + \vec{A} \cdot timeFactor - reductionFactor_2 \tag{4}$$

5   Gesture Design

We defined seven gestures that cover the full range of possible gestures to evaluate the gesture recognition algorithm. Hereby, it is implied, that single states are recognized as well as sequences of states and changes of the key values. These gestures also try to prove that a series of movements can make up a recognizable pattern. We show this for a realistic number of steps. The classification of the gestures follows along the attributes of movement and shape. We differentiate motion between continuous and partitioned movements. If a gesture is performed without breaks it has a continuous movement while partitioned gestures are made up of a series of continuous sub-gestures with sufficient breaks in-between. The second differentiation between gestures is related to the gesture form. Gestures have either curvy or angular paths describing their shape, see Fig. 2. Following the classification along the attributes movement and shape, the seven gestures will be described accordingly.

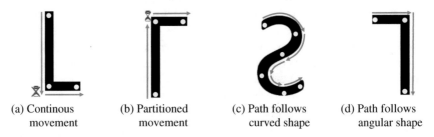

| (a) Continous movement | (b) Partitioned movement | (c) Path follows curved shape | (d) Path follows angular shape |

Fig. 2. Taxonomy of gestures along the attributes movement and shape.

As shown in Table 1, every key value has at least been used twice and in combinations with other values that made sense for those particular gestures. Additionally, the "Number of Steps" indicates how many different motions have to be performed in series for this gesture. We created gestures that have multiple steps, which demonstrates the capability of the system to create series of motions. In the following, we give detailed information of each gesture with the aim to transfer the knowledge so that the reader is able to create own gestures. For better readability, the following diagrams are simplified into a 2-dimensional abstraction of the real SMs. The *Circle Gesture* is defined with continuous movement and a curvy shape, see Fig. 3. This gesture seems to be intuitive and is supposed to be used when something has to be rotated. In order to perform this gesture, users have to perform a clockwise circular motion with their arm. The corresponding SM counts to transitions which is at least half of a circle. Start point of the circle can be at any point and due to the self-recovering nature of this particular SM, the recognition works always if the user does not stop circling. Axis and direction of turn can vary in the definition of the SM. The SM shown in Fig. 3. is not parametric.

Table 1. Key values and amount of states per gesture.

	Circle	Shake	Swipe	Hammer	Z	Lever	Ladle
Acc	✓	✓	✓	✓	✓		
Vel			✓	✓			
Grav			✓			✓	✓
‖Acc‖			✓	✓			✓
‖Vel‖			✓	✓	✓		✓
Time		✓	✓				✓
#Steps	5	4	2	2	4	3	6

The *Shaking Gesture* is defined as a continuous movement with an angular shape. This intuitive gesture is an analogy to shake things. An example would be shaking a dice cup. In order to perform this gesture, the executed motion is described by an alternating up and down of the user's arm. The *Shaking Gesture* can either be defined in vertical direction or horizontal direction. The corresponding state machine is depicted in Fig. 4. and shows the definition of this gesture in vertical direction by counting transitions between up and down. The state machine is parametrized in a way that each two sectors on opposite directions can be used. The *Swipe Gesture* is defined as partitioned gesture following a curvy path, see Fig. 5. This gesture us in analogy to the swipe touch gesture on smartphones that could find use in interactions, where something has to be moved into the direction of the swipe. The gesture can be unintentionally performed in a naturel interaction; therefore, the definition of the state machine is designed in a quite restrictive manner with the use of all key values. In order to avoid the performance of the motion by accident, initially the user has to hold his hand still for around 0.3 s, after that he has to move his hand in the wanted direction and hold the speed for a given number of frames. This gesture also detects false alarms which is caused by rotating the wrist, that is done by checking $Grav = Grav_{atStart}$.

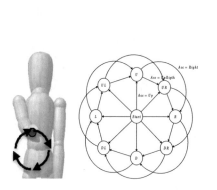

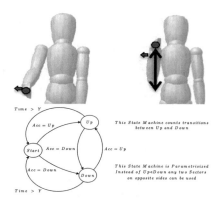

Fig. 3. Motion and SM of circle gesture: every arrow that points to state "L" has the condition $Acc = Left$.

Fig. 4. Motion and state machine of shaking gesture; the gray arrow indicates the direction of the watch.

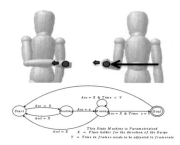

Fig. 5. Motion and state machine of swipe gesture uses all defined key values to avoid unintentionally gesture performance.

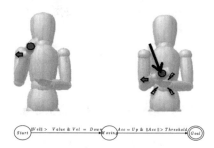

Fig. 6. Motion and state machine of hammer gesture demonstrates the usage of negative acceleration and velocity.

The *Hammer Gesture* follow the definition of a curvy and continuous gesture, see Fig. 6. It is performed by knocking the watch wearing hand in a hammer-swing-like motion onto the other hand. That way the de-acceleration is high enough to make a special pattern that we detect. It is supposed to be used for pushing buttons, smashing objects, or forging. The *Lever Gesture* was made to evaluate the usefulness of recognizing gestures just by the alignment of the watch to the gravity, see Fig. 7. No other key values are used for the gesture recognition. Therefore, we designed a partitioned gesture following an angular path with two stages in which the watch is rotated in two different directions using the gravity in those directions. First, the watch is rotated along the longitudinal axis and in the second state along the lateral axis. The Ladle Gesture is defined by partitioned movement following a curvy shape, see Fig. 8. The gesture demonstrates the combination of gravity gesture elements and the key values of velocity and time. The motion of the gesture is in analogy to scooping fluid and pouring it into another container. Due to the additional key values, merely rotating the wrist does not trigger the gesture recognition. The *Z Gesture* is defined as a continuous movement along an angular shape and was made to test the limits of the system by combining arbitrary motions into one recognizable gesture, see Fig. 9.

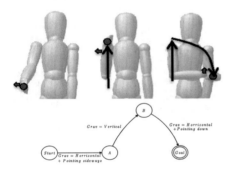

Fig. 7. Motion and state machine of two staged lever gesture only uses gravity as key value.

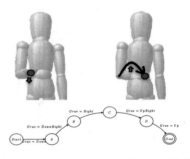

Fig. 8. Motion and state machine of ladle gesture combines gravity with velocity and time values.

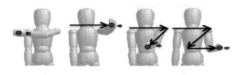

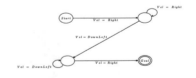

Fig. 9. Motion and state machine of Z gesture.

6   Discussion and Conclusion

The gestures we provided through the high flexibility of our system are easy to learn, effective, and user show a positive attitude towards using the technology. Primitive gestures as Swipe Left and Swipe Right seem simple, however it is challenging to design those gestures in a way, that they are easy to learn, easy to recognize, but not recognized while performing other gestures. More complex gestures, like Lever, Laddle, or Shaking incorporate less key values and are easier to design. Although, the circle gesture integrates up to nine potential states, not all of those have to be reached making the design and recognition of those gestures easier. Some limitations were discovered that have to be considered while designing gestures. It can be stated that gestures following continuous movement along angular shapes, like *Z Gesture*, are hard to learn and hard to recognize and should not be used. The reason is that a deceleration of the hand movement is easily recognized as movement into the opposite direction progressing the state machine into the next state too early. Combining arbitrary motions into one recognizable gesture is not possible in any case. Especially, for designing continuous gestures, reversing movements should be avoided. In a user study, we measured the effectiveness, which was measured as accuracy of the gesture recognizer describing the proportion of all measures, correctly classified. An average sensitivity rate of 90.64% for all performed gestures was achieved, with an average specificity rate of 99.46% and average accuracy rate of 98.36%. The best-performed gesture (shaking) had an accuracy of 99.52%, while the "weakest" gesture (swipe right) still had an accuracy of 97.40%. Compared to common technology like other smartwatches or electromyography armbands, the used device in this investigation uses less expensive sensors which can easily lead to inaccurate signals and measurements. Nevertheless, our approach is able to overcome this limitation and it led to satisfying results with low budget devices. Compared to optical tracking systems like 3D cameras, we could already prove in previous work [6] that the usage of smartwatches are promising alternatives to common gesture based interaction technology. Furthermore, our approach is able to identify more diverse gestures and even small movements like rotating the wrist, which would not be recognizable with those optical trackers. In this work, we presented a signal processing approach for enhanced multi-modal interaction interfaces, designed for smart- watches and smartphones for fully immersive environments that enhance the efficiency of interaction in virtual worlds in a natural and intuitive way. This work deals with the replacement of the common touch input gestures with actual body movement gestures. The challenge is described by missing sensor data, precisely gyroscope and magnetometer data, which together with acceleration is used to calculate orientation and motion dynamics of the device.

We present a transformation of the giving signal-processing from the smartwatch into arm movement gestures with the use of smoothing algorithms and gesture state machines. Based on six key values and statement sequences we are able to define precise arm movement gestures, exemplary demonstrated on seven different gestures. More gestures are conceivable and easily adoptable with our approach. The findings of the user study prove that the system as described in this work is able to recognize unique, primitive, and even complex gestures in an easy learnable way, while overcoming the missing sensor in low budget technology. The approach used performs quantitatively better results compared to the existing recognizer and allows more divers gestures incorporating different kind of states and key values. The tested gestures covered all key values and with this any kind of possible gesture types. The evaluation shows that complex gestures with many consecutive states are just as well designable as more primitive ones.

Acknowledgments. This research was funded by the German research foundation (DFG) within the IRTG 2057 "Physical Modeling for Virtual Manufacturing Systems and Processes".

References

1. Rupprecht, F., Hamann, B., Weidig, C., Aurich, J., Ebert, A.: IN2CO - a visualization framework for intuitive collaboration. In: EuroVis - Short Papers. ACM (2016)
2. Pausch, R., Proffitt, D., Williams, G.: Quantifying immersion in virtual reality. In: Proceedings of the 24th Annual Conference on Computer Graphics and Interactive Techniques, pp. 13–18. ACM (1997)
3. Bryson, S., Feiner, S., Brooks Jr., F., Hubbard, P., Pausch, R., van Dam, A.: Research frontiers in virtual reality. In: Proceedings of the 21st Annual Conference on Computer (1994)
4. König, W.A., Rädle, R., Reiterer, H.: Squidy: a zoomable design environment for natural user interfaces. ACM (2009)
5. Ball, R., North, C., Bowman, D.: Move to improve: promoting physical navigation to increase user performance with large displays. In: Proceedings of the SIGCHI Conference on Human Factors in Computing Systems, pp. 191–200. ACM (2007)
6. Rupprecht, F., Ebert, A., Schneider, A., Hamann, B.: Virtual reality meets smartwatch: intuitive, natural, and multi-modal interaction. In: Proceedings of the 2017 Chi Conference Extended Abstracts on Human Factors in Computing Systems, pp. 2884–2890. ACM (2017)
7. Bergé, L.-P., Serrano, M., Perelman, G., Dubois, E.: Exploring smartphone-based interaction with overview + detail interfaces on 3D public displays. In: Proceedings of the 16th International Conference on Human-Computer Interaction with Mobile Devices & Services, pp. 125–134 (2014)
8. Tregillus, S., Folmer, E.: VR-STEP: walking-inplace using inertial sensing for hands free navigation in mobile VR environments. In: Proceedings of the 2016 Chi Conference on Human Factors in Computing Systems, pp. 1250–1255. ACM (2016)
9. Sharples, S., Cobb, S., Moody, A., Wilson, J.R.: Virtual reality induced symptoms and effects (VRISE): comparison of head mounted display (HMD), desktop and projection display systems. Displays **29**(2), 58–69 (2008)
10. Freeman, E., Brewster, S., Lantz, V.: Do that, there: an interaction technique for addressing in-air gesture systems. In: Proceedings of the 34th Annual Conference on Human Factors in Computing Systems, CHI 2016. ACM (2016)
11. Wilkinson, G., Kharrufa, A., Hook, J., Pursgrove, B., Wood, et al.: Expressy: using a wrist-worn inertial measurement unit to add expressiveness to touch-based interactions. In: Proceedings of the Conference on Human Factors in Computing Systems. ACM (2016)
12. Arefin Shimon, S.S., Lutton, C., Xu, Z., Morrison-Smith, S., Boucher, C., Ruiz, J.: Exploring nontouchscreen gestures for smartwatches. In: Proceedings of the 2016 Chi Conference on Human Factors in Computing Systems, pp. 3822–3833. ACM (2016)
13. Houben, S., Marquardt, N.: WatchConnect: a toolkit for prototyping smartwatch-centric cross-device applications. In: Proceedings of the 33rd Annual ACM Conference on Human Factors in Computing Systems, pp. 1247–1256. ACM (2015)
14. Mace, D., Gao, W., Coskun, A.K.: Improving accuracy and practicality of accelerometer-based hand gesture recognition. In: Interacting with Smart Objects, vol. 45, pp. 45–49 (2013)
15. Mäntyjärvi, J., Kela, J., Korpipää, P., Kallio, S.: Enabling fast and effortless customisation in accelerometer based gesture interaction. In: Proceedings of the 3rd International Conference on Mobile and Ubiquitous Multimedia, pp. 25–31. ACM (2004)

16. Schlömer, T., Poppinga, B., Henze, N., Boll, S.: Gesture recognition with a Wii controller. In: Proceedings of the 2nd International Conference on Tangible and Embedded Interaction. ACM (2008)
17. Chu, H., Huang, S., Liaw, J.: An acceleration feature-based gesture recognition system. In: International Conference on Systems, Man, and Cybernetics, pp. 3807–3812. IEEE (2013)
18. Wagner, F., Schmuki, R., Wagner, T., Wolstenholme, P.: Modeling software with finite state machines: a practical approach. CRC Press (2006)
19. Wu, J., Pan, G., Zhang, D., Qi, G., Li, S.: Gesture recognition with a 3-D accelerometer. In: International Conference on Ubiquitous Intelligence and Computing. Springer (2009)

How Can AI Help Reduce the Burden of Disaster Management Decision-Making?

Mário Simões-Marques[1,2(✉)] and José R. Figueira[2]

[1] CINAV, Portuguese Navy, Alfeite, 2810-001 Almada, Portugal
mj.simoes.marques@gmail.com
[2] Instituto Superior Técnico, Universidade de Lisboa, Av. Rovisco Pais 1,
1049-001 Lisbon, Portugal
figueira@tecnico.ulisboa.pt

Abstract. Disaster management is a decision-making scenario where humans are faced with the assessment and prioritization of a large number of conflicting courses of action and the pressing need to take difficult trade-offs (*e.g.*, ethical, technical, cost-benefit) for selecting and assigning often very scarce resources in response to overwhelming humanitarian crises. The paper discusses the contribution of Artificial Intelligence methodologies for the development of Intelligent Systems that support decision-makers in the context of disaster management, providing examples of alternative methodologies for collecting and representing imprecise information, modeling the inference processes, and to convey naturalistically formulated recommendations and explanations to system users, also encompassing a User Experience perspective, addressing users' needs and requirements, the decision-making environment, equipment and task while using an Intelligent System that provides the support to their functions.

Keywords: Decision fatigue · Artificial Intelligence · Approximate Reasoning
THEMIS · User Experience

1 Introduction

Cognitive science frequently addresses thematics such as *bounded rationality, decision fatigue, analysis paralysis, decision avoidance, impulse decision*, im*paired self-regulation*, and *ego depletion* discussing the impacts and the burden on humans of decision taking. This burden is exacerbated when the decision process occurs under complex and stressful situations, processing big volumes of information, often shadowed by uncertainty, such as disaster management (DM) in catastrophes. In fact, DM is a decision-making scenario where humans are faced with the assessment and prioritization of many conflicting courses of action and the pressing need to take difficult trade-offs (*e.g.*, ethical, technical, cost-benefit) for selecting and assigning often very scarce resources in response to overwhelming humanitarian crises. Some authors note that making choices may imply a cognitive load/cost that impairs subsequent self-control, understood as the ability to keep focus on a long-standing goal by inhibiting immediate desires [1, 2]. For instance, Baumeister *et al.* suggest that

© Springer International Publishing AG, part of Springer Nature 2019
I. L. Nunes (Ed.): AHFE 2018, AISC 781, pp. 122–133, 2019.
https://doi.org/10.1007/978-3-319-94334-3_14

self-control can be viewed as a finite resource or energy, which can no longer be exerted by people that reached their natural limit, after engaging activities requiring effortful control, conducting to a state designated as ego depletion [3]. Narrowing the amount of reasonable decision options is a significant contribution to reduce the scope and number of decision-making situations that humans must face, hence reducing their burden and fatigue. The fact that the former US President Barack Obama or the former Apple CEO Steve Jobs reduced the variety of pieces in their wardrobes is commonly used to illustrate the importance of this contribution to help alleviate the burden of individuals that are subject to a high number of complex decisions along the day. In this paper, the authors discuss the contribution of Artificial Intelligence (AI) method-ologies for the development of Intelligent Systems that support decision-makers in the context of DM. The discussion encompasses also the User Experience perspective, addressing the disaster managers (the users) needs and requirements, the decision-making environment, equipment and task while using an Intelligent System that pro-vides the support to their functions. The paper will present study cases taken from the THEMIS (*disTributed Holistic Emergency Management Intelligent System*) project to illustrate the usage of AI methodologies in support of DM decision-makers.

2   Rationality and Artificial Intelligence

There is no commonly accepted definition of Artificial Intelligence. The definition proposed by the Oxford University is [4]:

"The theory and development of computer systems able to perform tasks normally requiring human intelligence, such as visual perception, speech recognition, decision-making, and translation between languages"

Russell and Norvig [5] analyzed several definitions and approaches adopted in AI identifying multiple dimensions that are generally addressed in literature. These authors noted that one of the facets where AI definitions differ regards the performance ref-erential used to measure success of intelligent systems, since some refer to '*human performance*' while others refer an '*ideal performance*', called rationality. In fact, this distinction between human and rational behavior, results from the mere awareness that human reasoning is not perfect, often being affected by error. Tversky and Kahneman refer that many decisions are based on beliefs concerning the likelihood of uncertain events, and that people rely on a limited number of heuristic principles, which are quite useful for reducing the complexity of tasks since they are highly economical and usually effective, but sometimes lead to severe and systematic errors [6]. These authors discuss the cognitive biases that stem from the reliance on judgmental heuristic, rec-ognizing that even experienced researchers are prone to the same biases when they think intuitively.

Furthermore, Herbert Simon noted that given an optimization goal, rationality (*i.e.,* the information-processing capacity) can face limits imposed by given conditions and constraints, some environmental (*e.g.,* risk and uncertainty, incomplete information about alternatives, complexity) and others intrinsic to the individual (*e.g.,* limited information-gathering and computing capacity) [7]. '*Bounded rationality*' is an expression that denotes these limitations. Simon introduces the terms *'satisficing'* and *'optimizing'* as labels for two broad approaches to ensure rational behavior in situations where complexity and uncertainty make global rationality impossible. The author states that in these situations, optimization becomes *approximate optimization* (*i.e., optimizing*), an approach where the description of the real-world situation is radically simplified until reduced to a degree of complication that the decision maker can handle with the heuristics used. Simon notes that the *satisficing* approach also seeks simplification but in a different direction, since it retains more of the detail of the real-world situation, while settling for a satisfactory, rather than an approximate-best, decision. Thus, *satisficing* requires the definition of some criterion (*i.e.*, an *'aspiration level'*) for determining that a satisfactory alternative has been found [7].

Bossaerts and Murawski [8] recently reviewed evidences showing that human decision-making is affected by computational complexity, noting that modern theories of decision-making persist ignoring such limitations. The authors refer that the rationality principle postulates that decision-makers always choose the best action available to them and does not consider the difficulty of finding the best option. These authors argue that to be more realistic, future theories of decision-making need to account both the resources required for implementing the computations implied by decision-making theory, and the resource constraints imposed on the decision-maker by biology.

Complementing these perspectives, some authors (*e.g.*, Vohs *et al.* [1], Inzlicht and Schmeichel [2]) note that making choices may imply a cognitive overload that impairs subsequent self-control (*i.e.*, the mental processes that allow people to override thoughts, emotions, or behaviors that compete with their overarching goals [3]). Baumeister *et al.* suggest that self-control can be viewed as a finite resource or energy, which can no longer be exerted by people that reached their natural limit, after engaging in activities requiring effortful control, conducting to a state designated as *ego depletion* [3].

Classical optimization provides theoretical foundations to support decision-makers in solving complex problems. However, the use of classical optimization is also faced with some challenges, for instance regarding the formulation of the problem, particularly when there is a significant number of goals to achieve, when they are conflicting, and when uncertainty is involved. Russell and Norvig warn that achieving perfect rationality (*i.e.*, always doing the right thing) is not feasible in complicated environments, because the computational demands are just too high, and the timeliness of the solution doesn't allow to do all the computations one might like [5].

The development of intelligent systems (IS), which apply AI methodologies is an attempt to reach, in a specific domain, a level of analysis and a performance, at least, comparable to (and desirably better than) that of a human expert. In fact, IS can engage in complex inference processes, necessary for evaluating alternative options and offering high quality conclusions and advice, and to offer explanations about the rationale that led to such conclusions [9]. The use of IS can contribute to circumvent some of the limitations of human decision-makers, therefore being very promising tools to support their decision-making tasks. The context of Disaster Management (DM) is quite demanding for decision-makers because, for coordinating the intervention of a variety of actors, they have to deal with large amounts of uncertain, incomplete and vague information [10]. The use of decision support systems (including IS) in DM is a recognized gap (referred, for instance, in [11]) which is addressed below, while describing some of the challenges facing THEMIS project.

3   Disaster Management and AI

According to the United Nations International Strategy for Disaster Reduction (UNISDR) *Disaster Management* means "*the organization, planning and application of measures preparing for, responding to and recovering from disasters*" [12]. Disaster management is a complex process for dealing with several families of natural (*e.g.*, geophysical, climatological, hydrological) and anthropogenic (*e.g.*, non-intentional, intentional) disasters [10].

Correia *et al.* discussed the Knowledge Engineering activities involved in the development of intelligent systems, focusing on methods for transferring the expertise from human experts to computers, implementing reasoning models based on the acquired knowledge, and for transferring knowledge back to humans, specifically in the context of disaster management [9]. This process, which is illustrated in Fig. 1, considered the model proposed by Turban *et al.* involving five major activities [13, 14]:

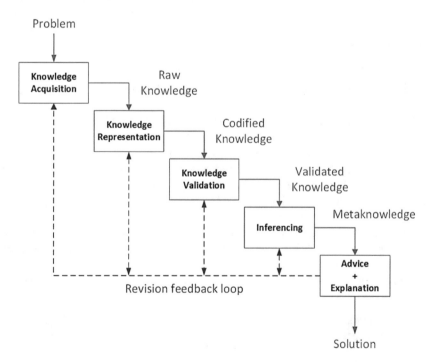

Fig. 1. The process of knowledge engineering (adapted from [13]).

- *Knowledge acquisition* (from experts and other sources) - involves the acquisition of knowledge (*e.g.*, from human experts and documents) specific to the problem domain or to the problem-solving procedures;
- *Knowledge representation* (in the computer and to the user) - involves encoding of the knowledge (*e.g.*, abstract concepts and their relations);
- *Knowledge validation* - (or verification) involves validating and verifying the knowledge until its quality is acceptable;
- *Inferencing* - is the reasoning capability that can build higher-level knowledge, for instance using heuristics;
- *Explanation and justification* (to the user) - delivers knowledge using adequate knowledge presentation and visualization formats, and involves an explanation capability (e.g., how a certain conclusion was derived by the intelligent system).

Under the scope of AI there are many study domains that address the interdependent issues that knowledge engineering handles related with the implementation of IS. Examples of alternative methodologies include Bayesian networks, Markov networks, case-based reasoning, ontologies, approximate reasoning (*e.g.*, fuzzy logics) and metaheuristics (*e.g.*, artificial neural networks, machine learning, genetic algorithms, swarm intelligence). The following section discusses in more detail the usage of fuzzy logics for collecting and representing imprecise information (*e.g.*, based on linguistic terms), for modeling the inference processes (namely, using approximate reasoning fuzzy rule-based approaches), which lead to naturalistically formulated recommendations and explanations to system users.

4   THEMIS – The Resource Assignment Example

THEMIS is the acronym for *disTributed Holistic Emergency Management Intelligent System*. This is an undergoing research and development project funded by the Portuguese Ministry of Defense aimed at supporting real time disaster management. A high-level conceptual perspective of the system is illustrated in Fig. 2 considering the scenario of a major disaster requiring international assistance, provided by multiple agencies. As shown in the figure THEMIS is meant as an intelligent system that accesses information from multiple sources (*e.g.*, users, sensors, crowdsourcing), provides situational awareness based a georeferenced common picture, shared among system users. The information about incidents is analyzed to assess response priorities and advise on the assignment of the available resources.

THEMIS general system requirements are to provide a common platform that supports:

1. the compilation of disaster incidents, integrating information from different sources;
2. the geo-referenced inventory of resources available for the operation (type, quantity, location);
3. timely decision-making through intelligent functionalities, adequate to the user profile, that assess the situation and recommend priorities of action, based on the current operational context, and the (quasi-) optimal allocation of resources;

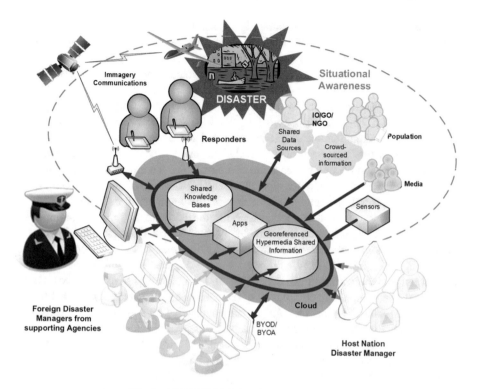

Fig. 2. THEMIS high-level conceptual view

4. knowledge repositories and analysis tools relevant to support response to different types of disaster, providing contextual advice, according to the scenario, characteristics of resources and user profile; and
5. preparedness and training, through simulations of humanitarian relief and disaster management operations.

The example discussed in here will focus particularly on the last component of the third requirement (resource assignment advice). The assignment problem is a very common problem whose purpose is to find a solution that minimizes the sum of costs (denoted as c_{ij} and corresponding, for instance, to a certain amount of time, distance or money) resulting from assigning a number of agents (denoted as i) to perform an identical number of tasks (denoted as j), subject to some constraints (*e.g.*, a task is performed by one and only one agent, the variable x_{ij} denoting the assignment of agents to tasks is binary), formulated as one or multiple objective functions of the type:

$$\min \sum\nolimits_{(i,j)} c_{ij}x_{ij}. \tag{1}$$

Operations Research (OR) offers "classical" approaches (*e.g.*, the Hungarian method) developed to compute optimal solutions of the assignment problem. However, computing an optimal and timely solution is not so trivial when the number of agents and tasks are not the same and is quite high, when the assignment feasibility criteria is not the same for all 'agent-task' pairs, or when there are multiple objectives to satisfy. In fact, as the number of agents and tasks increase, the number of possible solutions is a combinatorial explosion, making the problem potentially intractable using exact approaches. Note that if the number of agents and tasks equals n, the number of possible solutions correspond to the factorial $n!$ therefore getting to extremely large figures quite fast ($2! = 2$, $4! = 24$, $6! = 720$, $8! = 40.320$, $10! = 3.628.800$). Identically, when the combination of agents is considered to increase response capability (*e.g.*, joining a medical agent with a technical agent to rescue a trapped wounded victim) the problem becomes even more complex, since the number of potential teams is a k-combination of the number of agents available $C(n, k)$, where k is the size of the team. Despite forming teams tends to reduce the number of tasks performed, the number of agents and teams to consider for the assignment problem significantly increases (*e.g.*, 10 agents assigned individually can perform 10 tasks; assuming that the same 10 agents can operate individually or in teams of 2, the number of assignee possibilities is $C(10, 1) + C(10, 2) = 10 + 45 = 55$, and the number of tasks performed range between 5 and 10, depending on the number of teams formed). On the other hand, the number and complexity of the constraints also increases, since not all x_{ij} solutions are valid; note that assigning a team composed by agents i_1 and i_2, precludes the assignment of those agents, individually or integrated in any other team.

It is rather obvious that a human decision-maker is not able to mentally compute all possible solutions in the very plausible scenario of having to coordinate the assignment of much more than 10 agents. Even the usage of pen-and-paper decision aids will be of limited utility.

Classical OR also faces multiple challenges. For once, the timeliness of the computing outputs is paramount in disaster response, imposing limits on the "reasonable" amount of time spent to solve the assignment problem. A second issue is the formalization of the problem, since most data is imprecise (*e.g.*, victim or incident status, response time) and the criteria tend to be of a qualitative nature, rather than exact/quantitative. Even if a classical (*i.e.*, quantitative) formulation is adopted (such as the one suggested by expression (1)) the definition of costs c_{ij} requires embracing coherent and robust methods for translating and combining, objective and subjective, exact, imprecise and vague data and information, as humans do on their reasoning process. Finally, in these complex problems, the scalarization of multiple objectives (*i.e.*, the method to ensure simultaneous optimization of the various objectives) may also be challenging to define in a coherent and numerically efficient way.

Fuzzy set theory and Fuzzy logics [15–18], which are a generalization of the Classical set theory and Classical logics, is a quite consistent option to implement approximate reasoning processes suited to deal with the above-mentioned challenges facing the knowledge engineering activities required to design a DM intelligent system. Examples of use of fuzzy entities in such activities include:

- *Knowledge representation* - fuzzy sets, linguistic variables and fuzzy relations;
- *Inferencing* - approximate reasoning based on fuzzy rules and fuzzy aggregation operators;

- *Explanation and justification* - sentences in natural language, quantified or qualified using linguistic variables' terms and modifiers.

Considering this approach, the assignment problem can be based on a combination of criteria (multiple objectives), which, for instance, evaluates the *adequacy* (in this case a fitness or payoff measurement instead of a cost) of assigning each agent *i* to each task *j* pondering the relationship between them in terms of *agent availability, agent capability to perform task, nearness (distance), priority of the task.* Limitations of space do not allow to expand much on the theoretical concepts underlying Fuzzy set theory. Nevertheless, the basic principle is that a specific abstract concept can be represented by a fuzzy set, and the fitness of the elements of a given Universe of Discourse to such concept can be represented by a membership function that expresses quantitatively (in the interval [0, 1]) their degree of truth or adherence to the concept. A membership 0 means that is totally false that a certain element fits the concept (*i.e.*, it does not belong to the set), 1 means that is totally true that a certain element fits the concept (*i.e.*, is a full member of the set), while intermediate values reflect a partial degree of truth that an element fits the concept (*i.e.*, partial belonging to the set). The use of quantitative membership degrees is quite beneficial from a computational standpoint since it is possible to aggregate the terms of complex logical expressions using mathematical expressions that correspond to a wide variety of logical operations (*e.g.*, union, intersection, ordered weighted average, complement) and also modifiers (*e.g.*, concentration, dilation) which can be chosen in order to reflect the type of behavior that the system is expected to model.

Returning to the DM assignment problem, it can be formulated in natural language as "*agent i is adequate to perform task j if agent i available and agent i is capable to perform task j and agent i is near task j and task j is a priority*". This formulation has implicit an inference process that can be translated as an IF-THEN rule, where the degree of truth of conclusion "*agent i is adequate to perform task j*" (denoted as μ $adeq_{ij}$) results from the intersection of the degree of truth of several conditions (*availability, capability, nearness, priority,* denoted as $\mu\,av_i$, $\mu\,cap_{ij}$, $\mu\,near_{ij}$ and $\mu\,pri_j$) which are assessed based on the characteristics of- and relationships between *agent i* and *task j*. As already mentioned the IF-THEN rule can be computed numerically using the following expression, where $\cap$ reflects a fuzzy intersection operator (*i.e.*, *and* operator):

$$\mu_{adeg_{ij}} = \mu_{av_i} \cap \mu_{cap_{ij}} \cap \mu_{near_{ij}} \cap \mu_{pri_j}. \tag{2}$$

The membership functions can be continuous or discrete. Figure 3 illustrates two fuzzy sets. The one on the left is a continuous fuzzy set that conveys the concept of *nearness*; therefore, the degree of truth for small distances is high, decreases for intermediate distances, and is low for big distances. The one on the right is a discrete fuzzy set (in this case a linguistic variable) that conveys the concept of *availability*; where the different linguistic terms relate to membership degrees.

The most common operator for implementing a fuzzy intersection is the *min* function. Using this function, *agent i* is deemed as "inadequate" for assignment to *task j* ($\mu\,adeq_{ij} = 0$) if any of the conditions has a membership degree 0; *agent i* is deemed as

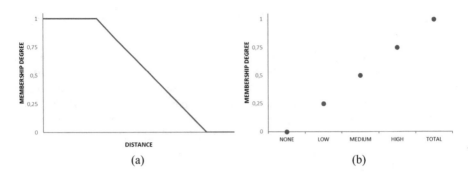

Fig. 3. Example of a continuous (a) and a discrete (b) fuzzy set

"totally adequate" for assignment to *task j* (μ $adeq_{ij}$ = 1) if all conditions have a membership degree 1; and *agent i* is deemed as "partially adequate" for assignment to *task j* (μ $adeq_{ij}$ ∈]0, 1[) if all conditions have a membership degree in the interval]0, 1].

However, there is a variety of fuzzy intersection operators, with different behaviors. Depending on the operator selected to implement the conditions aggregation, the result can either reflect the contribution of all conditions or just one. In fact, if the *min* function is used the value resulting from the aggregation always equals the worst membership value, not being influenced by any of the other membership values involved in the aggregation (*i.e.*, is nondiscriminant and noncompensatory). Another quite common operator is the *product*, where the result is influenced by all membership values involved in the aggregation, therefore potentially introducing some discrimination among different situations, but is also noncompensatory. Zimmermann and Zysno noted that people decisions often use compensatory procedures (between low and high degrees of membership) and suggested a class of hybrid compensatory operators (γ-operator), whose behavior is controlled by a γ compensation parameter [19]. Based on this rationale, other compensatory operators were proposed, such as Werners' compensatory "fuzzy and" operator, which is a convex combination of *min* and *arithmetical mean* functions [20]. This said, there is a certain degree of flexibility regarding the way of interpreting the logical "and" in the rule, reflecting a particular positioning in the range of optimistic to pessimistic judgements. Furthermore, is also possible to consider and incorporate some level of preference or importance for each condition of the IF-THEN rule, which can be conveyed by a weight assigned to it.

It is also worth to note that each of the conditions in the rule can result from the evaluation of previous rules. For instance, the *availability* criteria regarding *agent i* can result from the assessment and aggregation of conditions such as "*agent i is idle*" (*i.e.*, not assigned to a task) and "*agent i is on duty*" (*i.e.*, not in a resting period). Identically, the *capability* criteria can result from the assessment of conditions such as "*agent i is skilled to perform task j*" (*i.e.*, possesses the competencies to perform the task) and "*agent i is equipped to perform task j*" (*i.e.*, possesses the means/tools to perform the task). Therefore, the evaluation of a given rule may be part of a larger and more comprehensive inference process.

The use of the IF-THEN rules as presented here contributes to the selection of the most adequate agents to perform tasks and can be understood as corresponding to the assessment of *satisficing* conditions (refer to Sect. 2) in an approximate reasoning heuristic process. However, *per se* this approach does not ensure assignment optimality, it rather helps to significantly narrow the set of solutions to those that are worth of further consideration. The assignment solution will tend to optimality only if the added adequacy of agents to tasks is maximized, using the expression:

$$\max \sum\nolimits_{(i,j)} \mu_{adeq_{ij}} x_{ij}. \tag{3}$$

Nevertheless, as previously stated, computing all possible combinations of agents (including teams) and tasks may be very demanding and not compatible with the timeliness required from the decision-making process. Fortunately, in the AI toolkit there are methods (such as genetic algorithms) that, given a limited computation time, ensure quasi-optimal solutions, which can also contribute to take quite good decisions.

5 Some Remarks from a User Experience Perspective

According to ISO 9241-210 definition, User Experience (UX) includes all the users' emotions, beliefs, preferences, perceptions, physical and psychological responses, behaviors and accomplishments that occur before, during and after system use [21]. Simões-Marques *et al.* in [22] noted that, despite the goals of UX in the context of DM intelligent systems being conceptually like those of any other software application, the specificity and the complexity of the context of use, the variety of roles, or the criticality of the requirements present challenges in the analysis and assessment of UX that somehow differ from those of more common applications.

Common sense suggests that an intelligent system which reduces the burden of DM decision-makers and provides user interaction modes based on natural language will be appreciated by users. Nevertheless, there are several other features of a system design that contribute to its success or doom. Completeness, learnability, ease of use, intuitiveness are just some of them. THEMIS project team is quite committed in engaging prospective users along the design and validation process, to ensure high UX standards. For this purpose, a User centered design approach has been adopted and, after an initial characterization of the context of use and identification of users' and organizational needs, system requirements were derived, including for the decision support. The system is in a prototyping stage and the design concepts, namely regarding functionality and interaction modes, are being assessed, considering several types of equipment, according to user profiles. As shown in Fig. 2, disaster managers use fixed terminals while responders use mobile devices. Besides this, the knowledge activities are quite advanced, and the assessment and validation of inference models and explanation and justification features have already started, confirming the usefulness of the THEMIS intelligent system in support of disaster management.

6 Conclusion

The paper characterized Disaster management as a complex activity where decision-makers are faced with overwhelming humanitarian crises requiring the assessment and prioritization of many conflicting courses of action and the pressing need to make difficult trade-offs for selecting and assigning an adequate (*i.e.*, efficient and effective) response to incidents, optimizing the employment of the scarce resources available.

The limitations of human judgement and rationality were addressed, as well as their impact on decision-makers performance. The discussion identified the impossibility of human decision-makers mentally computing all alternative solutions that must be considered in a disaster management process. The paper also referred several of the challenges that may turn problematic the implementation of classical Operations Research problem solving methods.

The paper proceeded providing a perspective on the contribution that may result from applying Artificial Intelligence methodologies in the development of Intelligent Systems to support decision-makers in the context of disaster management. The paper referred different alternative methodologies for collecting and representing imprecise information, for modeling the inference processes, and for conveying naturalistically formulated recommendations and explanations to system users. An example was given on how to deal with the assignment problem, using a hybrid approach that combines fuzzy IF-THEN rules and optimization techniques.

It was mentioned that the analysis and discussion presented in this paper derived from work developed under the scope of the THEMIS project, which is an undergoing research and development initiative funded by the Portuguese Ministry of Defense, whose purpose is to create an intelligent system aimed at supporting real world disaster management activities.

To finalize, it was given a User Experience perspective on how the use of AI can help reducing the decision burden on disaster managers, highlighting the commitment of the THEMIS project team in engaging a representative group of users along the design and validation process, which is also guided by the set of users' needs and system requirements identified in the early stages of the project, and has to consider the specificity of the decision-making environment and equipment, as well as of the operational activities performed while using the system in support of disaster management.

Acknowledgments. The work was funded by the Portuguese Ministry of Defense and by the Portuguese Navy.

References

1. Vohs, K.D., Baumeister, R.F., Schmeichel, B.J., Twenge, J.M., Nelson, N.M., Tice, D.M.: Making choices impairs subsequent self-control: a limited-resource account of decision making, self-regulation, and active initiative. J. Pers. Soc. Psychol. **94**(5), 883–898 (2008)
2. Inzlicht, M., Schmeichel, B.J.: Beyond limited resources: self-control failure as the product of shifting priorities. In: Vohs, K.D., Baumeister, R.F. (eds.) The Handbook of Self-Regulation: Research, Theory, and Applications (3rd Edition), pp. 165–181. Guilford Press, New York (2016)

3. Baumeister, R.F., Vohs, K.D., Tice, D.M.: The strength model of self-control. Curr. Dir. Psychol. Sci. **16**(6), 351–355 (2007)
4. Oxford University Press. https://en.oxforddictionaries.com/definition/artificial_intelligence
5. Russell, S.J., Norvig, P.: Artificial Intelligence: A Modern Approach, 3rd edn. Prentice Hall Press, Upper Saddle River (2010)
6. Tversky, A., Kahneman, D.: Judgment under uncertainty: heuristics and biases. Science **185** (4157), 1124–1131 (1974)
7. Simon, H.A.: Theories of bounded rationality. In: McGuire, C.B., Radner, R. (eds.) Decision and Organization, pp. 161–176. North-Holland Publishing Company (1972)
8. Bossaerts, P., Murawski, C.: Computational complexity and human decision-making. Trends Cogn. Sci. **21**(12), 917–929 (2017)
9. Correia, A., Severino, I., Nunes, I.L., Simões-Marques, M.: Knowledge management in the development of an intelligent system to support emergency response. In: Nunes, I. (ed.) Advances in Human Factors and Systems Interaction. AHFE 2017. Advances in Intelligent Systems and Computing, vol. 592, pp. 109–120. Springer, Cham (2018)
10. Simões-Marques, M.J.: Facing disasters—trends in applications to support disaster management. In: Nunes, I.L. (ed.) Advances in Human Factors and System Interactions. Advances in Intelligent Systems and Computing, vol. 497, pp. 203–215. Springer, Cham (2017)
11. NAS: Facing hazards and disasters - understanding human dimensions. In: Kreps, G.A., Berke, P.R., et al. (eds.) National Academy of Sciences. National Academies Press, Washington, D.C. (2006)
12. UNISDR Terminology. https://www.unisdr.org/we/inform/terminology
13. Turban, E., Sharda, R., Delen, D.: Decision Support and Business Intelligence Systems, 9th edn. Prentice Hall Press, Upper Saddle River (2011)
14. Turban, E., Aronson, J.E., Liang, T.P.: Decision Support Systems and Intelligent Systems, 7th edn. Prentice-Hall of India, Inc., New Delhi (2007)
15. Zimmermann, H.-J.: Fuzzy Set Theory—and Its Applications. Springer, Netherlands, Dordrecht (2001)
16. Zadeh, L.: The concept of a linguistic variable and its application to approximate reasoning - I. Inform. Sci. **8**(3), 199–249 (1975)
17. Zadeh, L.: The concept of a linguistic variable and its application to approximate reasoning - II. Inform. Sci. **8**(4), 301–357 (1975)
18. Zadeh, L.: The concept of a linguistic variable and its application to approximate reasoning - III. Inform. Sci. **9**(1), 43–80 (1975)
19. Zimmermann, H.-J., Zysno, P.: Latent connectives in human decision making. Fuzzy Sets Syst. **4**(1), 37–51 (1980)
20. Werners, B.M.: Aggregation models in mathematical programming. In: Mathematical Models for Decision Support, vol. 48, pp. 295–305, Springer, Berlin (1988)
21. ISO 9241-210:2010: Ergonomics of human-system interaction – Part 210: human-centred design for interactive systems. ISO (2010)
22. Simões-Marques, M., Correia, A., Teodoro, M.F., Nunes, I.L.: Empirical studies in user experience of an emergency management system. In: Nunes I. (ed.) Advances in Human Factors and Systems Interaction. AHFE 2017. Advances in Intelligent Systems and Computing, vol. 592, pp. 97–108. Springer, Cham (2018)

Tackling Autonomous Driving Challenges – How the Design of Autonomous Vehicles Is Mirroring Universal Design

Susana Costa[1]([✉]), Nelson Costa[1], Paulo Simões[1], Nuno Ribeiro[2], and Pedro Arezes[1]

[1] ALGORITMI Centre, School of Engineering, University of Minho, Guimarães, Portugal
{susana.costa, ncosta, parezes}@dps.uminho.pt,
paulo.simoes@gmail.com
[2] BOSCH Car Multimedia Portugal, SA, Braga, Portugal
nuno.ribeiro@pt.bosch.com

Abstract. In the future, the world will be characterized by highly densely populations, with growing share of mobility-impaired/disabled persons, a critical problem regarding the sustainability of the metropolises, whose resolution may reside in autonomous vehicles. A broader range of users will be allowed a, so far, denied mobility in level 4 and level 5 SAE autonomous vehicles, a goal to be accomplished through Universal Design, a design which intends to be the closest possible to the ideal design. For such purpose, Human Factors and Ergonomics are key. Literature review and research have shown that there is evidence of application of the seven Universal Design principles in these new autonomous vehicles and that, given the nature and purpose of the Universal Design, with the increase of autonomy, there is natural increased evidence of Universal Design. A novel model for interaction of the Universal Design influencers is proposed.

Keywords: Autonomous driving · Universal Design · Human Factors
Human-systems interface · Autonomous vehicles

1 Introduction

The United Nations have anticipated that, by 2025 there will be more than 8 billion people living on Earth [1].

It is expected that, by 2050, the number of persons over the age of 80 years old will be increased by 446% over the year 2000, with estimates of population density increasing from 46.5 persons/km2 to 68.4 persons/km2 [1]. These figures highlight two major challenges of the upcoming world: (1) highly densely populated by (2) growing community of mobility-impaired/disabled persons.

This poses a future critical issue to be solved (which needs to be planned ahead): for the world to be sustainable, while encompassing the needs of fast mobility to succeed a competitive world, with increasingly stringent requirements, whilst a considerable portion of the society is composed by aged, impaired citizens, how can one

© Springer International Publishing AG, part of Springer Nature 2019
I. L. Nunes (Ed.): AHFE 2018, AISC 781, pp. 134–145, 2019.
https://doi.org/10.1007/978-3-319-94334-3_15

tackle the autonomous driving challenges? For sure, this portion of the society will also be reflected among the users of these new vehicles and, therefore, accounting for the characteristics of this demographic class will be key for broadening the range of users. But not only; when trying to accomplish Universal Design, all prospective users have to be considered. Notwithstanding, Ehrlich et al. (2006) stress that even taking into account elderly and disabled persons in the mainstream design process, some persons will be left out: "a 'tail' of people who are unable to use a given product" – in a reference to the 95th percentile - as the design of "all products and devices so that they are usable by all people" is not possible. In all truth, such ideal design does not exist [2].

The solution may rely in changing the current standard of an all-purpose vehicle into more flexible solutions, e.g., by replacing the traditional business model of car with distinct on-demand mobility solutions. Indeed, autonomous cars have the potential to overcome an array of transportation challenges by improving road safety, optimizing traffic flow, increasing efficiency in transportation, allowing new mobility models, available to broader groups of people, and providing additional comfort for drivers and passengers.

In 1985, an incipient concept was being brought up by Ron Mace, who first defined Universal Design as "the design of products and environments to be usable by all people, to the greatest extent possible, without the need for adaptation or specialized design" [3].

Indeed, Universal Design takes into consideration users from all backgrounds, life experiences, life stages, cultures and their capabilities, limitations and dimensions. As such, Universal Design is the opposite of designing for the typical user [4].

This approach may seem too exhaustive, time-consuming, even prohibitive, from an economic point of view. However, Universal Design has an inconspicuous presence in our daily lives (e.g., presence sensor-activated automatic sliding doors, side-walks with surface of functional topography, pedestrian walkways equipped with traffic lights with multimodal warnings). In fact, multimodality goes hand in hand with the Universal Design, because it allows, by making use of more than one sense, a redundancy of the information that one wants to transmit to the other, ensuring that it reaches him/her in an efficient manner. Even the typical user benefits greatly from multimodality.

The new vehicles, which are evolving towards increasing autonomy, are already taking advantage of these same principles, of which we are (or may be) oblivious.

Nowadays, it is quite common, while driving, to be assisted by visual, acoustic and even haptic feedbacks, all concurring to build driver's situation awareness. Obviously, this multimodality in presentation of information has many advantages. On the one hand, it helps to resolve the burden of driver saturation with information, by resorting to alternative sensing channels of communication. On the other hand, it also allows for redundancy in the transmission of information, if several modalities are simultaneously coding the same information. This is why high-level warnings (imminent danger warnings) are, usually, multimodal [5–8], and this is no more than the application of, at least, one principle of Universal Design.

For a better understanding of the richness of the Universal Design concept, some disambiguation must be made, starting with Universal design and accessible design, which are often used interchangeably, but are not synonyms. Even though the two concepts share essential design principles and strategies, accessible design is legally mandated whereas Universal Design is not. Erlandson states that "the legal mandates and design guidelines associated with accessible design require that designs be compliant subject to legal penalties [9]. Universal Design, on the other hand, is motivated more by global competition and the creation of more universally accessible and usable products and services demanded by international markets." [9].

It is often observed that the target users for whom a particular service or infrastructure is designed are not the only ones who benefit, prefer and privilege it, but also by the users who are the norm (e.g., the curb cuts, designed for wheelchair users that end up being used also by parents who walk strollers, cyclists, pedestrians who suffer from joints' pains). In fact, a more recent trend is that one way of achieving Universal Design will be to appeal to the hedonic characteristics of products and services. So, by creating a mainstream design, everyone will want to have (own, benefit), and not just the ones who need [2, 9].

Several authors have stressed the need to go beyond usability, focusing the advantage of designing for a better user experience [10–13]. The best design strategy, thus, seems to be the envisioning of the emotional response of the prospect user, the hedonic qualities' appeal like pleasure, luxury, exclusiveness and status.

In sum, the conjuncture and societal pressures and values (ageing society, technological evolution, laws, global markets and competitiveness, evolving conceptualizations of disability) are concurring to foment the mainstreaming of Universal Design. Figure 1 depicts this conceptual model [9].

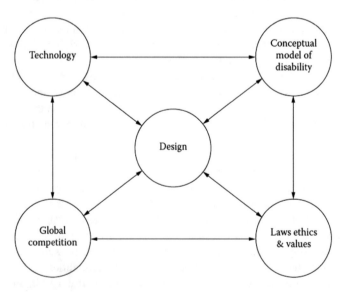

Fig. 1. Conceptual model depicting the interactions between technology, global competition, laws, ethics, and disability that influence the design process. (Retrieved from [9]).

Universal Design is, also, sometimes used interchangeably with two other terms: Inclusive design and design for all. It is not the focus of this paper to discuss to which extent these can be seen as synonyms, nevertheless it should be noted that their ultimate objective is common and that the definition may vary between countries.

The group responsible for developing the concept of Universal Design, which included Ron Mace, also established seven principles to be observed during the process of Universal Designing. This set of principles were structured by name, definition and guidelines [4, 14, 15]. Table 1 presents all seven principles of Universal Design.

It should be noted that, for being able to efficiently apply these Universal Design principles, one has to master issues like anthropometry, ergonomics, biomechanics and physiology, all sciences that the Human Factors and Ergonomics enclose. Anthropometric dimensions, for example, are important when the design has to meet functional ranges, and through ergonomic studies, the best modalities to convey different kinds of information can be assessed.

Even though user acceptance of this novel autonomous technology is yet to be completely clarified, one thing is certain, if autonomous driving is to gain acceptance by car users and become a reality in real world traffic driving environment, new generation autonomous vehicles should be shaped by Human Factors and Ergonomics principles.

This paper aims at showing how the design of increasingly autonomous vehicles is increasingly approaching Universal Design.

2 Methods

This work was carried through scoping review of the autonomous vehicles' characteristics that obey principles of Universal Design. It is a methodology which, according to Colquhoun [16], is "s a form of knowledge synthesis that addresses an exploratory research question aimed at mapping key concepts, types of evidence, and gaps in research related to a defined area or field by systematically searching, selecting, and synthesizing existing knowledge". It is a very popular methodology for systematizing scientific knowledge, with a steep growing expression among the scientific community the last 4 to 5 years.

3 Results and Discussion

As Erlandson emphasized, the design principles are arranged in a hierarchical structure, where "equitability imposes constraints on the other design principles in that they must be applied so that the designed entities are accepted by a broad spectrum of users." [9]. Indeed, "equitability forces the integration of the other Universal Design principles" and, also, "the designs must be age and context appropriate as well as aesthetically pleasing". This clearly bespeaks the huge gap between accessible design and Universal Design. In point of fact, accessible design is not concerned with equitability, but focused on "removing or preventing environmental barriers as prescribed by law,

Table 1. The seven principles of Universal Design (adapted from [4, 5]).

1. Equitable use

Definition	Guidelines
The design is useful and marketable to people with diverse abilities	1a. Provide the same means of use for all users: identical whenever possible, equivalent when not 1b. Avoid segregating or stigmatizing any users 1c. Make provisions for privacy, security, and safety equally available to all users 1d. Make the design appealing to all users

2. Flexibility in use

Definition	Guidelines
The design accommodates a wide range of individual preferences and abilities	2a. Provide choice in methods of use 2b. Accommodate right- or left-handed access and use 2c. Facilitate the user's accuracy and precision 2d. Provide adaptability to the user's pace

3. Simple and intuitive use

Definition	Guidelines
Use of the design is easy to understand, regardless of the user's experience, knowledge, language skills, or current concentration level	3a. Eliminate unnecessary complexity 3b. Be consistent with user expectations and intuition 3c. Accommodate a wide range of literacy and language skills 3d. Arrange information consistent with its importance 3e. Provide effective prompting and feedback during and after task completion

4. Perceptible information

Definition	Guidelines
The design communicates necessary information effectively to the user, regardless of ambient conditions or the user's sensory abilities	4a. Use different modes (pictorial, verbal, tactile) for redundant presentation of essential information 4b. Maximize "legibility" of essential information 4c. Differentiate elements in ways that can be described (i.e., make it easy to give instructions or directions) 4d. Provide compatibility with a variety of techniques or devices used by people with sensory limitations

5. Tolerance for error

Definition	Guidelines

(continued)

Table 1. (*continued*)

The design minimizes hazards and adverse consequences of accidental or unintended actions	5a. Arrange elements to minimize hazards and errors: most used elements, most accessible; hazardous elements eliminated, isolated, or shielded 5b. Provide warnings of hazards and errors 5c. Provide fail safe features 5d. Discourage unconscious action in tasks that require vigilance
6. Low physical effort	
Definition	Guidelines
The design can be used efficiently and comfortably and with a minimum of fatigue	6a. Allow user to maintain a neutral body position 6b. Use reasonable operating forces 6c. Minimize repetitive actions 6d. Minimize sustained physical effort
7. Size and space for approach and use	
Definition	Guidelines
Appropriate size and space is provided for approach, reach, manipulation and use regardless of user's body size, posture, or mobility	7a. Provide a clear line of sight to important elements for any seated or standing user 7b. Make reach to all components comfortable for any seated or standing user 7c. Accommodate variations in hand and grip size 7d. Provide adequate space for the use of assistive devices or personal assistance

guidelines, and standards". Regardless, an accessible design requirement can, also, be equitable.

This paper proposed to show how the design of increasingly autonomous vehicles is increasingly approaching Universal Design. For this purpose, an exposition of the various evidences gathered is made hereon.

Principle 1: Equitable Use - The design is useful and marketable to people with diverse abilities.

The concept of automated vehicles itself applies this principle. SAE Level 4 and Level 5 autonomous vehicles will allow for a wider range of users to benefit from an otherwise denied mobility. According to [4], "No user is excluded or stigmatized. Wherever possible, access should be the same for all; where identical use is not possible, equivalent use should be supported. Where appropriate, security, privacy and safety provision should be available to all." For instance, a bilateral transhumeral amputee can experience mobility in a level 4 SAE autonomous vehicle, in certain well-defined, limited road contexts. Since Level 4 SAE autonomous vehicles can be designed not to request, in any case, the driver to perform the dynamic task, this user can interact with the vehicle, which performs the dynamic driving task solely, by communicating, for instance, voice control inputs to the vehicle, or by making choices with just the eye gaze. When considering Level 5 SAE autonomous vehicles, this mobility becomes even more broadened, by allowing users to travel in all road (unlimited) contexts. This encompasses great autonomy in mobility for otherwise severely constrained persons (e.g., blind persons). Also, the concept of vehicle sharing is expected to minimize the gap between social strata, by providing economic solutions of mobility. Indeed, this is in line with what Erlandson has stated: "Non-stigmatizing and equitable imply being age and context appropriate, being aesthetically pleasing, being affordable, and having a broad market appeal".

Principle 2: Flexibility in Use - The design accommodates a wide range of individual preferences and abilities.

As Dix posed it, "the design allows for a range of ability and preference, through choice of methods of use and adaptivity to the user's pace, precision and custom" [4]. Increasingly, car manufacturers are aware of the variability among their users, in fact, not everyone likes to listen to the same radio station, the same type of music, not everyone likes to answer mobile calls while driving (even with recourse to legal channels such as bluetooth) and manufacturers have already realized that. From there, the concepts of personalization and customization in the automotive context were born. Indeed, allowing the user to predefine which radio stations he/she will be presented first is customizing the interface, which is the same as choosing the graphics/background colours of the centre stack display. Another thing, however, has to do with when a regular user of a vehicle makes their daily choices by giving priority to certain radio stations, information which the vehicle will apprehend. Here, machine reading plays a key role because it allows the vehicle to learn the user's preferences and, over time, to give priority to those same preferences (for example, by placing the radio stations most heard by the user at the top of the list of available radio stations, on the display, "betting" that the user's choice will fall preferentially on one of them and, thus, easing the human-machine interface. This is called personalization.

Multimodality is, also, an evidence of flexibility in use. For instance, if the vehicle has a multimodal warning consisting of both sound and visual icon, in some cases (low level warnings) the user may choose to silent the acoustic warning, resorting only to the (preferred) visual aid.

Another feature that depicts accordance to this principle is customization of the Human-Machine Interface (HMI) of the system, whereby the user may, for instance, predefine the HMI so as to not display infotainment (warnings, information) during highly demanding, complex driving tasks.

Principle 3: Simple and Intuitive Use - Use of the design is easy to understand, regardless of the user's experience, knowledge, language, skills, or current concentration level.

One of the main concerns of automated vehicles is the trust issues. Many researchers have focused on this matter and it seems that one major key to the success of the autonomous driving paradigm is the resolution of trust issues among prospective users, a concept strictly related to the coherence of the system and also annoyance. Indeed, another way of posing this principle could be make the design work in the expected way [15]. It seems reasonable to assume that, if the vehicle starts a frenzy of sounds, lights, colours and vibrations without nothing in its surroundings justifies this type of behaviour, it will leave the user with doubts about the efficiency of the system in detecting possible threats. Even without going to extremes, if one wishes to decrease the volume of the radio to request an indication on the street and, although pressing the correct interface, the sound instead of decreasing increases, it will cause stress, embarrassment to the user and the level of confidence in the whole system will suffer.

Similarly, the whole judgment about vehicle system will suffer if a little beep constantly alerts the user to things he/she considers insignificant, causing the user to fail to recognize value in the valences that the system supposedly offers. This lack of balance, which is often not anticipated by the designers/manufacturers, is responsible for the deactivation of these same features, which are classified as annoying and, as such, distracting rather than useful or beneficial to the driver. Likewise, one's confidence in the system will also be affected (in a more serious way) by the lack of a warning that should have been made by the system and that it failed to display. Clearly, if an event occurs without the driver having been advised in advance, confidence will be shaken because the design did not work as expected. In the advent of the new autonomous vehicles, some situations occurred (some fatal and wide-spread in the media), which put in question the future of the autonomous driving paradigm. Needless to say, since then, systems have been upgraded and technology has been refined so that, in fact, the design works as expected.

Principle 4: Perceptible Information - The design communicates necessary information to the user, regardless of ambient conditions or the user's sensory abilities.

According to this principle, as stated by Dix, redundancy of information is important, interpreting this principle as a need for information to be "represented in different forms or modes (e.g. graphic, verbal, text, touch) [4]. Essential information should be emphasized and differentiated clearly from the peripheral content. Presentation should support the range of devices and techniques used to access information by people with different sensory abilities".

Nowadays, it is quite common, when performing a reverse manoeuver in a vehicle, to be aided by both visual and acoustic cues, in the judgment of relative distances from another object(s) with which one can collide. It works well with both hearing-impaired and non-hearing-impaired drivers, because the two modalities are redundant for the same information. But if one could add an haptic feedback warning to the same context, the hearing-impaired would benefit from a really redundant warning, for they would be able to both see and feel it (thus, making use of two senses), whilst the non-hearing-impaired would have three senses activated by three redundant modalities. The relation between this evidence and the principle is evident, as the principle may also be rephrased as: "designs should provide for multiple modes of output" [15]. Also, it is not by chance that different sources of information are allocated in different places within the cockpit. In truth, the more important the type of information to be conveyed to the driver, the more close to his/hers line of sight within his/hers natural posture (which is why less important information, like radio station list is at a wider viewing angle - in the centre stack - and velocimeter, tell tales, warnings regarding dangerous conditions of the road and road signs appear on the Head Up Display (HUD), combiner or instrumental panel/dashboard).

Principle 5: Tolerance for Error - The design minimizes hazards and the adverse consequences of accidental or unintended actions.

The new paradigm of autonomous driving relies on an important functionality - the connected vehicles, whereby vehicles are connected to other vehicles, infrastructures, and all the surroundings involving the vehicle. Even though margin is left for foul play, information and warning systems are being developed to prevent the users of such vehicles from inadvertently making a mistake, like wrong-way driving, surpassing speed limits or other way disobeying mandatory road signs. This is in line with the postulate by Dix: "Potential hazards should be shielded by warnings" [4]. It is not, also, due to chance the location of the handle that opens the hood of the vehicles. Its dangerous potential when occurring during a journey is so high, that the reach of this handle tries to force it to only open when really desired and never inadvertently, which, again, fall within the interpretation of Dix "Potentially dangerous situations should be removed or made hard to reach" [4].

Principle 6: Low Physical Effort - The design can be used efficiently and comfortably and with a minimum of fatigue.

Being able to select any option or feature just using a voice command or the eye gaze are the epitome of this principle, where little to no effort is made. Within the scope of autonomous vehicles another relevant concept which has been (and still continues to be) thoroughly studied by researchers worldwide - workload, Indeed, in the attempt of taking full profit of all the features made available by state-of-the-art technologies, in order to get some advantage relatively to the direct competitors (with infotainment, here, taking a big share of blame), the driver begins to receive too many requests, overloading him/her with information and ending up doing the opposite of what it was supposed to (instead of enjoying a driving pleasant user experience, the system renders a more stressful and more difficult to manage task). This forces that alternative channels of communication be explored, so as to avoid workload, more comfortable and less physical demanding channels. These are also great examples in what concerns to mainstream features, as they may benefit greatly physically-impaired and illiterate users in the future autonomous cars, but may also be preferred to other more conventional modalities by typical users. Dix state that the design "should allow the user to maintain a natural posture with reasonable operating effort" and that "Repetitive or sustained actions should be avoided", which also reports to the location of the most requested and sought information – located in the dashboard or HUD/combiner, so as to prevent adoption of awkward, sustained postures [4]. Adjustment of the steering wheel, thus allowing more natural postures and, hence, less fatigue, is also evidence of principle 6 being attended.

Principle 7: Size and Space for Approach and Use - Appropriate size and space is provided for approach, reach, manipulation, and use regardless of the user's body size, posture, or mobility.

This principle has been attended to since, for instance, the seats of the vehicles are adjustable, thus considering the covering of different body dimensions and different body postures. As Dix affirmed "All physical components should be comfortably reachable by (...) users. Systems should allow for variation in (...) size and provide enough room for assistive devices to be used." [4].

It can also be seen in the adjustment of the steering wheel, making it possible to place it closer to those with shorter ranges of reach. In addition, vehicles have greater interior spaces, contributing greatly to obeying to this principle. Research is being conducted to make future autonomous cars the third living space of whomever uses them. Once again, their prototypes envision even larger interior spaces, with room for instalment of, for instance, small offices and gymnasiums. These wide conformations will also allow for persons who move in wheelchairs or other alternative means of mobility to enter directly into the vehicles (through lifting platforms), not depending on third parties to make their journey (nowadays it is a critical, physical demanding and non-autonomous process).

The essence of Universal Design forces that, as the level of autonomy increases, so does the evidence of Universal Design.

A conceptual model for Universal Design, based on Erlandson [9], depicting the interactions between technology, global competition, laws, ethics, disability, the social conjecture and acceptability and trust – influencers of the Universal Design - is proposed (Fig. 2).

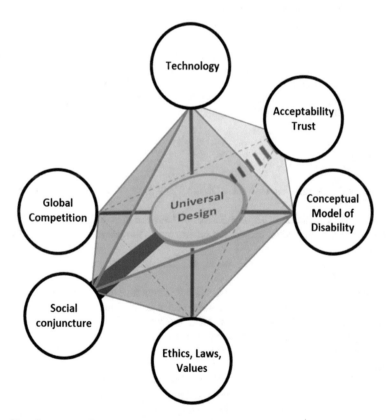

Fig. 2. Novel conceptual model for Universal design, based on [9], depicting the interactions between technology, global competition, laws, ethics, disability, the social conjecture and acceptability and trust – influencers of the universal design process proposed.

4 Conclusions

In the future, the world will be facing two major challenges: (1) highly densely populated by (2) growing community of mobility-impaired/disabled persons. The solution may rely on highly autonomous vehicles.

Evidence of Universal Design in these new autonomous vehicles can be found, among others, in the available modalities of communication between the driver and the vehicle, some of which can even be customized or personalized according to the driver's capabilities and/or taste.

In the future, a broader range of users will be allowed a, so far, denied mobility in level 4 and level 5 SAE autonomous vehicles. This will be accomplished through Universal Design, a design which intends to be the closest possible to the ideal design (which, in turn, is impossible to exist in its most pure form).

Naturally, with the increase of autonomy, there is increasing evidence of Universal Design of autonomous vehicles.

An innovative conceptual model for Universal Design, depicting the interactions between technology, global competition, laws, ethics, disability, the social conjecture and acceptability and trust is proposed.

Acknowledgments. This work has been supported by COMPETE: POCI-01-0145-FEDER-007043 and FCT – Fundação para a Ciência e Tecnologia within the Project Scope: UID/CEC/00319/2013 and by the European Structural and Investment Funds in the FEDER component, through the Operational Competitiveness and Internationalization Programme (COMPETE 2020) Project nº 002797; Funding Reference: POCI-01-0247-FEDER-002797.

References

1. Preiser, W.F., Ostroff, E.: Universal Design Handbook. McGraw Hill Professional, New York (2001)
2. Salvendy, G.: Handbook of Human Factors and Ergonomics. Wiley, Hoboken (2012)
3. Mace, R.: Universal design: barrier free environments for everyone. Des. West **33**(1), 147–152 (1985)
4. Dix, A., Finlay, J., Abowd, G., Beale, R.: Universal Design, pp. 365–394. Springer, Heidelberg (2004).
5. Bakowski, D.L., Davis, S.T., Moroney, W.F.: Reaction time and glance behavior of visually distracted drivers to an imminent forward collision as a function of training, auditory warning, and gender. Procedia Manuf. **3**, 3238–3245 (2015)
6. Gold, C., Körber, M., Lechner, D., Bengler, K.: Taking over control from highly automated vehicles in complex traffic situations: the role of traffic density. Hum. Factors **58**(4), 642–652 (2016)
7. Lu, Z., Happee, R., Cabrall, C.D., Kyriakidis, M., de Winter, J.C.: Human factors of transitions in automated driving: a general framework and literature survey. Transp. Res. Part F: Traffic Psychol. Behav. **43**, 183–198 (2016)
8. Häuslschmid, R., von Bülow, M., Pfleging, B., Butz, A.: Supporting trust in autonomous driving. In: Proceedings of the 22nd International Conference on Intelligent User Interfaces, pp. 319–329. ACM (2017)

9. Erlandson, R.F.: Universal and Accessible Design for Products, Services, and Processes. CRC Press, Boca Raton (2007)
10. Agarwal, A., Meyer, A.: Beyond usability: evaluating emotional response as an integral part of the user experience. In: CHI 2009 Extended Abstracts on Human Factors in Computing Systems, pp. 2919–2930. ACM (2009)
11. Uotila, M., Falin, P., Aula, P., Vaaranka, P.: Designing luxury: understanding the hidden values and pleasure factors of luxury and high-level design products. In: Digital Proceedings of Fashion Forward, the International Symposium on Fashion Marketing and Management (2005)
12. Hassenzahl, M., Tractinsky, N.: User experience - a research agenda. Behav. Inf. Technol. 25(2), 91–97 (2006)
13. Kang, S.R., Lee, E.: User experience: beyond usability. In: 6th Asian Design Conference (2003)
14. Christophersen, J.: Universal Design 17 Ways of Thinking and Teaching. Husbanken, Drammen (2002)
15. Story, M.F.: Principles of universal design. In: Universal Design Handbook (2001)
16. Colquhoun, H.L., Levac, D., O'Brien, K.K., Straus, S., Tricco, A.C., Perrier, K., Kastner, M., Moher, D.: Scoping reviews: time for clarity in definition, methods, and reporting. J. Clin. Epidemiol. 67(12), 1291–1294 (2014)

Assessment of Pilots Mental Fatigue Status with the Eye Movement Features

Liwei Zhang[1,2], Qianxiang Zhou[1,2(✉)], Qingsong Yin[1,2],
and Zhongqi Liu[1,2]

[1] School of Biological Science and Medical Engineering, Beihang University,
100083 Beijing, China
{zhangliweil, zqxg}@buaa.edu.cn, 632330893@qq.com,
lzq505@163.com
[2] Beijing Advanced Innovation Centre for Biomedical Engineering,
Beihang University, 102402 Beijing, China

Abstract. With the domain of aero technics, aviation mental fatigue research has been aimed at preventing the tragedy caused by fatigue flights. There are specific effects of the mental fatigue on behavioral performance in pilots, characterized as a decrease of reaction rate and distractibility of attention. Fourteen subjects completed an experiment with visual search tasks and simulated flight task. The eye movement data were recorded by the eye tracker with 120 Hz of the sampling rate. The results of the questionnaire showed that the fatigue value in the fatigue periods was higher compared with the awake periods. Moreover, the results of eye movement features showed that the PERCLOS and blinking frequency amplitude increases monotonically with increasing fatigue value during the task. Lastly, the conclusion can be made that the measurement of eye movement pattern can be used to assess and forecast pilots mental fatigue status.

Keywords: Aviation · Mental fatigue · Eye movement · PERCLOS

1 Introduction

Multiple factors contribute to the aviation metal fatigue, such as long-endurance flights, complex flight operation and poor weather conditions. Symptoms of aviation fatigue include drowsiness in the brain, lack of concentration, decreased of thinking ability, weakness of logical thinking, noticeable decline in memory, and a drastic reduction in work efficiency. As early as 1976, NASA's study indicated that about one-fifth of aviation accidents are related to flight fatigue [1]. According to a survey conducted by the U.S. Department, 4.1% of the aviation accidents that occurred in the United States between 1971 and 1977 were confirmed as being caused by pilots' mental fatigue, and 10.5% of the accidents could be related to mental fatigue [2]. In 2013, the flight 1354 of UPS company crashed at Birmingham-Shuttles worth International Airport. Accident investigations conducted by the United States National Transportation Safety Board (NTSB) showed that there were errors in the pilots' operation when the aircraft are approaching, and these errors were also consistent with the fatigue [3]. It was also

© Springer International Publishing AG, part of Springer Nature 2019
I. L. Nunes (Ed.): AHFE 2018, AISC 781, pp. 146–155, 2019.
https://doi.org/10.1007/978-3-319-94334-3_16

reported that three-quarters of the aviation accidents were caused mainly by the pilots themselves, and that mental fatigue caused their body functions and mental status to decline was an important factor [4]. Therefore, mental fatigue in flight had been one of the focuses of aviation medical and civil aviation safety research [5].

Fatigue is caused by many factors such as the psychology, physiology and the environment in which people live. At present, mental fatigue can be evaluated by objective and subjective ways. In the subjective, the assessment of fatigue status is made through the scores of sleepiness, fatigue and emotion scale. However, the subjective assessment method has different assessment criteria and is much affected by individual differences. There are also many shortcomings such as random graffiti by volunteers who want to hide or exaggerate their real feelings or can't endure fatigue [6]. In the objective, the fatigue state is assessed by the physiological of EEG signals, ECG signals, head & mouth features. Schmidt [7] and Zhao [8] reported that P300 of the ERP component was most affected by driver's mental fatigue during simulated driving. Qiu [9] used the method of face-specific area identification method to detect driver fatigue status. Furthermore, Australian scholar developed a Head Position Sensor (HPS) which identifies the driver's fatigue state through driver's head displacement [10]. In driving, visual information accounts for four-fifths of the total amount of data obtained by the brain. It is noteworthy that eye movement has a close relationship with cognition and becomes one of the hot spots to study mental fatigue. In general, the eyes closed more than 0.5 s during driving, indicating that driver is in fatigue or non-alert state. Percentage of Eyelid Closure Over the Pupil Time (PERCLOS) is the percentage of time the eyes closed during a specific period [11]. Some researchers suggest that PERCLOS can be priority used as a real-time, non-contact method to assess mental fatigue [12].

In the present study, the mental fatigue was evaluated by combining multiple features from multiple perspectives. The subjective fatigue and mental status questionnaire were used to estimate the fatigue value (FV), and then the features of blinking frequency (BF) and PERCLOS were extracted to establish a multi-index fusion mental fatigue assessment model. The model enabled non-contact monitoring of mental fatigue. It has crucial social significance and scientific value for improving flight safety.

2 Materials and Methods

2.1 Participants

Fourteen male volunteers from BeiHang University participated in this study. The volunteers were 23 years to 27 years of age, with an average of 24 years. All participants were right-handed and had a normal or corrected-to-normal vision, without a history of psychiatric disorders. The ethics committee of the BeiHang University approved this study. All participants knew the task before the experiments and signed the informed consent.

2.2   Task and Procedures

Simulated Flight Task. In the present study, the flight scenes of a simulated flight software were displayed on a 19-in computer screen (refresh rate: 60 Hz). Volunteers used the flight joystick to control the aircraft in a simulated flight task as shown in Fig. 1. The basic flight operations in the task were shown in Table 1.

Fig. 1. Simulated flight experimental

Table 1. Basic flight operations in a simulated Flight Task

	Flight operation
1	Takeoff
2	Climbing
3	Steering
4	Flat flight
5	Cruising
6	Landing

Target Search Task. The 1-h target search task was used to generate mental fatigue. The configuration interface of the task was shown in Fig. 2 in which ⊗ was the target, ⊗, ⊗ , ⊘, ⊕, ⊖, ⊖, ⊕ were the interferences. Each volunteer took 6 blocks × 100 trials of target search tasks, where every trial displayed 6 s. Then, the volunteers were asked to press the left button of the mouse as soon as possible, once they identified and acquired the target in the experiment. If they confirmed there was no target item on the screen, the right mouse button would be pressed. And if not sure, there is no button.

Fatigue Questionnaire. Mental status and fatigue questionnaire were administered to the volunteers. The mental status questionnaire was modified according to the "Fatigue Symptom Questionnaire" published by the Japanese Association for Industrial Health, which was developed to assess the fatigue and attention states under two conditions (see Table 2). The questionnaire was a four-item self-report inventory designed to

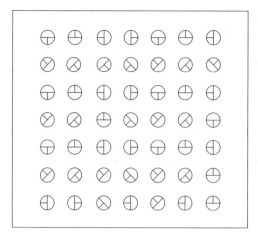

Fig. 2. The interface of target search task

measure transient or fluctuating affecting status on four different scales, mental clarity, attention concentration, sleepiness, and comprehensive assessment of fatigue. Participants were required to describe and score their feelings on a scale ranging from 1 (veryless) to 10 (extremely). In the mental status questionnaire, a low score indicates a good mental status.

Table 2. Mental status questionnaire

Scales	Score											
Mental clarity	Clarity	1	2	3	4	5	6	7	8	9	10	Confusion
Attention concentration	Concentration	1	2	3	4	5	6	7	8	9	10	Distraction
Sleepiness	Alertness	1	2	3	4	5	6	7	8	9	10	Sleepiness
Comprehensive assessment of fatigue	Non-fatigue	1	2	3	4	5	6	7	8	9	10	Fatigue

The subjective fatigue questionnaire (see Table 3) was divided into five levels of fatigue, excitement, alertness, light fatigue, moderate fatigue, and heavy fatigue. Participants were required to assess and score their feelings on a scale. A higher score indicates more fatigue.

Table 3. Subjective fatigue questionnaire

Fatigue scale	Symptoms of fatigue	Score
Excitement	Concentrated attention and interest in things	0
Alertness	Clear-headed and thinking	1–2
Light fatigue	Slightly tired	3–4
Moderate fatigue	Dizziness and fatigue	5–7
Heavy fatigue	Irritability and sleepiness	8–10

Procedures. Participants completed two visits (morning, afternoon) at Beihang University. After the setup for eye tracker, participants were seated 60 cm to 80 cm in front of an LCD monitor and given detailed task instructions. In the morning, subjects performed three blocks of simulated flight task with each block lasted 10 min, and in the afternoon, the subjects were asked to perform three blocks of simulated flight task after six blocks of target search task completed. Each participant attended a practice before the formal experiment to ensure that everyone was familiar with the tasks. In the formal experiment, the questionnaires were recorded every 10 min.

The desktop eye tracker (iView X RED, SMI company, Germany) was used in experiments with a sampling rate of 120 Hz, a tracking resolution of 0.03 deg, a gaze tracking accuracy of 0.4 deg and an average gaze error of less than 0.5 deg. The camera acquisition system with a frame rate of 30 frames/s was used to record the volunteer's facial video.

3　Results

3.1　Results of Fatigue Questionnaire Data

Five fatigue values were calculated in the awake and fatigue period, where fatigue value was the average of the questionnaire scores. The five fatigue values were named A1, A2, A3, B1 and B2, and corresponding to the following values: A1 = initial fatigue value in awake phase; A2 = initial fatigue value in fatigue phase; A3 = fatigue value after target search task; B1 = fatigue value in awake phase; B2 = fatigue value in fatigue phase. Three pairs of values were calculated with paired T-test, the results were shown in Table 4.

Table 4. Subjective fatigue value in different states

Group	Item	Mean	SD	t	p
1	A1	2.117	0.748	−12.76	0.000
	A3	6.858	1.747		
2	A2	4.083	1.278	−6.994	0.000
	A3	6.858	1.747		
3	B1	2.642	0.839	−10.266	0.000
	B2	6.167	1.490		

The results showed A2 and A3 had significant differences, indicating that the target search tasks induce the sense of fatigue. Moreover, the difference between B1 and B2 showed that in the simulated flight task of the fatigue period, the subjects experienced mental fatigue. Figure 3 showed the trend of subjective fatigue values of volunteers. The fatigue value was higher in the fatigue period compare to awake period.

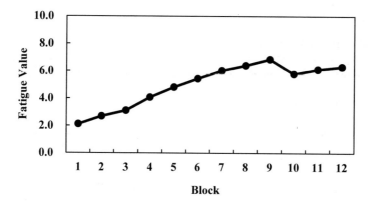

Fig. 3. The trend of the fatigue value (Block1–3 represent the simulated flight task in awake period, Block4–9 represent the target search task in fatigue period, Block10–12 represent the simulated flight task in fatigue period)

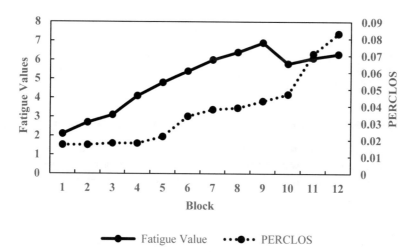

Fig. 4. Subjective fatigue values and PERCLOS in experiments (Block1–3 represent the simulated flight task in awake period, Block4–9 represent the target search task in fatigue period, Block10–12 represent the simulated flight task in fatigue period)

3.2   Results of PERCLOS

Figure 4 showed that PERCLOS was lower in the awake compare with fatigue period. Moreover, the PERCLOS increased monotonically with increasing fatigue value in the experiment.

3.3   Blinking Frequency

The trend of BF of volunteers in the whole experiment was shown in Fig. 5. It can be seen that the BF amplitude was the lowest in the awake phase of the simulated flight task. However, in the fatigue period, the value of the blinking frequency sharply increased.

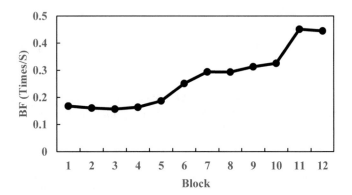

Fig. 5. Blinking frequency in experiments (Block1–3 represent the simulated flight task in awake period, Block4–9 represent the target search task in fatigue period, Block10–12 represent the simulated flight task in fatigue period)

3.4   The Coherence of Fatigue Value and Eye Movement Features

The correlations between the fatigue value and eye movement features were shown in Table 5. The results showed that fatigue value was significantly and positively associated with PERCLOS and BF. Similarly, PERCLOS was significantly associated with blinking frequency.

Table 5. Correlation analysis with eye movement features and fatigue Value

		BF	PERCLOS
Fatigue value	Correlation	0.605	0.722
	Significant	0.000	0.000
BF	Correlation	1	0.712
	Significant	-	0.000

3.5   Mental Fatigue Assessment Model

In summary, a mental fatigue assessment model based on eye movements features were established. The subjective fatigue values of simulation flight task in the awake period and fatigue period was divided into two categories, in which the fatigue value was higher than 5 as fatigue state (1) and less than 5 as the non-fatigue state (0). PERCLOS and BF as independent variables, the data were divided into two classes using the Bayesian algorithm. Ten volunteer samples were randomly selected to train the model, with the remaining four volunteer samples as validation. The classification results were shown in Table 6.

Table 6. Classification results of mental fatigue assessment model

Degree2				0	1	Summation
Initialization	No.	0		37	1	38
		1		5	17	22
	%	0		97.4	2.6	100.0
		1		22.7	77.3	100.0
Cross-validation	No.	0		37	1	38
		1		5	17	22
	%	0		97.4	2.6	100.0
		1		22.7	77.3	100.0

The classification results showed that in the initial grouping and cross-validation grouping, the accuracy rate of classification was 97.4%, the probability of the non-fatigue state being fatigue was only 2.6%.

According to the classification result, the classification function of non-fatigue state (0) is

$$f_1 = 3.906x_1 + 11.130x_2 - 1.134 \tag{1}$$

The classification function of fatigue state (1) is

$$f_2 = 9.282x_1 - 70.263x_2 - 6.055 \tag{2}$$

Where x_1 is the blinking frequency and x_2 is PERCLOS.

4   Discussion

Mental fatigue refers to a subjective feeling of lack of alertness and motivation in mind after a long period. The symptoms of mental fatigue include brain drowsiness, inability to concentrate, decreased the ability to think, logical thinking difficulties and unresponsiveness. In a monotonous, non-challenging environment, people can easily produce negative emotions such as slackness and irritability. Other studies have induced mental fatigue by setting a boring simulation driving environment [7, 15]. A target search task was designed to induce mental fatigue in this study. At the beginning of the study, subjective questionnaires were used to assess subjects' fatigue value. Afterward, the eye movement features verified that the mental fatigue status had significant differences in awake and fatigue period. Lastly, a mental fatigue assessment model was built according to PERCLOS and blinking frequency.

Simulated flight driving is commonly used to measure mental movements such as cognition, perception and motor response. Therefore, the results of questionnaires showed that fatigue value was higher in the fatigue period than in the awake period, which is consistent with another report [2].

The results showed that blinking frequency increased with increasing fatigue in simulated flight task. Moreover, as the sense of fatigue intensifies, the closing time of the eyes and the time elapsed in a single blink of an eye increased [16, 17]. The results were consistent with of Las' study, who was found that the rapid eye movements of volunteers disappeared and the frequency of small blink movements increased with the deepening of the degree of fatigue in fatigue driving [18]. The actions may be due to the body's self-feedback regulation. Meanwhile, the results in simulated flight task also showed that compared to fatigue period; the awake period was characterized by lower PERCLOS with a flat trend in line with the reference previous papers [4, 19]. Especially, the blinking frequency amplitudes positively correlated with corresponding PERCLOS, which was consistent with other findings [19].

There are still some potential limitations in the present study. First, the number of subjects was small, and individual differences can affect the final results. Second, less eye movement features were analyzed.

5 Conclusion

In the present study, the target search task accelerated mental fatigue, and mental fatigue value in fatigue period was higher than in awake period. Blinking frequency amplitude and PERCLOS increased with mental fatigue. Lastly, the mental fatigue assessment model was established by combine subjective fatigue values with eye movement features. It is noteworthy that the model has a non-contact ability to detect mental fatigue and its classification accuracy rate is 97.4%. In the future, we will study in the following aspects: (1) The mental fatigue model can perform multi-level classification; (2) Other eye movement features such as open eyes speed, fixation point position can be used in combination with PERCLOS to study pilot mental fatigue status. All the limitations should be considered in future studies of mental fatigue.

Acknowledgments. This research was funded by National Defense Basic Research Fund Project (A0920132003), Electronic Information Equipment System Research of Key Laboratory of Basic Research Projects of National Defense Technology (DXZT-JC-ZZ-2015-016) and Reform and Development of Beijing Municipal Institute of Labour Protection in 2017.

References

1. Gartner, W.B., Murphy, M.R.: Pilot workload and fatigue: a critical survey of concepts and assessment techniques. Contract, 20–31 (1976)
2. Xie, D.: A review of research on aviation fatigue flight. Public Commun. Sci. Technol. 5(9), 67–79 (2014)
3. Ni, H.: How NTSB Survey Fatigue Factors in Aviation Accidents [EB/OL] (2015). http://news.carnoc.com/list/332/332262.html

4. You, Y.: Research on Pilots' Fatigue Monitoring Technology Based on Machine Vision. University of Electronic Science and Technology of China, Chengdu (2011)
5. Yang, C., Luo, L., Tan, Z., et al.: The progress of international military aviation medical. Chin. J. Aerosp. Med.—Chin. J. Aerosp. Med. 22(1), 74–78 (2011)
6. Sirevaag, E.J., Kramer, A.F., Wickens, C.D., et al.: Assessment of pilot performance and mental workload in rotary wing-aircraft. Ergonomies 36(9), 1121–1140 (1993)
7. Schmidt, E.A., Schrauf, M., Simon, M., et al.: Drivers' misjudgement of vigilance state during prolonged monotonous daytime driving. Accid. Anal. Prev. 41(5), 1087–1093 (2009)
8. Zhao, C., Zhao, M., Liu, J., et al.: Electroencephalogram and electrocardiograph assessment of mental fatigue in a driving simulator. Accid. Anal. Prev. 45(1), 83–90 (2012)
9. Qiu, H.: Research on Fatigue Driving Detection Based on Face Detection. Guangdong University of Technology, Guangzhou (2008)
10. Zhou, Y., Yu, M.: Research on method of detecting drowsy driver. Chin. Med. Equip. J. 24(6), 25–28 (2003)
11. Hartley, L., Horberry, T., Mabbott, N., et al.: Review of fatigue detection and prediction technologies. Natl. Road Transp. Commiss. 2000, 4–10 (2000)
12. Wylie, C., Shultz, T., Miller, J.C., et al.: Commercial motor vehicle driver fatigue and alertness study technical summary. Commer. Driv. 837–13 (1996)
13. Yoshitake, H.: Methodological study on the inquiry into subjective symptoms of fatigue. J. Sci. Labour. 47, 797–802 (1971)
14. Thiffault, P., Bergeron, J.: Monotony of road environment and driver fatigue: a simulator study. Accid. Anal. Prev. 35(3), 381–391 (2003)
15. Minjie, W., Pingan, M., Zhang, C.: Driver fatigue detection algorithm based on the states of eyes and mouth. Comput. Appl. Softw. 30(3), 25–27 (2013)
16. Santamaria, J., Chiappa, K.H.: The EEG of drowsiness in normal adults. J. Clin. Neurophys. Off. Publ. Am. Electroencephalogr. Soc. 4(4), 327–382 (1987)
17. Lal, S.K.L., Craig, A.: Driver fatigue: psychophysiological effects. In: The Fourth International Conference on Fatigue and Transportation, Fremantle (2000)
18. Wang, L., Sun, R.: Study on face feature recognition-based fatigue monitoring method for air traffic controller. China Saf. Sci. J. 22(7), 66–71 (2012)

Effect of Far Infrared Radiation Therapy on Improving Microcirculation of the Diabetic Foot

Chi-Wen Lung[1,2,4], Yung-Sheng Lin[5], Yih-Kuen Jan[1,2,3],
Yu-Chou Lo[4], Chien-Liang Chen[6], and Ben-Yi Liau[7(✉)]

[1] Rehabilitation Engineering Lab, University of Illinois at Urbana-Champaign,
Champaign, IL, USA
[2] Kinesiology & Community Health,
University of Illinois at Urbana-Champaign, Champaign, IL, USA
[3] Computational Science and Engineering,
University of Illinois at Urbana-Champaign, Champaign, IL, USA
[4] Department of Creative Product Design, Asia University, Taichung, Taiwan
[5] Department of Chemical Engineering,
National United University, Miaoli, Taiwan
[6] Department of Physical Therapy, I-Shou University, Kaohsiung, Taiwan
[7] Department of Biomedical Engineering, Hungkuang University,
Taichung, Taiwan
byliau@hk.edu.tw

Abstract. Diabetes is associated with many severe complications, such as heart disease and nerve damage. Diabetics have elevated blood glucose levels that result in peripheral neuropathy. As a result, diabetics always feel painful about their lower limbs and may need to have lower limb amputation. The diabetic foot ulcers are hard to heal and require extensive medical treatments and follow ups. The purpose of this study was to evaluate the effect of infrared radiation therapy using a heating pad with carbon fibers (LinkWin Technology Co., Taiwan) on the diabetic foot. Ten diabetics and 4 non-diabetics were recruited for this study. The participants were assessed for their microvascular function before and after the interventions in each month for three consecutive months. The results showed that surface temperature increase by 2° after using the pad in the first and third months ($p < 0.05$); and blood flow increase on the plantar foot ($p < 0.05$) but not on the dorsal foot. The increase in blood flow may alleviate diabetic-related complications. The results of autonomic nervous testing indicated that the activity of sympathetic nervous increased after using the pad. In conclusion, the use of heating pad with far infrared radiation could improve blood flow and autonomic nervous system in diabetics, therefore improving alleviating symptoms of diabetes-related complications.

Keywords: Diabetes · Plantar temperature · Microcirculation
Autonomic nervous

© Springer International Publishing AG, part of Springer Nature 2019
I. L. Nunes (Ed.): AHFE 2018, AISC 781, pp. 156–163, 2019.
https://doi.org/10.1007/978-3-319-94334-3_17

1   Introduction

Diabetes mellitus is a metabolic dysfunction that controls blood glucose through insulin secreted by the human pancreas. Glucose enters the cells through the secretion of insulin. When the blood glucose concentration is high, insulin responds poorly to hypoglycemic reactions. It will lead to high blood sugar situation. Due to the body's lack of insulin secretion, or the body's immune system on the body of insulin-producing beta cells to attack, eventually leading to the body cannot produce enough insulin, causing protein, carbohydrate, and fat and other nutrients metabolism [1–4]. In the early stages of diabetes, most diabetics did not have obvious symptoms except for unpleasant sensations. It is estimated that there are currently 190 million people with diabetes in the world, and it should be improved that control of diabetes in medical care [5–9] Diabetes will be complicated by neuropathy, neuropathy is a high incidence of diabetes and a symptom, and neuropathy can be divided into (1) cerebral neuropathy, (2) sensory polyneuropathy, (3) exercise neuropathy; (4) autonomic neuropathy. About half of people with diabetes have high blood pressure. Controlling the deterioration of the disease and complications has become an important issue. Diabetic foot easily causes wounds, distal nerve abnormalities, and various degrees of peripheral vascular diseases, as well as related foot infections, ulcers, or deep tissue damage. Foot lesions are more common in patients with diabetes than in patients without diabetes [10] and foot ulcers affect not only the patient itself but also the quality of life of patients and the social costs.

For patients with poor blood circulation in the foot and easy cold in the feet [11], there is a clear improvement over a month or so as a sustained footbath. However, for people with diabetes mellitus, special attention should be referred to water temperature because these patients cannot normally feel the peripheral nerve can sense the environmental temperature, even if the water temperature is too high, peripheral neuropathy lead to numbness in temperature, easily burn, triggering very serious. Diabetic foot ulcers are also a common long-lasting symptom of diabetes mellitus. However, they are easily ignored because of many factors. They cause disturbance to patients' daily life and aggravates other physiological conditions. Therefore, the need for amputation in diabetic patients [12–15] can be reduced if an easy-to-use, noninvasive, and no side-effect-promoting foot blood circulation and soothing products are used. It can contribute to the better work-life of diabetic patients and reduce medical resources expenditure.

Therefore, the purpose of this study is to assess whether there is a real benefit to the circulation of the foot by far-infrared radiation. This study describes the use of far-infrared technology and constant temperature to assess whether plantar heating significantly contributes to the improvement of peripheral nerves and blood circulation in diabetic patients.

2   Methods

2.1   Participants

The subjects were separated into two groups. One group consisted of 10 diabetic patients with vascular disease symptoms and one group of 4 healthy non-diabetic patients. Both groups conducted far-infrared product evaluation, did not change the diet and lifestyle adjustments during the trial, and signed the consent form. These two groups of subjects had no open wounds and no amputation of the foot, they could walk without having to rely on aids. The diabetic group agreed to participate in the 3-month experiment, and the non-diabetic group agreed to participate in the 1-month experiment at least. The subjects recorded were tested for age, gender, height, weight, blood pressure, autonomic nervous activity, blood circulation in the foot, foot surface temperature.

2.2   Experimental Materials

The electrothermal material adopted in this study was provide by Link-Win Technology Inc, Taiwan, for experimental releasing of far-infrared textiles. The temperature of the material can be about 42°, the internal heating chip size is 18.5 cm long, 6.1 cm wide, 0.2 cm high, heating the host size 9.4 cm long, 6.7 cm wide and 1.8 cm high, each host uses four batteries. Far-infrared fiber materials release heat which can be absorbed quickly by the body, promoting subcutaneous temperature rise, body's microvascular dilatation and blood circulation (Fig. 1).

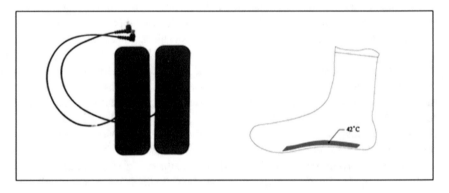

Fig. 1. Electric heating far-infrared heating chip and distribution.

When the human body feels pain, in addition to the physical sensation and psychological changes in mood, there are some changes in the physiological state including rapid heartbeat, increased blood pressure, increased adrenaline secretion, elevated blood sugar, slowed gastric activity. Therefore, in this study, ANSWatch was used for monitoring (ANSWatch monitor, TS-0411 manufactured by Taiwan Science Co., Ltd.). At the same time, the pulse signal and blood pressure were collected, and the autonomic nervous activity (autonomic nervous activity, sympathetic, parasympathetic, sympathetic/parasympathetic index) was analyzed (Fig. 2).

Fig. 2. ANSWatch monitor (TS-0411)

2.3   Experimental Design

Subjects wear electric far infrared socks. Before and after their use, measurements of heart rate, blood pressure, foot surface temperature, foot blood flow, and plantar pressure were taken and autonomic nerve analysis was carried out. Experimental socks were worn 30 min later, once again measurements of the subjects' heartbeat, blood pressure, foot surface temperature, foot blood flow were taken and autonomic nerve analysis was carried out. The patients were regularly while they wear the socks three times a week. Relevant data were measured every week and the subjects tracked for at least three months (Table 1).

Table 1. Subjects wear socks and the number of tests

	Wear times	Test times
Diabetics		
Per month	12 times	1 times
Total (3 months)	36 times	3 times
Non-diabetics		
Per week	3 times	
Per month	12 times	3 times
Total (1 month)	12 times	3 times

Note: The diabetic participants in the experiment were tracked three times per week, measuring the monthly record-related data for 3 months; non-diabetic patients were tracked 3 times a week, measuring records every two weeks.

2.4   Statistical Analysis

Statistical analysis was performed by using SPSS statistics 20 and the data was expressed as mean and standard deviation (x ± s). The significance test for the mean difference of each parameter was performed using the t-test, with $p < 0.05$ representing the significance of a statistically significant difference. We will also observed $p < 0.1$ that does not represent a statistically significant significance. The two groups were compared by paired t-test; one-way ANOVA was used to compare the three groups and the post-hoc LSD method was used for comprehensive comparison. Paired-sample t-tests were performed before and after each month, and three-way one-way ANOVA was used before and after the three-month experiment.

3   Results and Discussion

3.1   Foot Plantar Temperature

In diabetics, the first month using the Fluke-TiS20 thermal imager results show that there is a significant difference ($p < 0.05$) between the right foot temperature and the temperature of both feet. The average temperature of left foot, right foot and feet both increased significantly ($p < 0.05$). Average temperatures of non-diabetic patients increased about 6% before and after the first experiment, about 6% before and after the second experiment, and about 5% before and after the third experiment with significant differences (Fig. 3).

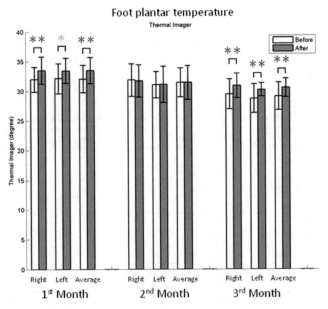

Fig. 3. Three months before and after the thermal imaging instrument foot temperature in diabetics. (*$p < 0.1$, **$p < 0.05$.)

3.2   Foot Blood Flow

In the second month before and after the experiment measured left foot blood flow increased significantly ($p < 0.05$), no significant changes in other values, blood flow before and after the experiment, all the first month without change in the second month left foot Ascension, all positions unchanged in the third month. In non-diabetic group, left flux first metatarsal head blood flow (Left Flux) about 11% before and after the first experiment increased about 3% before and after the second experiment, before and after the third experiment. The right lower extremity first metatarsal head blood. The flow rate (Right Flux) increased about 194% before and after the first experiment, about 10% before and after the second experiment, and about 40% before and after the third experiment (Fig. 4).

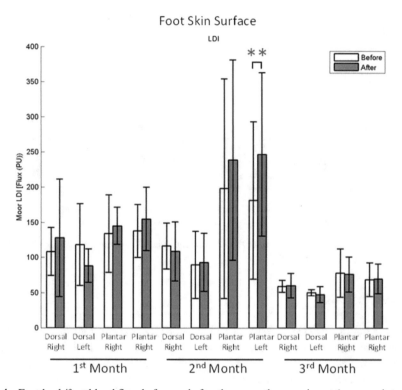

Fig. 4. Foot back/foot blood flow before and after three months experiment (mean and standard error)

3.3 HRV Analysis

In diabetic group, Sympathetic/parasympathetic differences were not significant before and after the first month ($0.05 < p < 0.1$). There was no significant change in the first month, the second month and the third month before and after the autonomic nerve activity test. In non-diabetic group, the total autonomic nerve activity (HRV) increased by about 121% before and after the first experiment and about 35% before and after the second experiment, with significant differences. Before and after the third experiment, the increase was about 99%. Sympathetic/Parasympathetic Balance Index (LF/HF) increased about 60% before and after the first experiment, and about 143% before and after the second experiment (Fig. 5).

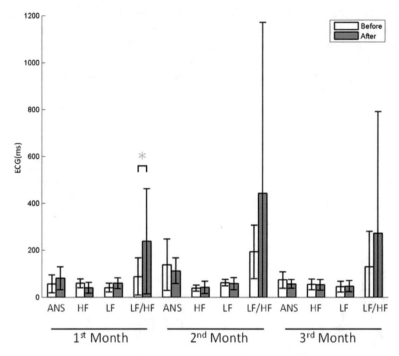

Fig. 5. Three months before and after the experiment of heart rate variability analysis in diabetic group. (mean and standard error)

4   Conclusions

We tested the use of electric heating pads to improve diabetic foot blood circulation, experimental results show that heating pads can promote lower limb blood circulation in diabetic patients. At the same time they will affect the autonomic nervous system, but may be affected by terminal neuropathy. The results show that the biological effects of infrared light can improve the blood circulation of the foot at a suitable temperature.

Acknowledgments. We are grateful for the Ministry of Science and Technology in Taiwan for financially supporting this research under contracts MOST 106-2221-E-241-002 and MOST 105-2221-E-241-006.

References

1. Adler, A.I., Boyko, E.J., Ahroni, J.H., Smith, D.G.: Lower-extremity amputation in diabetes. The independent effects of peripheral vascular disease, sensory neuropathy, and foot ulcers. Diab. Care **22**(7), 1029–1035 (1999)
2. Allet, L., Armand, S., Golay, A., Monnin, D., de Bie, R.A., de Bruin, E.D.: Gait characteristics of diabetic patients: a systematic review. Diab. Metab. Res. Rev. **24**(3), 173–191 (2008)

3. American-Diabetes-Association: Consensus development conference on diabetic foot wound care: 7–8 April 1999, Boston, Massachusetts. American Diabetes Association. Diab. Care **22** (8), 1354–1360 (1999)
4. Anjos, D.M., Gomes, L.P., Sampaio, L.M., Correa, J.C., Oliveira, C.S.: Assessment of plantar pressure and balance in patients with diabetes. Arch. Med. Sci. **6**(1), 43 48 (2010)
5. Armstrong, D.G., Lavery, L.A., Bushman, T.R.: Peak foot pressures influence the healing time of diabetic foot ulcers treated with total contact casts. J. Rehabil. Res. Dev. **35**(1), 1–5 (1998)
6. Armstrong, D.G., Lavery, L.A., Harkless, L.B.: Validation of a diabetic wound classification system: the contribution of depth, infection, and ischemia to risk of amputation. Diab. Care **21**(5), 855–859 (1998)
7. Ashry, H.R., Lavery, L.A., Murdoch, D.P., Frolich, M., Lavery, D.C.: Effectiveness of diabetic insoles to reduce foot pressures. J. Foot Ankle Surg. **36**(4), 268–271 (1997)
8. Boyko, E.J., Ahroni, J.H., Stensel, V.L.: Skin temperature in the neuropathic diabetic foot. J. Diab. Complicat. **15**(5), 260–264 (2001)
9. Cameron, N.E., Eaton, S.E.M., Cotter, M.A., Tesfaye, S.: Vascular factors and metabolic interactions in the pathogenesis of diabetic neuropathy. Diabetologia **44**(11), 1973–1988 (2001)
10. Faxon, D.P., Fuster, V., Libby, P., Beckman, J.A., Hiatt, W.R., Thompson, R.W., Topper, J. N., Annex, B.H., Rundback, J.H., Fabunmi, R.P., Robertson, R.M., Loscalzo, J.: Atherosclerotic vascular disease conference. Circulation **109**(21), 2617–2625 (2004)
11. Ikem, R., Ikem, I., Adebayo, O., Soyoye, D.: An assessment of peripheral vascular disease in patients with diabetic foot ulcer. Foot (Edinb.) **20**(4), 114–117 (2010)
12. Izuhara, M.: Relationship of cardio-ankle vascular index (CAVI) to carotid and coronary arteriosclerosis. Circ. J.: Off. J. Jpn. Circ. Soc. **72**(11), 1762–1767 (2008)
13. Jiang, Y.-D.: Incidence and prevalence rates of diabetes mellitus in Taiwan: analysis of the 2000–2009 Nationwide Health Insurance database. J. Formos. Med. Assoc. **111**(11), 599–604 (2012)
14. Lavery, L.A., Higgins, K.R., Lanctot, D.R., Constantinides, G.P., Zamorano, R.G., Athanasiou, K.A., Armstrong, D.G., Agrawal, C.M.: Preventing diabetic foot ulcer recurrence in high-risk patients: use of temperature monitoring as a self-assessment tool. Diab. Care **30**(1), 14–20 (2007)
15. Mayfield, J.A.: Preventive foot care in diabetes. Diab. Care **27**(Suppl. 1), S63 (2004)
16. Morales, F., Graaff, R., Smit, A.J., Bertuglia, S., Petoukhova, A.L., Steenbergen, W., Leger, P., Rakhorst, G.: How to assess post-occlusive reactive hyperaemia by means of laser Doppler perfusion monitoring: application of a standardised protocol to patients with peripheral arterial obstructive disease. Microvasc. Res. **69**(1–2), 17–23 (2005)
17. Namekata, T.: Establishing baseline criteria of cardio-ankle vascular index as a new indicator of arteriosclerosis: a cross-sectional study. BMC Cardiovasc. Disord. **11**(1), 51–61 (2011)
18. Petrofsky, J.: The use of laser Doppler blood flow to assess the effect of acute administration of vitamin D on micro vascular endothelial function in people with diabetes. Phys. Ther. Rehabil. Sci. **2**(2), 63–69 (2013)
19. Qiu, X., Tian, D.-H., Han, C.-L., Chen, W., Wang, Z.-J., Mu, Z.-Y., Liu, K.-Z.: Plantar pressure changes and correlating risk factors in Chinese patients with type 2 diabetes: preliminary 2-year results of a prospective study. Chin. Med. J. **128**(24), 3283–3290 (2015)
20. Robinson, C.C., Klahr, P.D.S., Stein, C., Falavigna, M., Sbruzzi, G., Plentz, R.D.M.: Effects of monochromatic infrared phototherapy in patients with diabetic peripheral neuropathy: a systematic review and meta-analysis of randomized controlled trials. Braz. J. Phys. Ther. **21** (4), 233–243 (2017)

Microinteractions of Forms in Web Based Systems Usability and Eye Tracking Metrics Analysis

Julia Falkowska[1(✉)], Barbara Kilijańska[2], Janusz Sobecki[1], and Katarzyna Zerka[3]

[1] Wroclaw University of Science and Technology, Wrocław, Poland
{julia.falkowska,janusz.sobecki}@pwr.edu.pl
[2] University of Wrocław, Wrocław, Poland
barbara.kilijanska@gmail.com
[3] Polish Academy of Science, Warsaw, Poland
katarzynazerka@gmail.com

Abstract. Microinteraction is a small function dedicated to a single task. Four elements can be specified in its structure: a trigger, rules, feedback loops, and modes. Triggers and feedback give visual hint to the user, thus the interaction might be investigated in eye tracking studies. We were aiming to investigate the usefulness of microinteraction model in eye tracking user experience studies and propose metrics that are able to capture the effect. The second goal was to validate influence of different triggers and instant feedback on the usability of web-based data-entry forms. Reported experiment results show shorter time ($p < .05$) needed to complete the form while placeholders are being used as triggers and visual in-line feedback applied. We proposed also eye tracking metrics useful in further studies on microinteractions.

Keywords: Usability · User experience · Microinteractions · Eye tracking

1 Introduction

To ensure proper user experience (UX) of digital products and systems is a motivation for many academic and commercial projects. However, the whole UX community is now seeking an answer for questions: how to capture the best quality of experiencing the interaction, as well as, what kind of framework to use to design it better.

Designing UX is a fascinating process. All may agree that you should have a vision, some ultimate goal, and see the big picture when starting this job. But, "the magic is all in the details" [1]. Apparently, paying attention to small task-related interactions shapes the impression that one has while interacting with the system. You could say that UX and ergonomics relay on microinteractions quality.

A microinteraction is a small function within a larger system that is dedicated to a single task [2]. It greatly affects usability of the system and the overall UX. To design good microinteractions it is crucial to know the context of usage and develop empathy with users. This knowledge can be gathered by doing research and validating the

I. L. Nunes (Ed.): AHFE 2018, AISC 781, pp. 164–174, 2019.
https://doi.org/10.1007/978-3-319-94334-3_18

microinteractions quality. Both qualitative and quantitative research are being done on the matter. The second is mostly done using A/B testing of a product feature on a very large sample of users. Both methods do not bring a full understanding why a certain microinteraction works well or not, and why it leaves a certain impression. To fully explore every single step of microinteraction, proposed by Saffer, you should know when and how attention and reactions are distributed. To examine that, we employed eye tracking methodology.

2 User Experience

User Experience (UX) can be defined as users' judgment of a product arising from their experience of interaction and qualities of the product, which engender the effective use of the product and user's pleasure [3]. This goes far beyond using the graphical user interface (GUI) itself. Nevertheless, in Human-Computer Interactions studies we are focused to assess and design this part the most.

We could say that the User Experience (UX) field aims at establishing the psychological reality of technology users [4]. The key is a focus on the user and his subjective experience. UX approach is a global one, taking into consideration more instrumental side of assessing human-technology interaction, like usability, but also exceeding it on usefulness, experienced emotions and hedonic qualities [4]. According to the International Organization for Standardization (ISO, 9241-210) user experience is defined as 'a person's perceptions and responses that result from the use or anticipated use of a product or a service' [5].

2.1 Usability

It is important to distinguish UX and usability. The official ISO 9241-11 definition of usability is: "the extent to which a product can be used by specified users to achieve specified goals with effectiveness, efficiency, and satisfaction in a specified context of use" [5]. So it is more a quality attribute of the user interface (UI), covering whether the system is easy to learn, efficient to use, stays in our memory, helps us achieve our goals in a pleasant manner [6].

Usability refers more to the instrumental aspect of the system, focusing mainly on improving the ease of usage. We can define systems usability by 5 components. Learnability refers to a situation when the user is using the system for a very first time and assesses the time it takes to learn how to accomplish basic tasks within this system. Efficiency is being measured by the time of task performance when the user is already familiar with the system. The usage efficiency after some period of time without contact with the system is memorability. While assessing usability the number of error, made by the user during task completion, is taken into consideration, as well as their severity (the more severe the more likely prevents from the task completion) and if the user can recover from errors. Satisfaction refers to the pleasure of task completion, which is a more narrow than the whole UX.

2.2   Utility

The next and very important attribute of system UX is utility, which refers to meeting users needs. Usability and utility together determine whether something is useful. Only now we can see the full picture, as it does not matter much if some system is easy if it does not what we want to accomplish thanks to it [7].

3   Microinteractions

A microinteraction is a small function within a larger system that is dedicated to a single task [2]. A microinteraction can be a whole product, a feature of a bigger system, or just a small part of a feature. The microinteraction takes a small moment that can be either engaging or dull. They are all around us, in every piece of the hardware or digital product. Every time we set up a home appliance, log in to our account in different ranges of products, check notification on your smartphone or set your alarm clock, you are engaged in a microinteraction. And as Charles Eames said they are not the detail, they are the design [1]. We pay little conscious attention to them, unless something goes wrong. Then only we realize of what importance they are to the experience we can provide to the users. In microinteraction' structure, four elements can be specified: a trigger, rules, feedback loops, and modes.

3.1   Trigger

A Trigger is an action that starts or signals a microinteraction. It can be manual, done by a user or system-initiated. The system starts a micro interaction when a specific set of conditions is met. A good example is when the phone mutes itself when it detects a meeting in your calendar. Basically, a trigger springs from users needs, answering the need of task accomplishment. This determines the affordances, accessibility, and persistence of the trigger.

Affordance is of a great importance to presented study. "...the term affordance refers to the perceived and actual properties of the thing, primarily those fundamental properties that determine just how the thing could possibly be used. [...] When affordances are taken advantage of, the user knows what to do just by looking: no picture, label, or instruction needed" [8]. The visual affordance constitutes the rule that if a trigger looks like a button, it should work like the one.

After the triggers initiate the microinteraction the sequence of behaviors is engaged. It also starts the rule of the functionality. The way the trigger works should be consistent over time to allow people to build a mental model of possible interactions. The system should also use all the data it has to provide smart defaults, or at least s good prompt.

As Saffer [1] suggests manual triggers are very often a collection of many elements, like form fields, or radio buttons. Logging-in, means text-entry fields to populate with users email and password, might be the most commonly known example of those.

3.2   Rules

Rules determine what happens, how our interactions behave and a frame for this behavior. To put it simply rules specify what user can and cannot do, how the microinteraction works. They are invisible. Users have to build the mental model of them while interacting with the system or product. The visual clue apart from affordance should be delivered as a feedback.

3.3   Feedback

Feedback lets people know what's happening, gives the system response. Might be expressed on different modalities, as a sound, haptic or visual indication. It can be anything that helps users to understand the rules. Our smartphones give us vibrations, LED light blinking or visual clues of all sort on the screen. Feedback should be easy to detect but also should not overwhelm the user. It is also one of the best channels to express system personality by design. Taking digital systems into consideration, feedback is, obviously, mostly visual.

The last components of microinteractions are loops and modes. They determine the meta-rules, allow to run microinteractions all and over again, specify the time whiting the microinteraction remains, how it repeats and what happens in a case of changed conditions. This is also a base for personalization, the learning that system gains with the user.

4   Eye Tracking in User Experience Studies

Usability and UX verification can be successfully enhanced with eye tracking, which is the process of determining where the user is looking [9]. Eye tracking also enables to measure different characteristics of the eye and its movements. Human eyes are moving all the time making jumps of attention few times per second. These jumps are called saccades and their purpose is to focus visual stimuli into the fovea, which is a very small area of the highest visual sensitivity on the retina. Foveal vision occupies only two degrees of the whole visual field and with the increasing distance from this point the image becomes more blurry and colorless. The actual information is perceived during fixation only.

Today eye tracking is more accessible than ever before, and its popularity is growing amongst user experience researchers [10] and very differentiated disciplines, from cognitive psychology to sports science [11].

4.1   Eyetracking Hardware

The first eye trackers were constructed at the end of the nineteenth century [11]. They were mechanical constructs, which was pretty much uncomfortable for the participants. Later on in the beginning of the last century the principle of the fovea reflection of the external light that was registered photographically was introduced by Dodge and Cline. Over 60 years ago a number of different eye tracking technologies have been developed, such as lens systems with mirrors, electromagnetic coil system, electrooculography

(EOG) that measured the variations of electromagnetic signal from participants eyeball muscles and very precise Dual Purkinje system.

Todays' eye trackers as Tobii TX-300 [12] that was applied in our experiment reported in this paper are using infrared light and a camera to determine both the participant pupil and cornea reflection positions. These two values after corresponding calibration with the screen displaying the sample stimuli enables to determine the exact position of the gaze point on the screen. However to free participants from the necessity to use chin rest device the head movement compensation is required. This is achieved by the eye position in the 3D space.

The eye trackers are characterized by the following values (in the brackets the actual values for Tobii TX-300 are given) [12]: monocular or binocular (both), sampling rates (60 Hz, 120 Hz, 250 Hz or 300 Hz), gaze precision that describes the special angular variation between samples ($0.01°$), gaze accuracy that describes the angular average distance from the actual and measured gaze points ($0.4°$), freedom of head movement measured as area height width in cm with at least visible eye by eye tracker (37×17 cm), and minimum and maximum operating distance (50–80 cm).

4.2 Eyetracking Measures

We can distinguish the following eye-tracking measures [11]: movement measures that concentrates of the whole variety of eye movements, position measures which corresponds only to where the participant is or is not looking, numerosity measures that pertain to the whole spectrum of the number, proportion or rate of any countable eye movement event, latency measures which give the values of the onsets between two events. However this is not the only typology of eye tracking measures. For example [9] Bojko distinguishes the following types of measures: mental workload measures, cognitive processing measures, target findability measures and target recognizability measures.

Nowadays we can distinguish far more than one hundred eye tracking measures. Here we would like to describe only the measures we will be using in the experiment described in this paper: average fixation duration, fixation count, time to first fixation and time from first fixation on target to target selection and total visit duration.

Average fixation duration belongs to cognitive processing measures [9]. It measures cognitive processing difficulty, the longer fixation indicates usually deeper cognitive processing caused for example by more effort of the information extraction. Fixation usually ranges from 100 ms to 500 ms. The average fixation duration for reading is 200–250 ms and for the scene viewing 280–330 ms [9].

Fixation count is usually used in presenting the accumulated by all participants results in a form of a heatmap [9]. Each fixation of each participant adds to the fixation count and then presented in the heatmap as appropriate color in the fixation area. In this case the time of the fixation is not taken into account.

Time to first fixation belongs to [9] the attraction measures, which is a superclass of area noticeability measures that are useful for visibility assessment of an object or area by telling how many people or how quickly it was noticed. Time to first fixation on the area of interest (AOI) measure is used when it matters how quickly an area was noticed. It should always be used with the percentage of participants who actually have looked at AOI. Shorter times indicate that the area has better attention-getting properties than areas of longer duration. The second measure belongs to target recognizability measures, which means that the user should understand that the target leads to the desired goal. The measure returns the time it takes from the moment of first fixation to the selection moment (i.e. clicking on it). Total visit duration at AOI measures the time spent on the observing the specific area by the participant. This usually indicates the participant's motivation and attention. The higher values of this measure for the specific AOI indicates more interest on this AOI than for those AOI's with the lower measure values.

5   The Experiment

When people are in a goal-oriented decision-making situation they actively search for something that meets their present needs. In consequence, they focus their attention, mostly, towards interesting visual objects. During this time our field of vision might shrink to 1° [13] only, what is around 1% of what we can typically see. Our eyes are searching then for familiar shapes. We assume that in the context of web pages human eyes capture triggers. When it is detected, we associate a trigger's affordance to it. A feedback is another visual clue. We predict that the feedback let human eyes go and look for another trigger, in the context of multi-step interaction of web based data entry form. The rules of forms are commonly understood by Internet users, even though visual hints may still help to accomplish the tasks quicker, with less errors. We predicted that the performance will be enhanced by the trigger that is an integral part of the text-entry field, instead of the label above, and visual, inline feedback, instead of the feedback given after the form is submitted.

In the summary, we present the study of two main goals. The main one is to investigate the usefulness of the microinteractions model in eye tracking user experience studies and propose metrics that are able to capture the effect. The second goal was to validate the following hypotheses:

1. Placeholders as triggers and inline visual feedback will enhance the usability of web based forms.
2. Placeholders as triggers will cause faster task completion.
3. Visual inline feedback will cause quicker proceeding to the next task.

We decided not to separate triggers and feedback changes, due to the available group of participants. That should be a matter for further studies that would contain more experimental groups.

5.1   Stimuli

To carry out the experiment we have created two sample web forms. These forms looked the same and contained exactly the same text-entry fields: first name, last name, email and password, additionally one question with four answers marked with radio-button, the last answer contains text-entry field. Both forms ended with a submit button. All the labels and description were in English.

The versions differed in terms of triggers and feedback used. Version 1 was the basic version, with the simplest and most common trigger, namely the label of the field was placed over it. The rules of password entry were stated below the field. In the basic version we used validation that checks the correctness after clicking the submit button, as the feedback. The button itself is all the time active, there is no visual clue whether the form is or is not finished successfully.

Version 2 had different triggers and feedback, in order to check the possible difference in performance. As the trigger in this version we used placeholders in text-entry fields. Also when someone picked the option "other" in our radio-button question, the text-entry field lit up. As a feedback we used inline validation with positive, green color feedback. The password field tips were lit green as soon as the user fulfilled one of the requirements. Also the submit button had less color saturation while the form was not yet completed.

5.2   Study Details

The study was conducted in March 2018 in the HCI laboratory at Wroclaw University of Science and Technology. Twenty students, 11 males and 9 females, participated in this study, all aged 20. Eye tracking data was recorded by Tobii TX300.

The study scheme:

- Eye tracker calibration.
- Instruction: Please fill in the form, all the fields are required.
- Completing one Form (randomized order).
- Competing neutral task: checking the weather on the internet.
- Completing one Form (randomized order).

6   Results

After quality analysis of the collected data we excluded the part of the faulty data sets covering one of the conditions for a certain participants. We included 11 data sets for Form 1 and 11 data sets for Form 2 for further investigation and treated the sample as independent. The sample contains data from 6 males and 5 females, all aged 20 and fluent in English.

6.1   Usability Metrics

We have used two usability metrics in our experiment: number of errors and the total time of task completion. Both metrics were based on the results of all the tasks (Table 1).

Table 1. The number of errors shown by validation of forms during the task completion

Form	Number of participants
1	5
2	3

Form 1 displayed errors after clicking submit button, Form 2 was providing inline instant feedback. All the errors, for both forms, were made while entering the password (Table 2).

Participants needed significantly less time to complete Form 2, $t(10) = 1.77$. $p = .046$.

Table 2. The average time of task completion

Form	Average time in sec.	St. deviation
1	77.35	28.31
2	59.51	17.65

6.2   Eyetracking Metrics

In eye tracking data analysis we have used Tobii Fixation Filter (velocity threshold: 15 pixels/window distance threshold: 15 pixels).

After qualitative analysis of the recordings we decided to set up a start point of data analysis on the moment of the first activation of the very first field in our forms (First name). It was done due to a habit of people to quickly scan the whole page before starting completing the task, what affects all the needed statistics. We also set the end point on the moment of first Submit button usage, which exclude error correction from eye tracking analysis.

We have defined areas of interest according to the microinteraction model and functions that they perform: the name of the field, the field, the field with the name, the question, the answer and the submit button (Figs. 1 and 2).

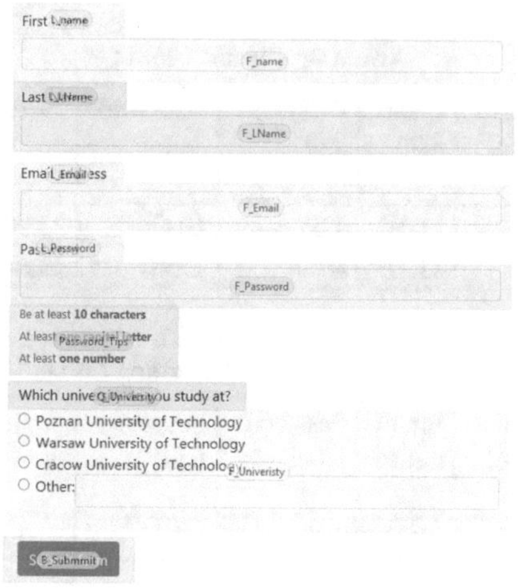

Fig. 1.   Area of interests: Form 1

Fig. 2. Area of interests: Form 2

For these AOI's we have determined the following metrics values: time to first fixation, time to first click (which we used in the metric presented later), fixation duration, fixation count and total visit duration (Tables 3 and 4).

Table 3. Fixation duration (sec)

Form	Last name label	Last name field	Email label	Email field	Password label	Password field	Password tips	Question	Answer
1	0.09	0.16	0.12	0.15	0.13	0.20	0.24	0.19	0.18
2	–	0.10	–	0.15	–	0.19	0.18	0.13	0.14

Table 4. Fixation count

Form	Last name label	Last name field	Email label	Email field	Password label	Password field	Password tips	Question	Answer
1	1.25	5.18	3.62	11.30	4.33	10.64	25.00	5.18	38.91
2	–	9.20	–	11.18	–	11.40	13.64	6.80	35.27

Table 5. Total visit duration (sec)

Form	Last name label	Last name field	Email label	Email field	Password label	Password field	Password tips	Question	Answer
1	0.11	0.82	0.42	1.74	0.55	2.13	5.91	1.01	7.34
2	–	0.93	–	1.65	–	2.21	2.54	0.91	5.09

We have not encountered significant differences between values of these metrics for Form 1 and Form 2 in presented metrics. Qualitative analysis may suggest though, that

Form 1 fixation duration may have a potential to behave as expected. The trigger contains the label and the field, so we should compare the sum of those two values in order to check the effectiveness of forms placeholders as triggers. Also Total Visit Duration on Password tips for Form 1 is higher (no significant), which encourage to investigate further on inline validation of password requirements (Table 5).

6.3 Trigger Effectiveness Metric

In order to examine the effectiveness of the trigger we proposed a metric, the difference between time to first field activation and time to first fixation on the trigger for each form element. To conduct the time analysis we agreed that time 0 for each participant is the moment of activation of the very first field (First name). Thus, it is excluded for this analysis. We counted the difference between the time to the text-entry field activation (the participants used either mouse clicks or tab option and we used the time of the click or the event respectively) and the time to first fixation on our trigger (Form 1 - label and text-entry field, Form 2 - text-entry field). For Form 1 we combined AOIs for the label and the field, to prevent the situation that scanning the form from below affects the metric (Table 6).

Table 6. The difference between time to first field activation and time to first fixation on the trigger for each form element.

Form	First name	Last name	Email	Password	Question
1	–	0.92	3.83	9.98	8.73
2	–	0.47	0.51	7.12	9.25

There was only one significant difference in the scores for Email field, Form 1 (M = 3.38, SD = 9.85) and Form 2 (M = 0.51, SD = 11.60), t(10) = 1.96, p = .033.

6.4 Experiment Findings

Basic usability metrics show, that Form 2 performs better in terms of the number of errors and the time needed to complete the task. We cannot conclude if this is an effect of the trigger, feedback or interaction between them.

Our proposed Trigger effectiveness metric (the difference between time to first field activation and time to first fixation on the trigger for each form element) was aiming to evaluate effectiveness of the trigger. Email field metric is statistically different according to the hypothesis. Qualitative analysis shows that two first fields (First name and Last name) were often scanned at the beginning of the task, most likely to familiarize with the task. Thus, in the future, we recommend to investigate longer forms, to be able to exclude elements placed at the beginning.

The feedback performance is visible in the data concerning Password field, as it was the most complex one. We analyzed eye tracking data up to the point of first submission and the data for password field are comparable. However, five participants spent additional, relatively long time, to correct validation errors. Thus, password instant validation contributed significantly to the total time of task completion.

7 Discussion

The results obtained in our experiment have several shortcomings. The study was made with the too small sample, especially for between-subject analysis that we conducted. Our goal was to minimalize variance among the participant groups and obtain more power of test that is appropriate for the chosen sample size thanks to within-subject design [14]. During data cleaning we had to exclude some of the data sets, thus the sample of the study was treated as independent.

We also designed the study aiming to test two microinteractions elements at once, having in mind that this is exploratory approach and this investigation should be repeated with more numerous experimental groups, to properly control all of the conditions. Microinteraction model seems to be promising as a framework for conducting a variety of user experience studies. The presented study shows how useful the eye tracking approach might be in validating different aspects of the model and interactions between its elements. Taking into consideration significant results for Email field and the total task completion time, the further studies with more power seem to be promising. We hope to encourage future research on the matter.

References

1. Saffer, D.: Microinteractions. O'Reilly Media Inc., USA (2014)
2. McDaniel, R.: Understanding microinteractions as applied research opportunities for information designers. Commun. Des. Q. **3**(2), 55–62 (2015)
3. Sutcliffe, A.: Designing for user engagement: aesthetic and attractive user interfaces. Synth. Lect. Hum. Cent. Inform. **2**(1), 1–55 (2009)
4. Van Der Linden, J., Amadieu, F., Van De Leemput, C.: User experience: a plural structure varying according to interaction types and social support. In: Human Computer Interaction (2017)
5. ISO 9241-210:2010(en): Ergonomics of human-system interaction—part 210: human-centred design for interactive systems
6. Nielsen Norman. https://www.nngroup.com/articles/definition-user-experience/
7. Nielsen. https://www.nngroup.com/articles/usability-101-introduction-to-usability/
8. Norman, D.A.: The Design of Everyday Things, p. 9. Doubleday, New York (1988)
9. Bojko, A.: Eye Tracking the User Experience. Rosenfeld, New York (2013)
10. Weichbroth, P., Redlarski, K., Garnik I.: Eye-tracking web usability research. In: Computer Science and Information Systems (FedCSIS) (2016)
11. Holmqvist, K., Nyström, M., Andersson, R., Dewhurst, R., Jarodzka, H., Van de Weijer, J.: Eye Tracking: A Comprehensive Guide to Methods and Measures. Oxford University Press, Oxford (2011)
12. Tobii TX-300 Eye Tracker. User Manual, Release 2, Tobii Technology (2014)
13. Eriksen, C., St. James, J.: Visual attention within and around the field of focal attention: a zoom lens model. Percept. Psychophys. **40**(4), 225–240 (1986)
14. Duchowski, A.: Eye Tracking Methodology: Theory and Practice. Springer, London (2007). Chap. 1

Design and Realization of Shooting Training System for Police Force

Bo Shi[(✉)]

Command and Tactics Department, Henan Police College,
Zhengzhou 450046, China
54216012@qq.com

Abstract. Shooting is one of the essential skills of the police; the safe use of weapons is a basic content of the training of police force. In recent years, shooting training system becomes a research hotspot. The cost of traditional military shooting training system is rather high, and one system can only provide for one person at one time. Basing on it, the Shooting Training System for Police Force was designed and realized in the paper. To begin with, Virtual Reality technology was discussed. Then, the system functions and structure of the Shooting Training System for Police Force were proposed in the paper. Furthermore, the teaching function of Shooting Training System for Police Force was analyzed in detail. Finally, the Shooting Training System for Police Force was realized. This system can be used for multi-people simultaneously training, and has great flexibility and low-cost advantages.

Keywords: Police shooting training · Virtual reality · Computer software
Image processing

1 Introduction

Shooting is one of the essential skills of the police, the safe use of weapons is a basic content of the training of police force. The current shooting course is mainly on basic training, understand the basic structure of common firearms, learning the basic knowledge of shooting, to grasp the common firearms' decomposition, combination and general troubleshooting through teaching, to grasp Various basic technologies through training, but live-fire can test the result of the shooting course training. Live-fire is a comprehensive assessment of the trainee's psychology, skills and overall quality. To meet the qualified standards though the assessment, you need to train hard in daily training, make full use of teaching resources, use of advanced high-tech simulation training system, to achieve the purpose of teaching and training [1].

2 Virtual Reality

Virtual reality (VR) is a computer-generated scenario that simulates a realistic experience. The immersive environment can be similar to the real world in order to create a lifelike experience grounded in reality or sci-fi. Augmented reality systems may also be

© Springer International Publishing AG, part of Springer Nature 2019
I. L. Nunes (Ed.): AHFE 2018, AISC 781, pp. 175–183, 2019.
https://doi.org/10.1007/978-3-319-94334-3_19

considered a form of VR that layers virtual information over a live camera feed into a headset, or through a smart phone or tablet device [2, 3].

Current VR technology most commonly uses virtual reality headsets or multi-projected environments, sometimes in combination with physical environments or props, to generate realistic images, sounds and other sensations that simulate a user's physical presence in a virtual or imaginary environment. A person using virtual reality equipment is able to look around the artificial world, move around in it, and interact with virtual features or items. The effect is commonly created by VR headsets consisting of a head-mounted display with a small screen in front of the eyes, but can also be created through specially designed rooms with multiple large screens [4].

VR systems that include transmission of vibrations and other sensations to the user through a game controller or other devices are known as haptic systems. This tactile information is generally known as force feedback in medical, video gaming and military training applications [5].

VR is used to provide learners with a virtual environment where they can develop their skills without the real-world consequences of failing. It has also been used and studied in primary education. For example, in Japan's online high school, VR plays a major role in education. Even the school's opening ceremony was a virtual experience for 73 of the students: they received headsets, which were connected to the campus hundreds of miles away – so they got to listen to the principal's opening speech without having to travel so far. According to the school's workers, they wanted to give the students a chance to experience VR technology, before having to use it live as part of their education. The specific device used to provide the VR experience, whether it be through a mobile phone or desktop computer, does not appear to impact on any educational benefit.

Thomas A. Furness III was one of the first to develop the use of VR for military training when, in 1982, he presented the Air Force with a working model of his virtual flight simulator the Visually Coupled Airborne Systems Simulator. The second phase of his project, which he called the Super Cockpit, was even more advanced, with high resolution graphics and a responsive display [6]. Furness is often credited as a pioneer in virtual reality for this research. The Ministry of Defense in the United Kingdom has been using VR in military training since the 1980s. The United States military announced the Dismounted Soldier Training System in 2012. It was cited as the first fully immersive military VR training system. Figure 1 is U.S. Navy personnel using a VR parachute training simulator.

Fig. 1. U.S. Navy personnel using a VR parachute training simulator

NASA has used VR technology for twenty years. Most notable is their use of immersive VR to train astronauts while they are still on Earth. Such applications of VR simulations include exposure to zero-gravity work environments and training on how to spacewalk. Astronauts can even simulate what it is like to work with tools in space while using low cost 3D printed mock up tools.

A head screen-wearing soldier sits at a gunner station while learning in a Virtual Training Suite. Flight simulators are a form of VR pilot training. They can range from a fully enclosed module to a series of computer monitors providing the pilot's point of view. By the same token, virtual driving simulations are used to train tank drivers on the basics before allowing them to operate the real vehicle. Similar principles are applied in truck driving simulators for specialized vehicles such as fire trucks. As these drivers often have less opportunity for real-world experience, VR training provides additional training time. Figure 2 is a head screen-wearing soldier sits at a gunner station while learning in a Virtual Training Suite.

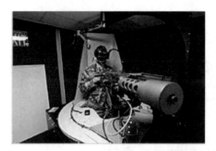

Fig. 2. A head screen-wearing soldier sits at a gunner station while learning in a Virtual Training Suite.

VR technology has many useful applications in the medical field. Simulated surgeries allow surgeons to practice their technical skills without any risk to patients. Numerous studies have shown that physicians who receive surgical training via VR simulations improve dexterity and performance in the operating room significantly more than control groups. Through VR, medical students and novice surgeons have the ability to view and experience complex surgeries without stepping into the operating room. On April 14, 2016, Shafi Ahmed was the first surgeon to broadcast an operation in virtual reality; viewers followed the surgery in real time from the surgeon's perspective. The VR technology allowed viewers to explore the full range of activities in the operating room as it was streamed by a 4 K 360fly camera.

3 Design of Shooting Training System for Police Force

Police shooting system is a new type of simulation shooting system that combines the computer multimedia technology with the acoustic automatic target-reporting technology. The training method is shoot the actual case scenario into teaching video or 3D animation video, controlled by computer, put it on the big screen through the projector, to produce the effects of actual scene environment and sound. Shooting trainee will practice the reaction ability of use firearms to shoot and the ability to deal with emergencies based on the various emergencies that displayed on the screen, and the moment of shooting and the location of target hit will be quickly recorded into the computer, to make judgments and statistics to the results finally, in order to improve trainee's level of shooting training.

3.1   System Functions

Random play the training program of real scenery image, bomb point recognition process, photoelectric bomb number statistics system, play the bomb point image, camera control and software detection system, personnel and database management.

3.2   System Structure

The system structure is Fig. 3.

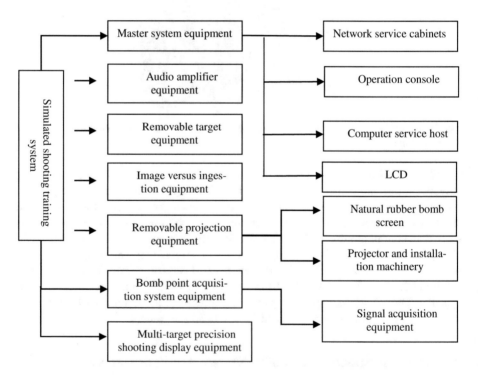

Fig. 3. The system structure of Shooting Training System for Police Force

4 The Teaching Function of Shooting Training System for Police Force

In order to achieve close actual combat, high simulation teaching with the unity of actual combat and training, the training content of interactive simulation teaching system be designed from different directions and angles. Figure 4 is the screenshots of the fixed target.

Fig. 4. The screenshots of the fixed target

4.1 Live-Fire Training System

In shooting teaching, the training content is edited into image video material through technical means, video materials of shooting training is processed, mark the bandits and hostages, occlusions and explosives, bunkers and non-bunkers in video, after computer processing, if the target is hit accurately, it will be shown qualified on the monitor by the edit system, and hostages and explosives were shown as unqualified. After training, the user can use the actual combat video training, and can quickly and accurately judge the trainee's shooting skill and actual combat ability. System provides training subjects with different content, there are various targets such as the chest ring target, half target,

Fig. 5. The picture of show hidden target

human target, body part target, face memory target, the movement pattern has fixed target, hide–show, up-down, move straight, appear randomly, and so on. According to the difference of training content, different target position and movement methods are adopted. Figure 5 is the picture of show hidden target and Fig. 6 is the picture of straight line moving target.

Fig. 6. The picture of straight line moving target

4.2   The Man-Machine Interaction Combating Shooting Training

Using high-tech simulation technology to reproduce an emergency that may occur in actual combat, the system breaks the traditional projection mode of dual-screen apart, adopts the dual-screen docking technology, and realizes the real scene from multiple angles. The screen is split into two screens when the trainees are combating shooting, and the dual-screen is rolling up into one screen when the trainees are simulated shooting or live-fire shooting. The system can realize interact live firing to trainees, criminals, thugs or terrorists holding weapons in the screen, use man-machine combating launchers to track training students automatically, carry on mutual shooting to the person in the screen through the laser signal. The system collects panoramic composite image video, and plays the scene according to the instructor's request, reduces the students' adaptability to program, and carries on the image confrontation training. The dynamic characters and unexpected events in the scene appear randomly, so we can realize human-computer interaction. According to the record of the trainee's shooting time and the shooting position, the students' shooting level can be judged effectively.

4.3   The Live-Action Video Combating Shooting Training

The shooting site is divided into A, B two areas, and the two sides reflect the actual scene synchronously through their respective screen. The system uses the cameras to transmit the scene to each other. The participants conduct tactical training on their own site, and carry on shooting up to the other person's projection scene, and the system analyzes the position of the bullet points through the computer and judges the trainees to be hit. Using real scene to shoot training, practicing two sides using terrain available, and combining with the tactical action, the system increase the actual effect and is very practical. So the system can be used in practical live firing or laser simulated shooting. The interactive live-fire simulation system is a comprehensive shooting training system which is developed by modern high technology. It is close to the actual combat. The design of the system is closely related to training requirements of police skills, achieving the unity of combat and training, trading gun speed, shooting speed and actual combat in particular, strengthening Sagittarius's ability to react quickly, quickly judge and deal with emergencies, so as to improve the teaching and training of shooting course.

5   Realization of Shooting Training System for Police Force

Based on semi-physical simulation technology, image acquisition and processing technology, 3D positioning technology, intelligent control technology, ballistics and other related theoretical technologies, virtual range shooting training system is one of the virtual shooting systems which are developed by adopting design viewpoint of system engineering and development thought of software engineering. And virtual range shooting training system is designed and developed on a certain platform.

The characteristics of the virtual range are not limited by time, climate and site. It has the ability to provide training and assessment for training personnel, also can provide various types and multiple range shooting according to training requirements. Only one controller can easily control the shooting process. And all shooting result is controlled by computer. The final result can be saved both in print and in database storage, avoiding the phenomenon of artificial modification.

The virtual range shooting training system simulates various types of targets through large screen projection, controls and simulates the state of the target using software. According to the requirements of the training course, the different parameters can be adjusted in real time to meet the requirements of various training subjects. According to the requirements, it can also modify the color, appearance and size of the target, adjust the distance and motion of the target, set weather conditions for shooting for representing a variety of meteorological conditions in the virtual scene, and change the appearance of the range. It can make full use of the remodeling of the virtual object to meet the training needs of training subjects.

Virtual shooting countermeasures system can provide virtual city streets, buildings, underground passageways and other kinds of scenes to help trainees to drill tactical movements and improve the speed and accuracy of shooting.

Through carrying out 3D modeling of scenes to local scenes such as city streets, shopping malls, buildings, underground tunnels, etc., making a building model that is basically the same as the actual situation, then texture mapping of the model according to background of actual photo and physical photos, we can get realistic virtual scenes. A virtual soldier with a certain tactical action capability is set up to imitate the enemy in the virtual scene, which can simulate the atmosphere of street fighting, trainers can shoot virtual soldiers through simulated firearms. At the end of the training, the system can be graded according to the performance of the trainees.

182   B. Shi

The above screenshot shows a virtual scene in a corner of an underground passageway. When the view is close to the enemy, the enemy emerges from the bunker and takes aim at the shooting. An enemy has been killed. Figure 7 is the picture of underpass.

The above screenshot shows a virtual scene in a corner of an underground passageway. When the view is close to the enemy, the enemy emerges from the bunker and takes aim at the shooting. An enemy has been killed.

Fig. 7. The picture of underpass

6 Conclusion

Building in an indoor environment, the police shooting training system occupy less space and not affected by the weather. Moreover, the target object of shooting is realized by computer software. So with great flexibility, the system can be replenished and updated continuously, no need to add new hardware. Highly automation of virtual shooting, so only a small number of security personnel can complete the process control of shooting training. The virtual shooting system can also be used for defense education. For the youth facing military training, there are both education meaning and entertainment, and there is no danger of live-fire shooting. Based on the virtual shooting platform, different virtual scene drivers and management programs are developed, finally the shooting system of a virtual shooting range and a virtual anti - firing system are realized, the paper achieve the expected goal.

References

1. Nieuwenhuys, A., Oudejans, R.R.D.: Training with anxiety: short- and long-term effects on police officers' shooting behavior under pressure. Cogn. Process. **12**(3), 277–288 (2011)
2. Negrotti, M.: Virtual reality. In: The Reality of the Artificial. Studies in Applied Philosophy, Epistemology and Rational Ethics, vol. 4, pp. 131–133. Springer, Heidelberg (2012)
3. Sisto, M., Wenk, N., Ouerhani, N., Gobron, S.: A study of transitional virtual environments. In: Proceedings of 2017 International Conference on Augmented Reality, Virtual Reality and Computer Graphics, pp. 35–49 (2017)
4. Knodel, M.M., Lemke, B., Lampe, M., Hoffer, M., Gillmann, C.: Virtual reality in advanced medical immersive imaging: a workflow for introducing virtual reality as a supporting tool in medical imaging. In: Computing and Visualization in Science, pp. 1–10 (2018)
5. Kitazaki, M., Hirota, K., Ikei, Y.: Minimal virtual reality system for virtual walking in a real scene. In: Proceedings of 2016 International Conference on Human Interface and the Management of Information, pp. 501–510 (2016)
6. Liu, Y., Hu, J., Cui, P.: An implementation approach for interoperation between virtools and HLA/RTI in distributed virtual reality system. In: Proceedings of 2015 6th International Asia Conference on Industrial Engineering and Management Innovation, pp. 293–299 (2016)

An Examination of Close Calls Reported Within the International Association of Fire Chiefs Database

James P. Bliss[(⊠)] and Lauren N. Tiller

Department of Psychology, Old Dominion University, Norfolk, VA 23529, USA
{JBliss, LTill002}@ODU.edu

Abstract. Close calls or near misses are common occurrences in the field of firefighting. As in many other domains, they represent a source of valuable information to inform personnel training and improve safety. Firefighting is uniquely complex because of the many possible causes of injury or death. Database repositories for close call narrative reports exist; however, they contain a wide variety of information that varies in usefulness. The authors applied a published taxonomy to categorize narrative reports within the www. firefighternearmiss.com online database. The authors analyzed 61 narratives from the database, categorizing them by situational variables, level of situation severity, and reported error frequencies. Results showed that the majority of narratives described close call events that were preceded by a sensory signal and attributable to human error. Notably, the vast majority of narratives included information from which actionable recommendations could be generated. The article concludes with suggestions for future database development.

Keywords: Firefighting · Close calls · Near misses · Severity
Database

1 Introduction

On average, firefighters in the United States respond to a fire call every 24 s [1]. To do so, they perform their work in an environment that includes a plethora of dangers. Toxic fumes, heat, falling and collapsing objects, chemical exposure, and even hysterical victims pose significant threats on the job. The National Fire Protection Association has compiled statistics concerning firefighter mortality for decades. Their findings indicate that 68 firefighters lost their lives while on duty in 2015. This figure represents a slight increase from the 2014 figure (64) but continues a trend of lower fatality rates for firefighters since 2008 [1].

Mortality statistics such as those reported above do not comprehensively reflect the dangerous nature of firefighting. Situations abound in which firefighters encounter risk without fatal consequences. Such "close calls" or "near misses" are an important and frequent aspect of occupational task performance because they often serve as harbingers of more critical events. The National Safety Council has estimated that 75% of all accidents are preceded by at least one near miss [2]. However, compiling accurate

I. L. Nunes (Ed.): AHFE 2018, AISC 781, pp. 184–194, 2019.
https://doi.org/10.1007/978-3-319-94334-3_20

statistics about the frequency of firefighting close calls is difficult. There are numerous databases for reporting firefighting close calls (such as www.firefighterclosecalls.com and www.everyonegoeshome.com); however, such reports are voluntarily submitted and contain diverse information about the events.

One factor that greatly complicates the estimation of close call frequency is the nature of the term itself. The terms "close call" and "near miss" are discussed within the firefighting community. However, there is little consensus regarding what the terms actually mean. In some situations, escaping death is considered a close call. In other situations, escaping serious injuries, hospitalization, and associated medical treatment costs is a close call. Further complicating the terminology challenges, close calls may refer to risks posed to the firefighters themselves or risks to fire victims or perhaps risks posed to bystanders observing the event as it unfolds.

Available literature from applied domains and the scientific community are of minimal help to resolve the definition problem. The medical community has studied the nature of close calls for many years. A survey of their membership by the Institute for Safe Medication Practices [3] has resulted in a definition that stresses a close call as "an error that happened but did not reach the patient." The Agency for Healthcare Research and Quality defines a close call as when "A patient is exposed to a hazardous situation but does not experience harm either through luck or early detection." [4]. Clearly, the emphasis of that definition is on avoiding personal harm. In contrast, one environmental organization defines close calls ("near misses") as "minor accidents or close calls that have the potential for property loss or injury" [5]. Because firefighters are responsible for protecting lives and property, the definition of a close call may be expected to be complex and inclusive. Indeed, Dodson notes that a close call is an "unintentional, unsafe occurrence that could have produced an injury, fatality, or property damage; only a fortunate break in the chain of events prevented the undesirable outcome" [6].

Firefighting is not unique with regard to close calls. Close calls are a frequent occurrence in many complex task situations, such as aviation [7], medicine, surface transportation, construction [8] and even sports [9]. The generality of the close call phenomenon has rendered it difficult to study, though some researchers have attempted to construct theoretical models to enable prediction, exploitation and mitigation of close call events [10]. In contrast to the notion of "luck" or chance as a driver of close calls, the goal of research is to leverage past information to increase safety.

One challenge surrounding close call research is the difficulty in applying conventional theoretical frameworks to close calls. As noted by Bliss [11], traditional tools such as Signal Detection Theory [12] offer minimal understanding of close calls. Sometimes referred to as "Near Hits" [13], close calls often feature event stimuli that do not fit neatly into the traditional categories of hit, miss, false alarm, or correct rejection. Given the ambiguity surrounding close calls, our research team has worked to understand them more clearly. Toward that end, the lead author published a taxonomy of close calls in 2013 [11] that differentiates several categories of close calls (see Table 1). In the same article, potential solutions were proposed to identify reasonable methods for determining close call severity. These include determining the number of lives at risk during a close call episode, calculating the monetary cost associated with

Table 1. Categories of close calls proposed by Bliss (2013)

Close call category	Definition
Un-Signaled Close Calls (UCC)	Events that occurred with no accompanying sensory signal
Signaled Close Calls (SCC)	Events that were preceded or accompanied by a sensory signal
Response-Driven Close Calls (RCC)	Events that occurred as the result of operator action(s)
Neglected Close Calls (NCC)	Events that occurred as the result of operator inaction(s)
Event-Driven Close Calls (ECC)	Events that occurred or continued spontaneously
Vicarious Close Calls (VCC)	Events that occurred to someone else observed by the operator
Disregarded Close Calls (DCC)	Events that occurred but were ignored by the operator

development of the episode, determining the time available until the close call degraded to a critical event, or tracking operational parameters central to the situation.

The current research represents part of an ongoing effort to apply the taxonomic structure advocated within Table 1 to a series of applied task domains. Recently, Tiller and Bliss [7] addressed this goal by conducting an examination of the Aviation Safety Reporting System (ASRS). They demonstrated the relevance of the close call categories described in Table 1 to a collection of extracted ASRS incident reports. The authors also used recommendations discussed within Bliss's [11] article to construct a close call severity scale for near-collision events. Ultimately, conclusions showed that close calls could be successfully stratified by severity as well as taxonomic category. Stratification by severity level is of special importance as organizations struggle to develop tools for interpreting close call anecdotes and leveraging the information within them to improve personnel safety.

This article describes our attempt to extend our investigation to the firefighting community. Since 2012, the International Association of Fire Chiefs (IAFC) has made a concerted effort to establish a database within which close call events can be made available to the firefighting community (www.firefighternearmiss.com). The overriding goal of establishing the database has been to assist local firefighting agencies as they focus their training efforts. The intent of this investigation is to provide structure to the reports within the database.

2 Focus of This Investigation

For the current research, the authors studied the occurrences of close calls reported within the IAFC Near Miss reporting database. Individual reports submitted by firefighters were interpreted and categorized according to Bliss et al.'s close call taxonomy [11]. The ultimate purpose was to facilitate mining of the data so users can better identify and stratify close call information.

3 Method

3.1 The International Association of Fire Chiefs (IAFC) Near Miss Online Database

The near miss online database is a repository that includes thousands of close call narratives that have been voluntarily and anonymously submitted by firefighters. The mission of the database is to help reduce possible injury or death to firefighters or emergency medical service providers by applying local lessons globally [14]. Because of the voluntary nature of the database entries, the reports represent soft data and are subject to self-reporting bias. Reports are openly available for review from http://www.firefighternearmiss.com.

3.2 Search and Analysis Procedure

We began our inquiry of the IAFC Near Miss online database by downloading records from the website on April 1, 2017. The authors originally selected 284 reports that were submitted in 2014, 2015, 2016, and 2017 using the "Browse Reports" option from the database home page options. Accessing the "Top Tags" report filter, narratives were selected from among the following seven categories of problems: "Equipment Malfunction," "Personal Protective Equipment (PPE)," "Equipment Failure," "Equipment Damage," "Improper Routine Maintenance," "Unseen Hazards," and "Communications." After removing the duplicates of some reports that were submitted under multiple tag identifiers, 174 unique reports remained. Complicating the procedure, it became apparent that the maintainers of the IAFC Near Miss database revised their close call report submission format in mid-January, 2015. Therefore, reports that were submitted prior to mid-January, 2015 did not contain sufficient event information to be analyzed in terms of taxonomic category and severity. Of the 174 reports, only 76 reports met the criteria for comprehensive analysis.

An additional complicating factor involved the nature of the incidents themselves. In some cases, firefighters submitted close call narratives that constituted an actual tragedy (one or more individuals suffered major injuries or death). Therefore, researchers examined the database again, removing those narratives so that only actual close call entries remained. After doing so, there were 61 remaining narrative reports. Of those, six events occurred in 2015, 42 occurred in 2016, and 13 occurred in 2017.

3.3 Categorization of Close Calls

After downloading the event narratives, the authors categorized each event according to the taxonomic categories proposed by Bliss [11]. The close call categories included Un-Signaled Close Calls (UCC), Signaled Close Calls (SCC), Response-Driven Close Calls (RCC), Neglected Close Calls (NCC), Event-Driven Close Calls (ECC), Vicarious Close Calls (VCC), and Disregarded Close Calls (DCC).

The first step was to categorize each close call event as either a SCC or UCC. An event was classified as an SCC if the event narrator indicated the existence of a sensory signal (visual, auditory, olfactory, gustatory, or tactile stimulus). Because of the diversity of events reported, there were numerous types of signals possible that could have alerted operators to impending danger. For example, stimuli that were not intended as alarms, alerts, or warnings may still have indicated danger (for example, a misplaced hose or broken piece of wood). If an event was classified as a UCC, the close call was not anticipated by the reporting firefighter or other personnel.

During the second step, events were categorized as Response-Driven (RCC), Neglected (NCC), or Event-Driven (ECC) close calls. RCC's occurred when the unfolding event progressed to a close call situation due to a specific action performed by the reporting firefighter or other personnel. As an example, in one event the driver of a fire engine proceeded through an intersection even though the engine was not guaranteed the right of way. As a result, the engine narrowly avoided a collision with a civilian vehicle. Conversely, events were categorized as Neglected Close Calls (NCC's) if the reporting firefighter or other personnel failed to follow standard procedure either during the event or prior to the event. As an example, during a house fire call, a firefighter reported surveying the ground for powerlines but failed to examine the overhead powerlines also. This caused him to step on a downed wire (but receive no injury). Another example of negligence that contributed to a close call was a visibly worn tire on an emergency vehicle that resulted in tire failure (but no serious injuries).

Some reports indicated ECC's, which described events that progressed into close call situations without active intervention or interruption by the reporting firefighter or other emergency personnel. For example, one fire engine suffered brake failure during descent on a steep hill grade (but no firefighters were injured). For the last category, Vicarious Close Calls (VCC), narrators recounted close call situations they observed as bystanders rather than participants.

Each near miss event narrative was analyzed by isolating the following specific provided information: date of occurrence and date published, database tag, title of report, type of event, type of injuries, number of people with injuries, number of fatalities, and contributing factors. As noted by Gnoni and Saleh [15], one of the most important aspects of close call investigation is the perform a root cause analysis. Toward that end, we focused closely on contributing factors. Contributing factors could have included faulty decision making, human error, improper communications, ineffective training, lack of resources, poor situation awareness, improper individual action, equipment failure, operator fatigue, and/or degraded structural or physical conditions of building. Reports also included the reporter's subjective analysis of the event and an indication of whether unsafe acts were performed.

3.4   Determination of Close Call Severity

After determining the close call categories represented by each event narrative, the authors estimated the number of human lives placed at immediate risk had the close call evolved into a critical situation. The number of lives at risk was sometimes difficult to determine because of the presence of bystanders, unknown occupants in buildings, or vague or incomplete event reporting by the narrative author. In such cases, the two authors discussed the narrative in depth, coming to an agreement regarding the number of lives purportedly at risk.

Once all narratives were analyzed to determine the number of lives purportedly at risk, the authors constructed a severity taxonomy that included five levels. If the events reported jeopardized one person's life, the event was categorized as Level 1 close call (L1); if two lives were placed at risk, the event was categorized as a Level 2 close call (L2); if three lives were placed at risk, the event was a Level 3 close call (L3); if four lives were placed at risk, the event was a Level 4 close call (L4); if five or more lives were placed at risk during the event, the event was categorized as a Level 5 close call (L5). Our intent was to develop a severity scale for firefighting close calls that was objective and generalizable across the diverse range of scenarios represented within the narrative database.

4 Results

4.1 Data Coding Procedure

All narratives were jointly coded by the two experimenters. As noted above, the authors first determined whether each narrative described a close call. If the reported event resulted in significant injury (including lost work time or hospitalization) or death for one or more individuals, we excluded that narrative from further analysis. Of the 76 initial reports reviewed, 15 described conditions that resulted in significant injury or death. Therefore, our final set included 61 narrative reports.

Analysis of the close call types represented within each narrative yielded the data displayed in Fig. 1. Only three of the 61 narratives included only one type of close call situation; 58 narratives (95%) warranted at least two classifications (for example, a signaled close call that impacted someone besides the narrative author would be classified as both SCC and VCC). As shown in Fig. 1, the most common category of close call was the Signaled Close Call (N = 43, or 71% of narratives). By comparison, Event-Driven Close Calls, where the situation unfolded without direct interaction or intervention, was less common (N = 14, or 23% of narratives). Because all narratives were reported as close calls within the IAFC database, we did not apply the DCC (Disregarded Close Call) category label to any narratives.

Examining the number of lives at risk within each narrative revealed a bimodal pattern. Most of the narratives included either one life at risk or five (or more) lives at risk (L1 = 19; L2 = 11; L3 = 8; L4 = 5; L5 = 18; see Fig. 2). Figure 3 includes a more granular display of the same results. This is particularly useful for those events that represented the highest level of severity, because of the variability of lives at risk within that category (five or more). Generally, it appears that most close call events threatened fewer lives (the modal category was Level 1 severity). However, Fig. 3 shows that there was considerable variability in the high severity group. In addition, four cases in the high severity group required the experimenters to estimate the number

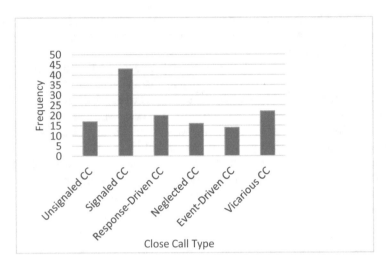

Fig. 1. Frequency of event narratives for each close call category.

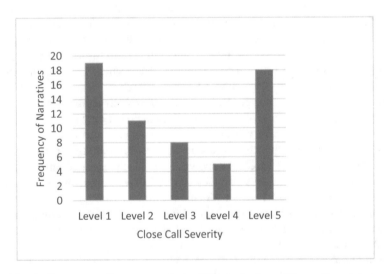

Fig. 2. Frequency of event narratives within each close call severity category.

of lives at risk because the narrative was unclear. For example, the author may not have specified how many bystanders were present or how many occupants were in an involved vehicle.

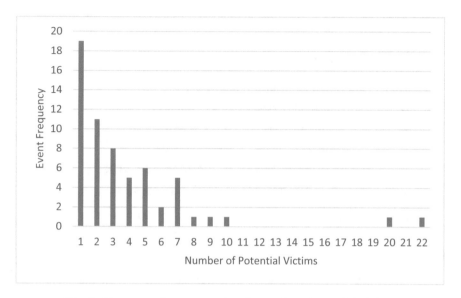

Fig. 3. Frequency of event narratives by number of potential victims.

A basic root cause analysis was performed by examining the cited event cause within each narrative (see Fig. 4). Approximately half (30, or 49%) of the narratives identified "human error" as a causal factor, 25 narratives (41%) identified "situation awareness," and 23 narratives (38%) identified "decision making." Eight narratives (13%) called out "training issues" and seven (11%) listed "problematic communications" as a contributing factor.

Further analysis of the root causes revealed specific causes associated with most events. In the vast majority of cases the causes pertained to the development of the initial unsafe operational environment. However, within five narratives the author made a direct reference to conditions that prevented the unsafe condition from resulting in injury or loss of life. The causes reported within the narratives reflected both errors of commission (N = 15 narratives) and errors of omission (N = 26 narratives). Errors of commission typically concerned firefighters following unsafe procedures such as opening the door of a home that is on fire without realizing there could be a flashover situation. Errors of omission often involved firefighters neglecting to follow established safety procedures, such as donning all required personal protective equipment. In cases where the narrative author addressed conditions that prevented further situation escalation, a general theme was that adherence to prior training had prevented further injury or loss of life.

Another important element of the analysis was whether each narrative included specific actionable information to prevent future similar occurrences. Of the 61 narratives reviewed, 42 (69%) specifically mentioned an alternative solution or actionable recommendation.

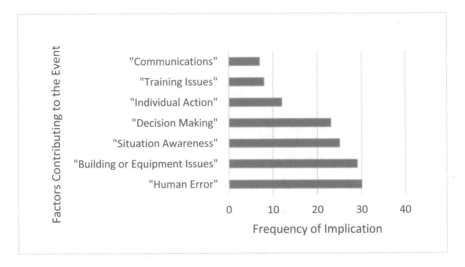

Fig. 4. Frequency of factors contributing to the close call events.

5 Discussion and Conclusions

The findings reported here constitute an important bridge between theoretical treatments of close calls and applied domains within which they frequently occur. Tiller and Bliss [7] noted that close call events within the Aviation Safety Reporting System followed the categories established in Bliss's [11] taxonomy. Similarly, the events noted within the IAFC database seemed well represented by the categories, with one exception (Disregarded Close Calls). The DCC category was relevant for aircraft pilot narratives because the ASRS database included a wide variety of incidents and accidents, attributable to a host of causes. In contrast, the IAFC database was specifically designed to include only close call narratives. Therefore, no close call events would have been disregarded by the operator or narrative author.

Another distinguishing factor of the IAFC database seems to be the frequent occurrence of a sensory signal preceding or accompanying the close call event. By comparison, only 21% (24 of 117) narratives taken from the ASRS database were classified as SCCs. One possibility for this difference is the existence of a limited number of discrete alarm systems in aircraft cockpits. In-flight dangers are typically signaled by the Engine Indicating and Crew Alerting System, an overspeed siren, a SELCAL chime, or a similar stimulus. In firefighting, formal alarm systems are scarce. However, environmental cues at the site serve as signals. Furthermore, they occur across almost all sensory channels.

Other differences exist between the aviation and firefighting narrative databases. The ASRS database is well organized and relevant to a variety of aviation incidents and accidents. Because it was designed with researchers in mind, we were able to focus our investigation closely on near-miss situations. For the current investigation, however, the narratives within the IAFC were more diverse and targeted toward the firefighting community. This rendered our investigation more challenging. In both cases, however, we were able to successfully construct a severity system that cleanly differentiated levels. For aviation, the severity system was predicated on absolute distance between aircraft. In firefighting, we elected to estimate lives at risk. Though it was often difficult to settle on a precise number from the narratives, our eventual approach seems practical and informative.

Of most direct benefit to the firefighting community was our examination of root causes. Unfortunately, the terminology used by the narrative authors reflects a level of generality that serves more to confuse than elucidate. The most common root cause identified was "human error." However, such a general term does little to inform training. It is unclear whether the error was sensory, cognitive, or action oriented; whether it was perceptual or motor; whether it reflected a "mistake" or a "slip" [16]; whether it was an error of omission or commission [17], or whether it was intentional or accidental. Similarly, broad causal terms such as "situation awareness" and "decision making" are difficult to interpret without additional information.

Fortunately, clear recommendations are possible to ensure that narratives uploaded to the IAFC database serve the applied firefighting community and the researcher community equally well. First, we encourage those who maintain the database to include a definition of "close call" and "near miss" on the home page of the web site. Doing so should reduce the number of narratives submitted that constitute actual tragedies. Second, we believe that the narrative submission form should include specific fields for descriptive input. Specific and helpful options might include requesting causes behind the situation escalation, actions that resulted in situation de-escalation or recovery, and indications of the particular types of errors that were observed (at a minimum, errors of omission or commission). Third, given the ultimate training goal of the information gathered, we recommend that narrative authors include specific lessons learned so that the firefighting community might consider changes to their curricula or refresher training.

In addition to the recommendations above, we encourage the research community to undertake further taxonomic efforts with close call databases. Though some data are protected from public access, other domains encourage examination. With continued examination of close calls, theoretical models may be refined to enable clear explanation and effective melioration of disasters.

References

1. Haynes, H.J.G.: Fire Loss in the United States (Final Report). National Fire Protection Agency, Quincy (2017)
2. National Safety Council: Near miss reporting systems (2013). http://www.nsc.org/ WorkplaceTrainingDocuments/Near-Miss-Reporting-Systems.pdf. Accessed 19 Feb 2018
3. Institute for Safe Medication Practices: ISMP survey helps define near miss and close call (2009). https://www.ismp.org/newsletters/acutecare/articles/20090924.asp. Accessed 19 Feb 2018
4. Agency for Healthcare Research and Quality: Adverse events, near misses, and errors. Patient Safety Primer, June 2009. https://psnet.ahrq.gov/primers/primer/34/adverse-events-near-misses-and-errors. Accessed 19 Feb 2018

5. Farmer Environmental Group: Safety topic: near misses! (2018). http://farmereg.com/safety-topic-near-misses1/. Accessed 17 Feb 2018
6. Dodson, D.W.: Fire Department Incident Safety Officer, 3rd edn. Jones & Barlett Learning, Burlington (2015)
7. Tiller, L., Bliss, J.: Categorization of near-collision close calls reported to the aviation safety reporting system. In: Proceedings of the Human Factors and Ergonomics Society Annual Meeting. Human Factors and Ergonomics Society, Santa Monica (2017)
8. Golovina, O., Teizer, J., Pradhananga, N.: Heat map generation for predictive safety planning: preventing struck-by and near miss interactions between workers-on-foot and construction equipment. Autom. Constr. **71**, 99–115 (2016)
9. Vanpoulle, M., Vignac, E., Soule, B.: Accidentology of mountain sports: an insight provided by the systemic modelling of accident and near-miss sequences. Saf. Sci. **99**, 36–44 (2016)
10. Dillon, R.L., Tinsley, C.H., Burns, W.J.: Near-misses and future disaster preparedness. Risk Anal. **34**, 1907–1922 (2014)
11. Bliss, J.P., Rice, S., Hunt, G., Geels, K.: What are close calls? A proposed taxonomy to inform risk communication research. Saf. Sci. **61**, 21–28 (2013)
12. Green, D.M., Swets, J.A.: Signal Detection Theory and Psychophysics. Wiley, New York (1966)
13. Wikipedia: Close calls (2018). www.wikipedia.com
14. International Association of Fire Chiefs: Firefighter Near Miss 2016 Annual Report (Final Report). National Firefighter Near Miss Program, Fairfax (2018)
15. Gnoni, M.G., Saleh, J.H.: Near-miss management systems and observability-in-depth: handling safety incidents and accident precursors considering safety principles. Saf. Sci. **91**, 154–167 (2016)
16. Norman, D.A.: The Psychology of Everyday Things. Basic Books, New York (1988). ISBN 0465067093
17. Kern, T.: Flight Discipline. McGraw-Hill, New York (1998)

Applications in Healthcare
and Patient Safety

Analysis and Improvement of the Usability of a Tele-Rehabilitation Platform for Hip Surgery Patients

Hennry Pilco[1]($\boxtimes$), Sandra Sanchez-Gordon[2], Tania Calle-Jimenez[2],
Yves Rybarczyk[3,4], Janio Jadán[5], Santiago Villarreal[3],
Wilmer Esparza[3], Patricia Acosta-Vargas[3], César Guevara[5],
and Isabel L. Nunes[6,7]

[1] Escuela Politécnica Nacional, Quito, Ecuador
hennry.pilco@epn.edu.ec
[2] Department of Informatics and Computer Science,
Escuela Politécnica Nacional, Quito, Ecuador
{sandra.sanchez,tania.calle}@epn.edu.ec
[3] Intelligent & Interactive Systems Lab,
Universidad de Las Américas, Quito, Ecuador
{yves.rybarczyk,santiago.villarreal,wilmer.esparza,
patricia.acosta}@udla.edu.ec
[4] CTS/UNINOVA, DEE, Nova University of Lisbon,
Monte de Caparica, Portugal
[5] Universidad Tecnológica Indoamérica, Ambato, Ecuador
{janiojadan,cesarguevara}@uti.edu.ec
[6] Faculty of Science and Technology, Universidade NOVA de Lisboa,
2829-516 Caparica, Portugal
{y.rybarczyk,imn}@fct.unl.pt
[7] UNIDEMI, Campus de Caparica, 2829-516 Caparica, Portugal

Abstract. The Tele-Rehabilitation platform for hip surgery allows patients to carry out part of their rehabilitation at home, without the need to travel long distances to a rehabilitation centre. A lack of usability that may prevent effectiveness, efficiency, and the satisfaction of patients, may lead to problems of confusion, error and delay, or even abandonment of the physical therapy. To perform the usability analysis, a set of heuristics were selected relating to aspects such as navigation, visual clarity, coherence, prevention of errors, user guidance, online help and user control. A cognitive walkthrough technique was also applied. With the results of the analysis, the design and implementation of improvements were performed. The web interfaces of the Tele-Rehabilitation platform were evaluated once again and compared with the baseline to ensure there was an improvement in usability.

Keywords: Telemedicine · Tele-rehabilitation · Hip surgery
Web user interface · Usability evaluation · Heuristic evaluation
Cognitive walkthrough evaluation

© Springer International Publishing AG, part of Springer Nature 2019
I. L. Nunes (Ed.): AHFE 2018, AISC 781, pp. 197–209, 2019.
https://doi.org/10.1007/978-3-319-94334-3_21

1 Introduction

The project "Tele-rehabilitation system for patients after hip replacement surgery" provides a web platform for physical tele-rehabilitation, for patients who underwent hip arthroplasty surgery. The goal of this platform is to allow patients to carry out part of the rehabilitation at home and communicate through the web their recovery evolution to the physiotherapist [1]. However, web interfaces for patient and physiotherapist have only been through a basic usability evaluation. Therefore, according to the ISO 9241-11 standard, which deals with ergonomic requirements, it cannot be determined whether compliance is being made with "the degree to which a product can be used by specific users to achieve specific objectives effectively, efficiency and satisfaction in a specific context of use" [2].

In this research, we present an overview of our ongoing project and how we are applying Usability Evaluation Methods during the software development cycle of the platform. Unlike other similar studies, our platform is focused on patients who have suffered a hip fracture. Hip fracture has a great worldwide incidence, mainly in people over 65 years of age. In 1990 it had an incidence of 1.66 million, however, there are studies that estimate that its incidence will surpass 6 million in 2050 [3–5]. These same studies reveal that post-fracture mortality (after one year) has a higher incidence in patients older than 85 years. In the United States, 200,000 hip fractures occur annually. This study begins with a literature review. Then, we select two evaluation methods and perform a first usability evaluation whose results allowed us to obtain a list of improvements to be developed. These improvements were implemented and evaluated for a second time. This way, and with the results obtained from the two evaluations, we were able to demonstrate the process of improving the usability of the platform step by step, before it completes its development cycle.

2 Literature Review

This section presents previous relevant published research that is used for this study and obtained through a systematic literature review (SLR). The protocol used in this SLR study starts specifying the research questions, search strings, search sources, inclusion and exclusion criteria, selection process, extraction of relevant data, synthesis of the data and ends with the selection of the relevant primary studies.

2.1 Research Questions

Due to the need of establishing a baseline for usability evaluation of the Tele-rehabilitation platform, we proposed three central research questions to get different approaches, recommendations, and methods, which the authors of the previously published researches have used in similar evaluations. Table 1 describes the research questions used in this SRL and their motivation.

Table 1. Research questions.

RQ	Research questions	Motivation
1	What are the usability requirements that are currently applicable to web interfaces?	Identify current usability requirements applicable to web interfaces
2	What are the benefits and/or limitations of usability evaluation applied to web interfaces?	Obtain references on difficulties encountered in similar usability evaluation projects applied to web interfaces Obtain references on usability attributes that improve the results of the usability evaluation and the quality of the software product
3	What inspection methods are applicable in the evaluation of usability of web interfaces?	Obtain comparative references between different methods of inspection on web interfaces Obtain references about inspection methods and their main attributes in the evaluation of usability

2.2 Search Strings

The search strings were composed for a set the terms derived from the research questions and combined the terms: ("telerehabilitation" AND "summative evaluation"), ("summative evaluation" AND "web user interface"), ("usability evaluation" AND "web interfaces"), ("user interface" AND "cognitive evaluation"), ("web user interface" AND "heuristic evaluation"), ("web user interface" AND "cognitive walkthrough evaluation"), ("cognitive walkthrough evaluation").

2.3 Search Sources

The next step in the review protocol was to choose search sources, which are scientific databases that guarantee to provide a high level of quality in the research publications. The scientific articles obtained from these databases were used as a source of preliminary information for the literature review: Scopus, The ACM Digital Library, IEEE, ResearchGate, Scientific Electronic Library Online, Elsevier, Springer, Taylor & Francis.

2.4 Inclusion and Exclusion Criteria

The inclusion and exclusion criteria aim to identify the most relevant scientific articles, reduce the number of searches obtained and select those that provide adequate answers to the previously defined research questions. The inclusion criteria (IC) and exclusion (EC) sets are detailed in Table 2.

Table 2. Inclusion and exclusion criteria.

Id	Criteria	Description
IC-1	Full-Text	Find the full text of an article
IC-2	Approach	Usability heuristics and cognitive walkthrough in specific domain
EC-1	State of being in force	Exclude articles of the last 10 years were reviewed
EC-2	Other heuristics	Paper or theses that contain proposed for the other aspects

2.5   Review Results Analysis

For the search of relevant studies, we used three specialized search engines. In this way, of the universe of studies found, only those that belong to the defined search sources were considered. The details are shown in Table 3.

Table 3. Search results.

Source engine	Results	Pre-selected	Not relevant	Selected
SCOPUS	30	22	17	5
GOOGLE SCHOLAR	500+	18	3	15
RESEARCHGATE	230	6	1	5

Finally, we performed a quantitative analysis of the selected documents to verify that they are within the research domain and determine what information will be taken as reference for the preparation of the usability evaluation methods for this study. Summary of the results obtained by using the Atlas Ti tool [18] was Usability (1914 words), Heuristic (1430 words), Evaluation (1351 words), Cognitive (332 words), Nielsen (264 words). According to the quantitative analysis, the selected published researches are semantically inside this study domain and show that usability, heuristic, cognitive and Nielsen are the words with more relevance. However, this was not the most relevant factor in selecting the published researches, it only helped to sort and prioritize the final selection of 25 relevant published researches.

3   Usability Analysis

In [3, 5, 6, 8], the authors recommend complementing the heuristic evaluation with another usability evaluation method, like cognitive walkthrough, because this combination found usability problems not identified by heuristic evaluation and vice-versa. (Figure 1 shows the usability evaluation implemented in our study).

Heuristic evaluation is a usability inspection method that identifies usability problems based on usability principles or usability heuristics. In this way it is possible to measure/evaluate usability [10]. The heuristic evaluation was proposed by Nielsen

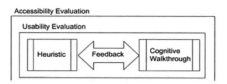

Fig. 1. Usability evaluation method

and Molich [10, 11] and involves usability experts who inspect a product interface based on heuristics and identify usability problems. These problems are associated with usability heuristics, then the frequency, severity and criticality of each problem are evaluated [10].

The cognitive walkthrough is a usability evaluation method in which one or more evaluators work through a series of tasks, ask a set of questions from the perspective of the user, and check if the system design supports the effective accomplishment of the proposed goals [7]. The International Organization for Standardization in the standard ISO/IEC 9126-1 defines usability for a software product. The ISO 9241-11 suggests the possibility of using specific metrics for measuring performance [12]. The ISO/IEC 25010 defines usability with one of eight characteristics of the software product quality [7]. Supported by these three standards, we developed 5 additional specific heuristics for the usability evaluation of the tele-rehabilitation platform according to the study of Rusu et al. [13]. Many methodologies have been proposed to develop sets of usability heuristics. For this study, we used a combination of 10 Nielsen heuristic principles and 5 specific usability heuristics based on the methodology proposed by Rusu et al. [13]. This is because it: (1) presents six clearly defined steps; (2) includes a standard template for specifying heuristics; (3) includes validation methods; and (4) can be applied iteratively [14]. For carrying out the heuristic evaluation and the cognitive walkthrough, a set of templates was developed for three selected web interfaces of the platform. In these templates, we included the steps for completing a task, we recorded the user actions and the potential user problems. Table 8 show the rationale used to select the three web interfaces. Finally, the results of this initial usability analysis constituted our baseline and we obtained our first list of usability problems to implement improvements and perform a second analysis.

3.1 Baseline Definition

Establishing a baseline based on usability attributes to be measured is an important part of any usability evaluation. This is because it determines the current status of web interfaces and identifies the characteristics based on recognized heuristics principles for specific applications, which in our case is web interfaces. Therefore, for this study, it is important to start the usability analysis with Nielsen's 10 heuristics principles [10] and complement with five specific heuristics. Tables 4 and 5 show the definitions of these heuristics.

According to the 10 heuristic principles and the 5 specific usability heuristics, we made a heuristic evaluation template based on the following four-point scale: (0) heuristic not applicable; (1) not fulfilled; (2) partially fulfilled; and, (3) fully fulfilled [11]. This scale will help us in the results section, not only to assess whether the interfaces comply with a certain heuristic or not, but also to identify their degree of severity. The greater the degree of severity, the greater priority for the development of improvements.

The degree of severity is determined according to the following agreement:

1. High Severity: When there is coincidence of criteria of the three experts.
2. Severity Medium: When there is coincidence of criteria of two experts.
3. Low Severity: When there is no majority.

Table 4. Nielsen 10 heuristic principles [10].

Id	Heuristic principles	Definition
HP1	Visibility of system status	The system should always keep users informed about what is going on, through appropriate feedback within a reasonable time
HP2	Match between system and the real word	The system should speak the users' language, with words, phrases and concepts familiar to the users. Make the information appear in a natural and logical order
HP3	User control freedom	"Emergency exit" to leave the unwanted state without having to go through an extended dialogue. Support undo and redo
HP4	Consistency and standards	Users should not have to wonder whether different words, situations, or actions mean the same thing
HP5	Error prevention	Even better than good error messages is a careful design which prevents a problem from occurring in the first place
HP6	Recognition rather than call	Instructions for use of the system should be visible or easily retrievable whenever appropriate
HP7	Flexibility and efficiency of use	Allow users to tailor and speed up frequent actions
HP8	Aesthetic and minimalist design	Dialogues should not contain information which is irrelevant or rarely needed
HP9	Help users recognize, diagnose, and recover from error	Error messages should be expressed in plain language (no codes), precisely indicate the problem, and constructively suggest a solution
HP10	Help and documentation	Any such information should be easy to search, focused on the user's task, list concrete steps to be carried out, and not be too large

Table 5. Show 5 specific usability heuristics [13].

Id	Specific heuristic	Definition
SH1	Clarity	The web interfaces should be easy to understand and operate, without the need for a manual. They should use clear graphic elements, text and language
SH2	Feedback	The web interfaces should deliver appropriate feedback on users' actions and keep them informed on the jobs' progress [13]
SH3	Shortcuts	The web interfaces should implement shortcuts so that the flow of actions made by the user is short and fast
SH4	Physical constraints	Screen should be visible at a range of distances and in various types of lighting; the distance between targets (e.g. icons) and the size of targets should be appropriate; size should be proportional to distance [13]
SH5	Extraordinary users	The web interfaces should be suitable for people with disabilities

For the cognitive walkthrough, we used a performance measurement testing technique and an interview as the inquiry method for the users [14]. For the template, we defined the users' objectives, steps to complete tasks, and questions to evaluate in the interview for each interface selected. We define four simple questions to evaluate: is the control for the action visible; will the interface allow the user to produce the effect the action has; will users succeed in performing this action; and will users notice that the correct action has been executed successfully? A four-point scale was defined for the questions in the cognitive walkthrough template: (0) not applicable; (1) partially fulfilled; (2) partially with fulfilled assistant, and (3) not initially.

3.2 Usability Evaluation

For this study, the usability evaluation was performed for two phases. The first phase was performed for the three web interfaces selected, to acquire a first list of improvements to develop. The second phase was performed in the questionnaire page with the developed improvements, to acquire a second list of improvements to develop in future works. In Table 6, we describe the web interfaces selected and the justifications. To carry out the evaluation the following steps were followed:

1. Select three heuristic evaluation experts with a total of 41 years of experience in software development and 17 years of expertise in usability. The authors agree that three is the optimal number of evaluation experts [8, 10, 15].
2. Select two cognitive walkthrough evaluators with backgrounds in web design and usability testing [9].
3. Perform the usability evaluation using heuristics and cognitive walkthrough templates on the questionnaire page. (Figure 2 shows the questionnaire page before the evaluation).
4. Report results.

Table 6. Web interfaces and selection criteria used.

Id	Interface	Rationale
UI1	Login Page http://telerehabilitation. cedia.org.ec/accounts/login/?next=/ patient/	This interface is important because it is main gate to the Tele-rehabilitation system. It also provides security of the resources assigned to the user and user identification for progress monitoring purposes
UI2	Questionnaire page http:// telerehabilitation.cedia.org.ec/extra/ questionnaire/	This questionnaire must be completed by the user before doing the rehabilitation exercise. It defines whether the user is sufficiently healthy to practice
UI3	Exercise assessment page http:// telerehabilitation.cedia.org.ec/ assessment/	This interface is a sample of the set of interfaces that show the exercises to the user and provides instructions

Fig. 2. The questionnaire page before the evaluation.

3.3 Results

Heuristic Evaluation. In Table 7, we show a summary of the result the first heuristic evaluation for the questionnaire interface.

Heuristics with high severity are highlighted in red, those with medium severity are in yellow, and those with low severity are in green. In the opinion of most experts, the questionnaire interface does not meet any heuristic. Also, two heuristics are not applicable to the questionnaire interface.

Cognitive Walkthrough. For the first cognitive walkthrough evaluation, the experts found 19 usability problems while users used the questionnaire interface. Table 8 shows where users encounter more problems in completing a task.

Table 7. Heuristic evaluation result. Phase I.

Heuristic	Not applicable	Not fulfilled	Partially fulfilled	Fully fulfilled
Visibility of system status	0	3	0	0
Match between system and the real word	0	1	2	0
User control freedom	0	1	1	1
Consistency and standards	0	2	1	0
Error prevention	0	2	1	0
Recognition rather than call	0	2	1	0
Flexibility and efficiency of use	3	0	0	0
Aesthetic and minimalist design	0	2	1	0
Help users recognize, diagnose, and recover from error	0	2	1	0
Help and documentation	0	3	0	0
Clarity	0	1	2	0
Feedback	0	3	0	0
Shortcuts	3	0	0	0
Physical constraints	0	1	2	0
Extraordinary users	0	3	0	0

Table 8. Percentage of users encountering a problem in completing a task.

Question	Percentage	
Is the control for the action visible?	45%	of the users
Will the interface allow the user to produce the effect the action has?	55%	of the users
Will users succeed in performing this action?	73%	of the users
Will users notice that the correct action has been executed successfully?	55%	of the users

4   Refinement

To implement the improvements, we applied the following process: (1) Sort and pri-oritize the results by severity; (2) Select a suitable methodology for developing improvements; (3) Make a Change Control Request Form and record the improvements to do; (4) Re-evaluate and compare results; (5) Test and deploy the new improvements. (Figure 3 shows the schema).

Fig. 3. Schema of the process from implementation of the improvements.

4.1   List of Improvements

The list of the proposed improvements was grouped according to features defined in the template and arranged based on severity rating (High, Medium). Table 9 shows the most relevant list of proposed improvements, ordered and prioritized. This only includes improvements with a severity rating of High.

Table 9. Questionnaire web interface severity rating.

Id	Severity rating	Proposed improvements
I1	High	Implement messages/animations that confirm that "the form has been sent correctly"
I2	High	Validate the consistency of the information entered
I3	High	Implement guide notifications and help for the user
I4	High	Implement messages/animations/graphics that helps the user to correctly enter the information
I5	High	Change the list to checkbox control to enter the information
I6	Medium	Increase the size of the text
I7	Medium	Implement a help section

This is only a sample of the improvements to be made, but it is sufficient for a second usability evaluation. In the section future work, we propose more improvements to develop. (Figure 4 shows mock-up example).

Fig. 4. Shows the questionnaire page mock-up with the developed improvements. This mock-up was used for the second usability evaluations.

4.2 Second Evaluation

Heuristic Evaluation. According to the proposed improvement list, we built a model of the new questionnaire interface and submitted it to the second evaluation, obtaining the following results presented in Table 10.

In opinion of the experts there are still usability problems, so it will be necessary to develop a new list of more precise improvements to be implemented.

Cognitive Walkthrough. For the second cognitive walkthrough evaluation, while users used the questionnaire interface, the experts found 20 usability problems. Table 11 shows the percentage of users who have problems completing assigned tasks.

Table 10. Heuristic evaluation result. Phase II.

Heuristic	Not applicable	Not fulfilled	Partially fulfilled	Fully fulfilled
Visibility of system status	0	0	0	3
Match between system and the real word	0	0	3	0
User control freedom	0	1	2	0
Consistency and standards	0	0	1	2
Error prevention	0	0	1	2
Recognition rather than call	0	0	1	2
Flexibility and efficiency of use	3	0	0	0
Aesthetic and minimalist design	0	0	3	0
Help users recognize, diagnose, and recover from error	0	0	1	2
Help and documentation	0	1	1	1
Clarity	0	0	2	1
Feedback	0	0	2	1
Shortcuts	3	0	0	0
Physical constraints	0	0	3	0
Extraordinary users	0	1	0	2

Table 11. Percentage of users who have problems in completing tasks. Phase II.

Question	Percentage	
Is the control for the action visible?	100%	of the users
Will the interface allow the user to produce the effect the action has?	67%	of the users
Will users succeed in performing this action?	100%	of the users
Will users notice that the correct action has been executed successfully?	67%	of the users

5 Conclusion

Once the usability evaluation was carried out using the methods of heuristic inspection and cognitive evaluation, we found a set of usability problems that required improvements that were implemented and evaluated again. This allowed us to verify the improvement in the level of usability, fulfilling the goal of this study. In the literature reviewed, we did not find a method or process for the development and iterative implementation of improvements. Therefore, the first contribution of this study is to propose a cyclical and orderly process for the development of improvements. The Tele-rehabilitation platform for hip surgery patients was developed using Django CMS, a content management system that allows users without programming knowledge to create, edit, and publish web content. However, extending, personalizing, or adding new functionality to the Django CMS is complex, and it requires time to understand the code. This made the development and implementation of the improvements defined in the results of the usability evaluation difficult. We have discovered that executing inspection methods on an application, in our case a web application, helps to analyse whether the tools selected for its development provides enough flexibility to evaluate and implement usability improvements. Finally, for future work we plan to perform a literature review to select a suitable methodology to develop improvements in an agile way, and to perform a new iteration of usability evaluation regarding specific attributes based on public health standards and protocols. Furthermore, before performing tests with end users, we will carry out a combined evaluation of the heuristic and cognitive methods, called heuristic walkthrough [16], following the methodological approach proposed in [17].

Acknowledgments. This research has been partially supported by the Consorcio Ecuatoriano para el Desarrollo de Internet Avanzado (CEDIA).

References

1. Rybarczyk, Y., Deters, J., Cointe, C., Arián, G., Esparza, D.: Telerehabilitation platform for hip surgery recovery. In: Conference Second Ecuador Technical Chapters Meeting (ETCM). IEEE (2017)
2. International Organization for Standardization: ISO 9241-171. Ergonomics of human-system interaction – Guidance on software accessibility (2012). https://www.iso.org/obp/ui/#iso:std:iso:9241:-171:ed-1:v1:en
3. Negrete-Corona, J., Alvarado-Soriano, J.C., Reyes-Santiago, L.A.: Fractura de cadera como factor de riesgo en la mortalidad en pacientes mayores de 65 años. Acta Ortopédica Mexicana 28(6), 352–362 (2014)
4. Elena, G.: Evaluación de la estancia hospitalaria en prótesis de cadera (2011)
5. Third AJRR Annual Report on Hip and Knee Arthroplasty Data (2016). http://www.ajrr.net/images/annual_reports/AJRR_2016_Annual_Report_final.pdf
6. Afonso, A., Lima, L., Perez, M.: A avaliação da usabilida de de interfaces Web. A Investigação do sítio Web da secretaria de uma escola do Ensino Superior. Comput. Sci. Eng. 25–32 (2012). https://doi.org/10.5923/j.computer.20120001.04
7. Gulati, A., Sanjay, K.: Critical analysis on usability evaluation techniques. Int. J. Eng. Sci. Technol. (IJEST) 4, 990–997 (2012). Computer Science and Engineering Department
8. Xiao, L., Yan, X., Emery, A.: Design and evaluation of web interfaces for informal care providers in senior monitoring (2013). https://doi.org/10.1002/meet.14505001034

9. Lim, C., Hae-Deok, S., Lee, Y.: Improving the usability of the user interface for a digital textbook platform for elementary-school students. Educ. Technol. Res. Dev. **60**, 159–173 (2012)
10. Nielsen, J., Molich, R.: Heuristic evaluation of user interfaces. In: Proceeding CHI 1990 Proceedings of the SIGCHI Conference on Human Factors in Computing Systems (1990)
11. Alotaibi, M.B.: Assessing the usability of university websites in Saudi Arabia: a heuristic evaluation approach. In: 10th International Conference on Information Technology: New Generations (2013)
12. Ivanc, D., Vasiu, R., Onita, M.: Usability evaluation of a LMS mobile web interface. ResearchGate (2015)
13. Rusu, C., Roncagliolo, S., Rusu, V., Collazos, C.: A methodology to establish usability heuristics. In: The Fourth International Conference on Advances in Computer-Human Interactions (2011)
14. Quiñones, D., Rusu, C.: How to develop usability heuristics: a systematic literature review. Comput. Stand. Interfaces **53**, 89–122 (2017)
15. Schaarup, C., Hangaard, S., Hejlesen, O.: Cognitive walkthrough: an element in system development and evaluation – experiences from the eWALL telehealth system. In: Conference on ENTERprise Information Systems/International Conference on Project MANagement/Conference on Health and Social Care Information Systems and Technologies, CENTERIS/ProjMAN/HCist (2016)
16. Mahatody, T., Sagar, M., Kolski, C.: State of the art on the cognitive walkthrough method, its variants and evolutions. Int. J. Hum.-Comput. Interact. **26**(8), 741–785 (2010)
17. Kushniruk, A., Monkman, H., Tuden, D., Bellwood, P., Borycki, E.: Integrating heuristic evaluation with cognitive walkthrough: development of a hybrid usability inspection method. Stud. Health Technol. Inform. **208**, 221–225 (2015)
18. ATLAS.ti: Qualitative analysis tool (2018). http://atlasti.com/product/what-is-atlas-ti/

Educational Resources Accessible on the Tele-rehabilitation Platform

Patricia Acosta-Vargas[1]([⊠]), Wilmer Esparza[1],
Yves Rybarczyk[1,2], Mario González[1], Santiago Villarreal[1],
Janio Jadán[3], César Guevara[3], Sandra Sanchez-Gordon[4],
Tania Calle-Jimenez[4], Jonathan Baldeon[1], and Isabel L. Nunes[5,6]

[1] Intelligent & Interactive Systems Lab,
Universidad de Las Américas, Quito, Ecuador
{patricia.acosta,wilmer.esparza,yves.rybarczyk,
mario.gonzalez.rodriguez,santiago.villarreal,
jonathan.baldeon}@udla.edu.ec
[2] CTS/UNINOVA, DEE, Nova University of Lisbon, Monte de Caparica,
Portugal
[3] Universidad Tecnológica Indoamérica, Ambato, Ecuador
{janiojadan,cesarguevara}@uti.edu.ec
[4] Escuela Politécnica Nacional, Quito, Ecuador
{sandra.sanchez,tania.calle}@epn.edu.ec
[5] Faculty of Science and Technology,
Universidade NOVA de Lisboa, 2829-516 Caparica, Portugal
{y.rybarczyk,imn}@fct.unl.pt
[6] UNIDEMI, Campus de Caparica, 2829-516 Caparica, Portugal

Abstract. This research is part of a telemedicine platform project to guide and accompany the patient online, during rehabilitation after total or partial hip replacement surgery. The study proposes to apply the Accessibility Guidelines for educational content in accordance with the Web Accessibility Initiative (WAI) accessibility guidelines. The main functionalities of the tele-rehabilitation platform involve the execution of rehabilitation movements, remote communication with health professionals, and therapeutic education of the patient during the recovery process. This article discusses the guidelines that the teaching-learning resources for elderly patients must meet to generate inclusive and easily accessible resources. The present study takes into consideration specific parameters relevant to the design of educational resources, with the aim of providing more accessible and inclusive educational guidelines for elderly patients.

Keywords: Accessibility elderly patients · Educational resources
Tele-rehabilitation platform · WAI · WCAG 2.0

1 Introduction

Nowadays, it is necessary to consider the different levels of education, especially for elderly patients and those with disabilities. Therefore, the educational resources of the Tele-rehabilitation platform must provide instant and ubiquitous access to all types of services and content, including documents and multimedia resources.

© Springer International Publishing AG, part of Springer Nature 2019
I. L. Nunes (Ed.): AHFE 2018, AISC 781, pp. 210–220, 2019.
https://doi.org/10.1007/978-3-319-94334-3_22

The accessibility known as universal accessibility is the degree to which all people can use an object, visit a place or access a service, regardless of their technical, cognitive or physical capabilities [1]. Accessibility describes techniques, methods, and theories to make media, in its multiple forms, more accessible to people with disabilities. Which implies that resources are accessible and inclusive for all people as possible. In general, the educational resources of the platform should include subtitles in the videos for people with hearing problems, audio description for blind or low vision audiences, alternative access to printed media.

The United Nations Convention on the Rights of Persons with Disabilities [2] "Recognizes the importance of access to the physical, social, economic and cultural environment, to health and education and to information and communication, so that people with disabilities can fully enjoy all human rights and fundamental freedoms". Although several countries in the world have introduced regulations that try to put them into practice, these guidelines are not applied entirely, in such a way that the educational resources do not comply with the suggested standards.

The number of people who have undergone surgery for Total Hip Arthroplasty (THA) has increased significantly in the last 10 years, and it is estimated that it will continue to increase [3]. THA is a surgery that consists of replacing the femoral head and the acetabulum of the hip joint. This surgery is usually performed in older adults, because of a degenerative joint disease or progressive wear of the joint. However, the demographics of patients who choose to undergo a THA have progressively become younger [4]. There are two types of THA that are practiced, either in elderly patients with low demands for activity (cemented), or in young patients with high demands (uncemented) [5]. Younger individuals may have higher functional and participation goals than older adults, which may modify the structuring of rehabilitation protocols. In any of the cases, what is intended after surgery is to relieve pain, restore normal function and improve quality of life [5].

In this study, the following question was considered: Are multimedia resources accessible to all users? Multimedia resources are now shared on websites as learning resources, but not all multimedia resources are accessible and inclusive.

A multimedia resource is considered accessible if its content is available to all users, regardless of their disability or context of application [6]. It is important that the educational resources of the platform are accessible even for people who use a screen reader. A screen reader is a software application that interprets the resource and presents the user with a text-to-speech synthesizer or a Braille line. This research analyzes accessibility problems with multimedia resources, especially those related to the video.

Educational processes can be oriented at different stages of rehabilitation and include: preventive, curative and maintenance processes. These processes depend on the pathology to be treated and whether it produces permanent or temporary dysfunction [7]. An example of temporary dysfunction is complete or partial hip replacement. In this type of intervention, the education of the patient is extremely important to avoid harmful movements and guide the rehabilitation process, which in turn optimizes the recovery process and promotes the empowerment of the patient [8].

This therapeutic education of the patient serves to: (i) instruct the general consequences of the procedures and their probable risks, and (ii) guide the patient through the rehabilitation process in all stages. A good orientation prevents the patient from performing movements that are not indicated, avoids fear, reaffirms assurance in functional tasks and motivates the patient to complete the rehabilitation program.

The application of the guidelines provides the necessary conditions for the resources of the rehabilitation platform to be accessible and fully usable by any type of patient, especially elderly people.

Considering that the patient must perform the exercises in standing position and at a distance from the computer, inclusive resources are proposed to guide the patient in their learning process.

The research applies the creation of multimedia learning objects to become a didactic material with accessible content. One of the main elements that have proven to be useful and efficient in education as a mean of transmitting knowledge is the implementation of multimedia teaching materials, including the use of videos, which have become a very important support material in the process of teaching-learning.

The method used to evaluate access to educational resources is as follows: (i) identify the type of resource, (ii) review access to resources in accordance with the WAI guidelines, (iii) implement the WCAG 2.0 registration recommendations, (iv) the results, (v) analyze the results.

This study presents a method to evaluate multimedia educational resources, as a case study will be applied to analyze the videos. This study presents a method to evaluate educational resources, the research carried out can serve as a starting point for future work related to learning, as well as to support the rehabilitation process using a therapeutic platform.

The multimedia accessibility policies propose [9] that: "All the multimedia elements as audio or video, produced or published must be accessible at the time of publication". Multimedia accessibility proposes a simple text transcription, which is why it is necessary to consider a transcription in audio-only recordings, to meet all the success criteria suggested by WCAG 2.0 at levels A and AA according to Media Guide 1.2 based on time.

The rest of this article is structured as follows: Sect. 2 provides details on the background and related work, Sect. 3 presents the accessibility in educational resources, Sect. 4 presents the method used and the case study, Sect. 5 presents the results and discussion, and finally, Sect. 6 presents the conclusions and future work of this research.

2   Background and Related Work

Recent advancements in the use of information and telecommunications technology in the healthcare industry has allowed telemedicine to emerge as a critical component for healthcare providers to supply services to patients. Telemedicine allows to evaluate, diagnose and treat patients from a distance and cost effectively [10].

A study from Rybarczyk and Vernay describes a Web application that aids patients and health professionals to improve the assessment of motor disabilities and needs for daily activities [8]. The platform allowed patients and professionals to get useful information about the patient's physical and psychological condition without consuming too much time and resources in the process.

In 2015, a research related to the study of patients with brain injury used a tele-medicine service, based on the Kinect camera [11]. Another study [12] measured the service and usability levels of a tele-rehabilitation system in patients with brain injury obtaining satisfactory results. A 3D virtual sensor from Microsoft Kinect was the core of the tele-rehabilitation system demonstrating the feasibility of use for patients with brain damage.

In 2016, the ePHoRt project was launched with the aim of developing a web-based system for the remote monitoring of rehabilitation exercises in patients after hip replacement surgery. An experiment was carried out to evaluate the validity and accuracy of the system using a low-cost motion capture device [7]. The results show that the Kinect is a suitable hardware for the therapeutic purpose of the project. Using the Dynamic Time Warping technique, it was possible to discriminate between correct and incorrect execution of the movements performed by the patient [13].

Post-surgical rehabilitation protocols for THA are well known and their correct practice has a positive impact on the prognosis of patients. The restoration of functionality plays an important role in the process of reinsertion of the patient to daily life activities. In this sense, it is considered that the total functional performance, under favorable conditions, should be achieved in the first 6 months of the postoperative period [14].

Therefore, the development of an appropriate rehabilitation program is of paramount importance. Rehabilitation programs begin between the second and the third postoperative day and, in general, extend intensively between three and four weeks. Patients should follow postoperative surgical precautions to allow tissue healing and prevent the risk of dislocation up to 6 to 12 weeks after THA [15]. Obviously, the amount of physical therapy sessions a patient receives has an impact on early recovery. In general, the number of therapy sessions is subject to some restriction, by reduced personnel or by the cost of treatment. Therefore, functional recovery can be prolonged, which has an impact on the costs of health services.

One of the current technological possibilities that could increase the number of treatment sessions, without increasing the costs, is the use of tele-rehabilitation. Tele-rehabilitation is the use of information and communication technologies as an instrument to ensure a remote rehabilitation service and continuity of care [16, 17]. Being a communication tool, tele-rehabilitation can also be a channel of education for the patient.

Previous studies of conventional educational programs for patients such as conferences, videos, websites, phone calls, found that structured education had an effect on postoperative pain, anxiety and functional recovery [18]. In addition, educational programs were effective in reducing hospitalization days, readmissions, surgical re-interventions and, therefore, costs of the health service [19]. However, there are no studies that evaluate a tele-rehabilitation educational program, its accessibility, and the design of educational resources, with the objective of providing more accessible and inclusive educational guidelines for patients.

3 Accessibility in Educational Resources

Multimedia resources [20] include texts, images, graphics, animations, video and sound to present or communicate certain information. At present, there are different types of disabilities, such as the auditory, cognitive, neurological, physical, and speech or visual, to access and understand the media. For example, people who are blind [21] could not access visual information in videos, for which they require alternative audio information.

Consequently, the proposed accessibility requirements for multimedia elements are based on the Accessibility Guidelines for Web Content 2.0 (WCAG 2.0). The approach makes the requirements have a greater impact on the final accessibility of a multimedia resource [22]. The accessibility requirements for a video to meet the accessibility level of AA are:

> *Priority 1 (Compliance A):* textual transcription, subtitles, appropriate flashes and flashes, reproduction control, possibility of being handled through the keyboard and the appropriate tabulation order.
> *Priority 2 (Compliance AA): subtitles* and audio description.

The incorporation of video elements in an accessible website is a complex process, which involves the investment of a large amount of time and effort.

The first thing that is applied is the textual transcription, this alternative is based on text and can be represented through any sensory movement such as vision, hearing or touch and interpreted in different ways for support products when reading aloud.

According to the W3C [21] resources that include multimedia, as the videos contain the application of alternative content technologies, such as:

> *Described video:* contains a description of visual elements to make resources accessible to people who are blind or visually impaired. The descriptions include actions, gestures, scene changes.
> *Text video description:* text video descriptions are provided as text files. It should be noted that some users who use screen readers, for example, those who are elderly or have learning difficulties, can slow down the speed of speech.
> *Extended video descriptions:* video descriptions are generally provided as a recorded voice.
> *Clean audio:* clean audio is aimed at audiences with hearing problems and consists of isolating the audio channel that contains spoken dialogues and non-verbal information.
> *Content navigation by content structure:* most people are familiar with the advancement of content backward or forwards. However, in cases where the content is based on time, moving forward or backward is not effective.
> *Captioning:* the subtitles must consist of be a textual representation of the audio, it must express exactly what is in the audio track.
> *Enhanced captions/subtitles:* enhanced subtitles are programmed text keys that help with more information, provide multiple subtitle tracks that will improve the understanding of content for users.

Sign translation: sign language shares the same concept as subtitling, presents voice information in an alternative format. Recognizing that not all devices are capable of handling multiple video transmissions, this is a requirement that must be included for browsers where the hardware is compatible.

Transcripts: synchronized subtitles are recommended for people with hearing disabilities, although for some users they are not viable.

4   Method and Case Study

In the evaluation of educational resources, the WCAG 2.0 guidelines for videos and the Photosensitive Epilepsy Analysis Tool (PEAT) were applied to identify the most common accessibility problems that are presented in educational resources. In this preliminary work, five random samples were taken that are part of the material for therapeutic education of the patient.

The method used to evaluate the learning resources, in this case the videos, consists of five sequential phases, as shown in Fig. 1.

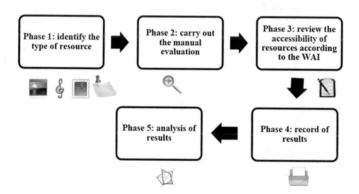

Fig. 1. Evaluation of multimedia resources

Phase 1: identify the type of resource, in this phase the educational resources are reviewed, for this study five videos of the educational material were randomly selected.

Phase 2: carry out the manual evaluation of the selected multimedia material, in accordance with the criteria of WCAG 2.0. For this scenario, a table with the requirements of an accessible video was generated.

Phase 3: review the accessibility of the resources according to the WAI guidelines, in this phase the information is validated with the checklist generated in phase 2. In addition, the PEAT tool was applied that works automatically to verify compliance specifically with the guideline 2.3 of the Web Content Accessibility Guidelines (WCAG) 2.0.

Phase 4: record of results, in this phase the results are recorded in a spreadsheet, in this case, Microsoft Excel was used.

Phase 5: analysis of results, in this phase the statistical tool R and Microsoft Excel was applied to perform the data analysis and to identify compliance with WCAG 2.0.

In addition to the manual evaluation with WCAG 2.0 guidelines, the Photosensitive Epilepsy Analysis Tool[1] (PEAT) of the University of Wisconsin Trace Center was applied, which allows developers to identify content on their sites that can induce seizures in people with seizures photosensitive disorders. PEAT captures a recording of web content or software and evaluates it for sequences of luminance flashes and red flashes that can cause seizures in people who are photosensitive.

Table 1 shows the resources evaluated manually, contains the video identifier, the URL where the resource is hosted, the descriptive video, the description of the text video, the extended descriptions of the videos, the clean audio, the navigation of content by structure of content, subtitles and subtitles improved, translation of signs, transcriptions, video quality, total score and the percentage obtained.

Table 1. Manual evaluation of the video resource

Id	URL	Dv	Tvd	Evd	Ca	Cncs	C	Ecs	St	T	Q	Tl	%
V1	https://youtu.be/hIuonL37Rec	0	0	0	0	100	0	0	0	0	95	195	19.5
V2	https://youtu.be/y9I7tsspA3E	0	0	0	0	100	0	0	0	0	85	185	18.5
V3	https://youtu.be/Y4lNPMwUHNw	0	0	0	0	100	0	0	0	0	80	180	18.0
V4	https://youtu.be/p-KZs7NRXl8	0	0	0	0	100	0	0	0	0	90	190	19.0
V5	https://youtu.be/_fzZH3FcQa8	0	0	0	0	100	0	0	0	0	85	185	18.5

Id = Identifier, URL = Uniform Resource Locator, Dv = Described video, Tvd = Text video description, Evd = Extended video descriptions, Ca = Clean audio, Cncs = Content navigation by content structure, C = Captioning, Ecs = Enhanced captions/subtitles, St = Sign translation, T = Transcripts, Q = Quality, Tl = Total, % = percentage

Figure 2 shows a screenshot of the Web application. It is observed that the patient performs rehabilitation exercises. It should be noted that this resource is part of the patient education section, the educational resources will be hosted on the platform located at http://telerehabilitation.cedia.org.ec/.

Table 2 records the results obtained in the evaluation with PEAT, contains the video identifier, length of material, luminance flash failures, red flash failures, extended flash warnings, result status and percentage.

___
[1] http://trace.umd.edu/peat.

Fig. 2. Screenshot of the resource to evaluate

Table 2. Evaluation with the photosensitive epilepsy analysis tool

Id	Lm	Lff	Rff	Efw	Rs	%
V1	000:10.20	0	0	0	P	100.0
V2	000:15.15	0	0	0	P	100.0
V3	000:15.21	0	0	0	P	100.0
V4	000:02.14	40	0	0	F	66.7
V5	000:04.22	0	0	0	P	100.0

d = Identifier, Lm = Length of material, Lff = Luminance flash failures,
Rff = Red flash failures, Efw = Extended flash warnings, Rs = Result status,
P = Passed, F = Failed.

5 Results and Discussion

When performing the analysis, it is observed that no video complies with the WCAG 2.0 guidelines, that is, they are not inclusive. The defects are related to the absence of the video describing the description of the text video, the extended video descriptions and the clean audio.

In Fig. 3, the relationship between the analysis performed manually and with the PEAT tool is observed. In the mosaics it is observed that videos one, two, three and five pass the test made with the PEAT tool while video four has faults related to the luminance flash that is not accessible for people with seizure problems, video four complies with 66.7%.

The multimedia material of the sample has an average compliance of 93.3% in relation to guideline 2.3 of the WCAG 2.0. On the other hand, manually analyzing the videos with the parameters of the described video, description of text videos, extended descriptions of videos, clean audio, content navigation by structure of content, subtitles, improved subtitles, translation of signs, transcriptions and quality, it is observed that video one has a higher percentage of compliance with 19.5% followed by video four with 19%, then video two and video five with 18.5% video three finally achieves 18% compliance. Among the five videos evaluated an average of 18.7% is reached, which is very low, which implies that the videos for the therapeutic education of the patient are not accessible or inclusive.

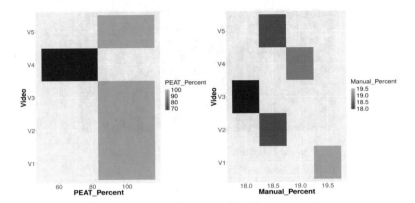

Fig. 3. Analysis manually and with the PEAT tool

Table 3 aggregates the analysis of the results and contains the video identifier, the percentage of PEAT, the percentage of the manual evaluation and the uniform weighted average between the PEAT and the manual evaluation. From the results of Table 3, a descriptive statistic is obtained with the mean corresponding to 56.0, with a median of 59.3, and a standard deviation of 7.4. Considering the average of PEAT and Manual Tests performed, all videos fall below 60% of accessibility compliance. Table 3 records the analysis of the results, contains the identifier, the percentage of PEAT, the percentage of the manual evaluation and the average between the PEAT and the manual evaluation.

Table 3. Evaluation with the photosensitive epilepsy analysis tool

Id	% PEAT	% Manually	%Average
V1	100.0	19.5	59.8
V2	100.0	18.5	59.3
V3	100.0	18.0	59.0
V4	66.6	19.0	42.8
V5	100.0	18.5	59.3

From the results of Table 3, a descriptive statistic is obtained with the following data: the mean corresponds to 56.0, the typical error is 3.3, the median is 59.3, the mode is 59.3, the standard deviation corresponds to 7.4, the sample variance is 54.4, kurtosis is 5.0, the asymmetry coefficient is −2.2, the range is 16.9, the sum is 280.1, the number of variables evaluated was 5.0, the maximum corresponds to 59.8, the minimum is 42.8 and the level of confidence with a 95.0% certainty corresponds to 9.2.

6   Conclusions and Future Work

The results of this study show that the evaluated multimedia resources evaluated did not reach the acceptable level of accessibility. Therefore, it is necessary to rectify the errors to comply with the level of accessibility recommended by the W3C. It is necessary to include accessibility measures in the development of educational materials through the application of a checklist to correct the problems identified. The results obtained in the evaluation can serve as a point of reference to compare with educational resources in future implementations of the tele-rehabilitation platform. The results obtained can provide ideas on how to develop and improve educational resources to make them more accessible and inclusive. This research can serve as an input for future projects with a greater number of evaluators. Educational resources can offer several alternative formats to meet the needs of several users. This study has limitations, but it can serve as a lesson learned for future work by implementing educational resources. Future research may propose better methods for using multimedia resources when applying the guidelines proposed by WCAG 2.0. With respect to videos and sound recordings, in both cases, a transcription of the dialogues and a description of the sounds should be provided. In the case of videos, a video description of the image must also be provided. However, inappropriate use of multimedia elements may cause a barrier to user access.

Acknowledgments. The authors thank CEDIA for partially funding this study through the project "CEPRA XI-2017-15 Telerehabilitación". They also thank MSc. Alejandro Galvis and MSc. Luis Gallardo for his collaboration in the evaluations carried out.

References

1. Acosta-Vargas, P., Luján-Mora, S., Acosta, T., Salvador-Ullauri, L.: Toward a Combined Method for Evaluation of Web Accessibility (2018)
2. United Nations: Convention on the rights of persons with disabilities. Treaty Ser. **2515**, 3 (2006)
3. Kurtz, S.M., Ong, K.L., Lau, E., Bozic, K.J.: Impact of the economic downturn on total joint replacement demand in the United States: updated projections to 2021. J. Bone Jt. Surg. - Am. **96**, 624–630 (2014)
4. Ravi, B., Croxford, R., Reichmann, W.M., Losina, E., Katz, J.N., Hawker, G.A.: The changing demographics of total joint arthroplasty recipients in the United States and Ontario from 2001 to 2007. Best Pract. Res. Clin. Rheumatol. **26**, 637–647 (2012)
5. Salavati, A., Duan, F., Snyder, B.S., Wei, B., Houshmand, S., Khiewvan, B., Opanowski, A., Simone, C.B., Siegel, B.A., Machtay, M., Alavi, A.: Optimal FDG PET/CT volumetric parameters for risk stratification in patients with locally advanced non-small cell lung cancer: results from the ACRIN 6668/RTOG 0235 trial. Eur. J. Nucl. Med. Mol. Imaging **44**, 1969–1983 (2017)
6. Brady, E., Zhong, Y., Bigham, J.P.: Creating accessible PDFs for conference proceedings. In: Proceedings of the 12th Web All Conference - W4A 2015, pp. 1–4 (2015)
7. Rybarczyk, Y., Kleine Deters, J., Cointe, C., Esparza, D.: Smart web-based platform to support physical rehabilitation. Sensors **18**(5) (2018)

8. W3C: Web Content Accessibility Guidelines. https://www.w3.org/TR/WCAG20/
9. Baruch, H., Ehrlich, J., Yaffe, A.: Splinting–a review of the literature. Refuat. Hapeh. Vehashinayim. **18**, 29–40, 76 (2001)
10. Mendes, P., Rybarczyk, Y., Rybarczyk, P., Vernay, D.: A web-based platform for the therapeutic education of patients with physical disabilities. In: 6th International Conference of Education, Research and Innovation (ICERI 2013), pp. 4831–4840 (2013)
11. Holden, M.K., Dyar, T.A., Dayan-Cimadoro, L.: Telerehabilitation using a virtual environment improves upper extremity function in patients with stroke. IEEE Trans. Neural Syst. Rehabil. Eng. **15**, 36–42 (2007)
12. Kizony, R., Harel, S., Elion, O., Weiss, P.L., Feldman, Y., Shani, M., Kizony, R., Weiss, P. L., Harel, S., Kizony, R., Obuhov, A., Zeilig, G.: Tele-rehabilitation service delivery: journey from prototype to robust in-home use. In: International Conference on Virtual Rehabilitation, ICVR 2015, pp. 178–182 (2015)
13. Rybarczyk, Y., Kleine Deters, J., Gonzalo, A.A., Esparza, D., Gonzalez, M., Villarreal, S., Nunes, I.L.: Recognition of physiotherapeutic exercises through DTW and low-cost vision-based motion capture, pp. 348–360. Springer, Cham (2017)
14. Contributors, P.: Total Hip Replacement. https://www.physio-pedia.com/index.php?title=Total_Hip_Replacement&oldid=181448
15. Van Der Weegen, W., Kornuijt, A., Das, D.: Do lifestyle restrictions and precautions prevent dislocation after total hip arthroplasty? A systematic review and meta-analysis of the literature. Clin. Rehabil. **30**, 329–339 (2016)
16. Kairy, D., Lehoux, P., Vincent, C., Visintin, M.: A systematic review of clinical outcomes, clinical process, healthcare utilization and costs associated with telerehabilitation. Disabil. Rehabil. **31**, 427–447 (2009)
17. Forbes, R., Mandrusiak, A., Smith, M., Russell, T.: Identification of competencies for patient education in physiotherapy using a Delphi approach. Physiotherapy **104**, 232–238 (2017)
18. Louw, A., Diener, I., Butler, D.S., Puentedura, E.J.: Preoperative education addressing postoperative pain in total joint arthroplasty: Review of content and educational delivery methods. Physiother. Theory Pract. **29**, 175–194 (2013)
19. Pelt, C.E., Gililland, J.M., Erickson, J.A., Trimble, D.E., Anderson, M.B., Peters, C.L.: Improving value in total joint arthroplasty: a comprehensive patient education and management program decreases discharge to post-acute care facilities and post-operative complications. J. Arthroplasty **33**, 14–18 (2018)
20. W3C: Multimedia Accessibility FAQ. https://www.w3.org/2008/06/video-notes
21. W3C: Media Accessibility User Requirements
22. W3C: Web Content Accessibility Guidelines (WCAG) 2.1. https://www.w3.org/TR/WCAG21/

Design of an Architecture for Accessible Web Maps for Visually Impaired Users

Tania Calle-Jimenez[1]([⊠]), Adrián Eguez-Sarzosa[1],
and Sergio Luján-Mora[2]

[1] Department of Informatics and Computer Science,
Escuela Politécnica Nacional, Ladrón de Guevara E11-253 y Andalucía,
Quito, Ecuador
{tania.calle,adrian.eguez}@epn.edu.ec
[2] Department of Software and Computing Systems, University of Alicante,
Carretera de San Vicente del Raspeig, Alicante, Spain
sergio.lujan@ua.es

Abstract. This study presents the design of a conceptual software architecture for the development of accessible web maps for visually impaired users. The conceptual software architecture proposed has three components: the database component is responsible for storing the web maps, the accessible web server component is responsible for converting a web map into an accessible web map, and the user side component is responsible for presenting the accessible web map to the user. A web application has been developed that shows the feasibility of implementing the proposed architecture. This application displays accessible web maps that provide additional information for visually impaired users. In addition, the application offers some alternatives for better interpretation of each map element. Visually impaired users receive feedback from the map displayed on the interface using a screen reader. The application has accessibility features based on WCAG 2.0 that facilitate the interaction of the visually impaired users.

Keywords: Architecture accessible · Web accessibility · Maps
Visually impaired users · WCAG 2.0

1 Introduction

Software architecture is important since the approach that is used to structure a web application has a direct impact on the capacity of the system to satisfy its quality attributes [1, 2]. One important quality attribute is usability, which has to do with how easy it is for users to operate the system. On the other hand, according to the ISO 25000 standard, accessibility is a part of usability and represents an important component in the development of web applications [3]. Improving web accessibility means that users with dis-abilities will be able to make better use of applications. To solve the problems of accessibility, the Web Content Accessibility Guidelines (WCAG 2.0) can be used. WCAG 2.0 [4, 5] is the standard that is most often legally adopted, via the ISO/IEC 40500 standard, in those countries that have legislation on web accessibility. There are three levels of accessibility compliance in WCAG 2.0: A, AA, and AAA [6].

© Springer International Publishing AG, part of Springer Nature 2019
I. L. Nunes (Ed.): AHFE 2018, AISC 781, pp. 221–232, 2019.
https://doi.org/10.1007/978-3-319-94334-3_23

This study presents the design of a conceptual software architecture for the development of accessible web maps for visually impaired users. The conceptual software architecture proposed has three components: the database component is responsible for storing the web maps, the accessible web server component is responsible for converting a web map into an accessible web map, and the user side component is responsible for presenting the accessible web map to the user. A web application has been developed that shows the feasibility of implementing the proposed architecture. This application displays accessible web maps that provide additional information for visually impaired users. In addition, the application offers some alternatives for better interpretation of each map element. Visually impaired users receive feedback from the map displayed on the interface using a screen reader.

The application has accessibility features based on WCAG 2.0 that facilitate the interaction of the visually impaired users. The conversion from regular map to accessible map is made using the Scalable Vector Graphic format (SVG) [7]. This format permits the inclusion of metadata that can be extracted in the browser to make the map accessible. The use of the SVG format allows the image to be scaled and resized without losing resolution, contrary to what normally happens when changing the size of traditional images. Scaling is important for visually impaired users because they may wish to expand the text in the browser or use assistive technologies, such as screen magnifiers and tactile graphic devices. The accessibility of the user interfaces of the application can be further improved using attributes of the technical specification Accessible Rich Internet Applications (ARIA) [8]. ARIA allows the assigning of functions, properties and information to code snippets to improve user interface controls. To validate the web accessibility of the web application, it was necessary to perform a series of tests to check the degree of compliance with WCAG 2.0.

2   Literature Review

Finding a satisfactory alternative to allow blind and visually impaired people to explore geographic information is an active field of research. There have been several practical solutions for accessible online maps developed by different authors. However, although some of them have already been implemented, most of them are still prototypes. The most relevant investigations are presented below.

The authors in [9] present a web service for blind people that allows them to select and generate tactile maps automatically. The selection of the maps is adapted to the requirements of blind users. The audio content of a map allows a previous feedback to check if the selected map corresponds to the information needs of the user. Changes can be initiated before printing the map, which saves 3D printing time. The user can select the characteristics to be included in the tactile map. In addition, the map representation can be adapted to different zoom levels and supports multiple printing technologies. Finally, the authors use an evaluation with blind users to refine their approach.

The authors in [10] developed a multi-sensory interactive map for visually impaired children. The authors conducted a formative study in a specialized institute to understand children's educational needs, their context of care and their preferences regarding interactive technologies. The results (1) outline the need for tools and methods to help children acquire spatial skills, and (2) provide four design guidelines for educational assistive technologies. The prototype enables collaborations between children with a broad range of impairments, proposes reflective and ludic scenarios, and can be customized by caregivers.

The authors in [11] developed a prototype that is located at Mapy.cz. Blind users can automatically adjust the maps in this prototype, which can then be read through touch, after being printed on microcapsule paper, opening up a whole new perspective on the use of tactile maps. Blind users can select an area of their choice in the Czech Republic and initiate the production of a tactile map, including the preparation of the map.

The authors in [12] present an accessible map visualization prototype that is designed with WCAG 2.0. This prototype was designed based on an architecture that helps mitigate the visual barriers of online geographical maps. As a case study, the prototype displayed a series of maps of Ecuador. In addition, the prototype was tested in different browsers and evaluated by accessibility tools; hence, the degree of accessibility of the prototype has been determined within a real-life environment.

As can be seen, there have been several research projects that present different prototypes, such as tactile relief maps, which can be used by blind and visually impaired people. These maps are used to acquire a mental representation of space, but the prototypes also have significant limitations, such as a limited amount of information. One solution to this problem could be interactive maps, which can overcome these limitations. However, the usability of these types of maps has not yet been evaluated, meaning it is still unknown whether interactive maps are equivalent to or even better solutions than traditional 3D maps.

While it is true that there have been several accessible prototypes developed for blind and visually impaired people, these studies do not present a specific software architecture for the development of accessible maps. As a result, there are still accessibility barriers in online maps.

3   Architecture Design

The design of the architecture is an important phase in any software development process; it can determine the success or failure of the software. Designing the software architecture implies adhering to a set of system restrictions, such as limitations of technology and code structuring, among others. The purpose of software architecture design is to understand the structure of the system. The result is a description of the components of the system and of the interfaces between the components [1].

The authors explain in [2] that software architecture is represented by the structure of the system, and how its components work together. Figure 1 illustrates the conceptual architecture of the software, where each section of the diagram represents a component. The components are composed of blocks; the boxes inside the blocks indicate that they have subcomponents. The arrows indicate the information that receives or requests another component, that is, the functionality that passes from one component to another in the direction of the arrows [1].

The conceptual software architecture proposed has three components:

- The databases are responsible for storing web maps.
- The accessible web server is responsible for converting an online map into an accessible online map.
- The user side is responsible for publishing the accessible online map when the user makes the request.

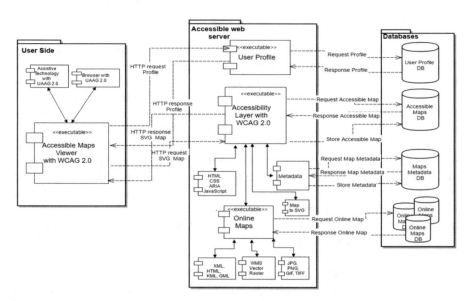

Fig. 1. Architecture. The three processes are: User Side *(left square)*, Accessible web server *(centre square)* and Databases *(right square)*.

3.1 User Side Component

This component is responsible for the functionality of the accessible user interface that is displayed in a browser. The user requests a user profile and receives a response from the database in the interface. The user side component has three subcomponents. The blind users interact with the first subcomponent—the assistive technologies—through the browser, which is the second subcomponent. Both of these subcomponents must be compatible with the User Agent Accessibility Guidelines (UAAG) [13]. The third subcomponent is the accessible maps viewer, which presents the map for navigation, with the help of the two previous subcomponents. The desired accessibility features for browsers, media players and other assistive software, used by people with disabilities to interact with computers, are described in UAAG [13]. These guidelines require the browser to present the alternative text, which functions with the screen reader software.

The accessible map viewer is the interface that provides information to the user, enabling them to request a map. This subcomponent works as follows:

1. User selects the user profile designed for blind and visually impaired users.
2. User makes the request of the map accessible to the accessibility subcomponent, which asks for the online map accessible to the database.
3. With the map accessible, user can navigate the map and the screen reader will describe thc mctadata contained in the map.
4. The user selects the user profile designed for blind and visually impaired users.
5. The user submits a request for an accessible map to the accessibility subcomponent, which requests the accessible online map from the database.
6. With the accessible map, the user can navigate the map and the screen reader will describe the metadata contained in the map.

The assistive technology subcomponent also helps improve the functioning of websites for people with disabilities [14, 15]. Blind users often use text-to-speech software, also known as screen readers, and Braille text hardware. Users with vision problems use screen magnifiers. For this prototype, ChromeVox was used as a screen reader, to describe the internal information. This information is stored in the metadata of the online map for the blind users. In addition, an accessibility toolbar was developed that changes the size of the source and the map, so that users with vision problems can better visualize the attributes of the map [12, 16].

3.2   Accessible Web Server Component

This component is responsible for ordering, receiving and storing accessible online maps. The component is initiated when the user makes a request for an accessible online map to the accessible web server, which is responsible for requesting information from the database. If the database contains that information, it is returned to the server and the server returns the request to the user.

On the other hand, if no accessible map is available, the database returns a message stating that the requested map has not been found. The accessible layer will then perform a search on the map website, which in turn makes the request to the databases that are online. The map is returned to the accessible layer. Once the map is in the accessible layer, it is sent to the SVG component, which transforms the map from its specific format to the SVG format.

Once the SVG map is obtained, it is returned to the accessible layer, which is responsible for sending the map to the metadata subcomponent. This subcomponent is responsible for adding the metadata to the map and returning it to the accessible layer. Finally, the accessible layer requests that the map be stored and displayed to the user.

The user profile subcomponent makes the request to the database and looks in the profile. For example, if a user is blind, this subcomponent makes the request and the database searches and responds with an accessible online map that is displayed in a browser for blind people. Alternatively, if the user is visually impaired, the map and the textual information will be enlarged in the browser, or the user will select the size of the letters and the spacing.

The accessibility layer subcomponent is responsible for converting any online map into an accessible online map and displaying it in a browser. This conversion is possible because the web application has been developed using technology and programming languages that help mitigate accessibility barriers. Some of the barriers faced by blind and visually impaired users when accessing geographic content include image maps, maps that are non-operable using a keyboard, maps with small text, and maps using non-distinguishable colours [17].

The accessibility layer is responsible for requesting the accessible map from the database. If the database has the map stored, it responds by attaching the map. Otherwise, the accessibility layer searches for the map on the web, obtaining a map in a different image format. The accessibility layer converts the map into a map in SVG format [18–20]. Once the map is in SVG format, the accessibility layer includes additional information in the metadata using the metadata component. Subsequently, the metadata and the map are stored in the database. This data can be viewed in a browser at the request of the user. Finally, the screen reader reads the information stored in the metadata. The accessible map and its metadata are stored in the database for future requests.

Metadata is an important component of geographic information. Because of this, the metadata is stored as an inherent part of the geographic information or as a separate document [18]. The metadata subcomponent stores additional information about a map and adds it to the SVG format through a web system interface. In this way, the screen reader is able to describe the metadata of the SVG map.

3.3   Database Component

This component constitutes the information store of the proposed conceptual architecture. The component stores the user profile that helps the user to obtain the necessary configuration options, making the system work in the way the user chooses. That is, a profile is a custom environment for a specific user.

The database also stores the metadata of the map and provides the management of transactions and the storage of accessible online maps. The database stores the SVG map along with the metadata and features necessary for the map to be understood by the blind and visually impaired users through the interface.

4   Accessible Prototype Development

The goal of this architecture is to improve the usability and accessibility of maps to blind and visually impaired users. Based on the proposed architecture, a prototype has been developed that displays accessible maps online. This prototype was developed with the pattern Model View Controller (MVC). This programming scheme is based on a layer model, and with the aim of differentiating between processes, users and data, while providing a coupling between these elements. Figure 2 shows the architecture of the application with the pattern MVC. The application was developed from a Single Page Application (SPA) because it is a web application. For communication with the data layer, the sails.js framework was used as a backend, so that the application would save the data in the view. The node.js framework was also used to per-form the basic operations, such as creating, updating and deleting, as well as storing the entire application. Finally, the view layer uses the angular framework as a frontend to separate the layer seen from the model layer.

The model layer connects to the database that is located in the framework sails.js and node.js. This layer sends the necessary instructions for the database to store the geographic information, metadata and data dictionaries, with which the application operates. The model layer is responsible for managing the operations on the information, in addition to updating the information so that it appears in the view. The information requests that are sent to the model layer are made through the controlling layer.

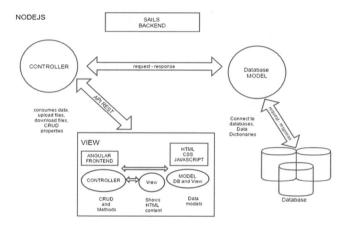

Fig. 2. Model-view-controller pattern. The image explains the MCV patter used to the prototype and their interaction with the programmer languages.

The visible layer is developed with the angular framework and represents the frontend. The layer seen, when being developed with SPA, is also an MVC, with the objective that the information is stored in the view and then sent to the model layer. In addition, the view layer contains the programming code that is displayed as the user interfaces. For the web application, the view contains the SVG format with which the maps were developed.

SVG format interacts with HTML and CSS for presentation, and JavaScript to handle the events that have been implemented in the application. The HTML code is complemented with JavaScript code to build the logic behind the web application and define the behaviour of the maps.

The controlling layer responds to and verifies the requests provided by the user, communicates with the model and selects the data that will be presented in the view. The controlling layer is composed of functions and this layer uses the Rest API and is responsible for performing the basic operations of creating, deleting and updating.

4.1   Accessible Map

One of the main problems was how to design the accessible maps since maps are usually made in an image format. To solve this problem, the SVG format was used, a format that is compatible with HTML (as mentioned in the previous sections). SVG images do not lose visual quality when the size is changed, so they can be displayed in different screen resolutions without loss of information.

The SVG format allows metadata to be included, which can later be extracted in the browser to make the map accessible. The use of the SVG format allows the image to be scaled and resized without flickering or pixilation, contrary to what can happen when resizing traditional images. Scaling is important for users with low vision, who may wish to expand the text in the browser or use assistive technologies, such as screen magnifiers and tactile graphic devices, which usually have low resolution.

To improve the accessibility of the user interface attributes, ARIA was implemented in the geographic content. ARIA allows the assigning of functions, properties and information to code snippets to improve user interface controls. Once the application is implemented, people with visual disabilities can use a screen reader and a keyboard as a means of navigation.

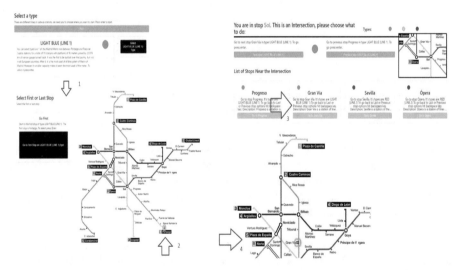

Fig. 3. Madrid metro 1955–1966 map. This image shows how the metro map can be navigated by blind and visually impaired people.

Figure 3 shows part of a map of the Madrid Metro. This interface starts with the selection of a 'type': that is, the line that the user wants to select. For this example, the first stop was chosen. The interface shows the map of the Madrid Metro with the first stop (Portazgo) highlighted. Once the user is at the defined stop, the stop can be changed with the arrow keys, from left to right; to return to the beginning the backspace key is used. Thus, the user can navigate through the map using the keyboard.

Whenever the user reaches an intersection, a new interface is displayed, which allows the user to select the stop on the line that they want. For a blind or visually impaired user, the screen reader guides them, reading out the activities that the user performs step-by-step.

In addition, the intersection interface—showing the image of the intersection and which stops the user can choose—is presented. The screen reader reads all the contents of the interface, regarding each attribute, while the user advances with the keyboard. For users who can see, an image of the intersection is highlighted to select the stop.

4.2 Testing

Automated accessibility tools can only identify with certainty a limited number of barriers in web content. A human being must interact with an accessibility review to make decisions about possible problems that automated tools cannot identify: for example, any check that requires the evaluation of meaning, such as whether a link text accurately describes the purpose of the link, or if an image's alternative text describes the content of the image in a comprehensible manner.

For this reason, accessibility tests were carried out in a real-life environment with the participation of an experimental group of 17 users. These users were blindfolded during the execution of the tests. The users had previous experience in the use of technology, such as computers and mobile devices, since they were engineering students and knew the tools. However, the students had not previously used screen readers, which implied a difficulty when manipulating the application.

Table 1 presents the results of the tests with the experimental group. There are nine questions based on the following parameters: use, understanding, interaction,

Table 1. Test results with experimental group.

Question	(−3)	(−2)	(−1)	(0)	(1)	(2)	(3)
How easily did you learn to use the website?	1	1	1	1	3	7	3
How understandable did you find the structure of the website?	1	1	1		5	7	2
How efficient did you find the navigation in relation to your aims on the website?	1		2	1	5	6	2
How easy did you find the process of achieving a task on the website?	2		2	1	5	5	2
How predictable did you find the controls when interacting with the website?	2	1			5	7	2
How do you feel about the controls for interacting with the website?	2	1		2	6	6	
How do you feel your accessibility requirements were supported by the website?	2		2	2	1	6	4
How do you experience this website with regards to your accessibility requirements?		1		3	5	6	2
How adaptable is the website to your interaction requirements?			1	1	4	6	3

experience and accessibility. These questions were assessed on six levels of difficulty: very difficult (−3), moderately difficult (−2), difficult (−1), neutral (0), easy (1), moderately easy (2) and very easy (3).

As can be seen, the majority of users rated the website as easy to use—that is, in the range 1 to 3—since the experimental group had no major problems with the use of the technology. However, there were also some students who rated it difficult to use (−3 to −1), due to their lack of practice using a keyboard, screen reader and browser without the ability to see.

The following feedback describes the improvements suggested by users after testing the application. The feedback was the most interesting result of the testing process, and helped to improve the final version of the accessible online map application [21]:

- The users explained the need to implement keyboard shortcuts with clear explanations that would enable switching from one metro line to another, without needing to travel the entire map.
- The users stated that it would be necessary to practise using the application several times, since the application would be easier to understand with continuous use.
- The users mentioned that the pronunciation of the screen reader confused them. This impaired their navigation and led to their selecting wrong directions.

5 Conclusion

Nowadays it is increasingly important that digital platforms are accessible to all people, regardless of their physical ability or age. The goal of digital inclusion is vital since many traditional services and basic functions are now presented online. Accessibility allows websites to be easily understood, read and navigated. The most important thing is that the content is available to people with different types of disability, such as visual, physical, auditory, speech, cognitive, language, learning and neurological disabilities. Web accessibility also makes the content more useful to the elderly, whose skills can decrease with age.

The WCAG 2.0 guidelines help blind and visually impaired people to be able to locate themselves on a map by means of voice commands transmitted by the screen reader. Most of the studies highlighted in the literature review show no interest in applying these accessibility guidelines to the development of geographic platforms, since maps are generally visual and therefore not easily accessible. This shows a current weakness in the geographical development of the web.

The developed architecture mitigates the main accessibility barriers to online maps. As result of this architecture, a web application was developed to create accessible maps and allow these maps to be navigated by blind and visually impaired people, using the keyboard. Users get a detailed description of the characteristics of the map with the use of a screen reader.

The use of web technologies can greatly improve the accessibility of a website. This study shows how the use of ARIA, HTML, CSS, SVG and JavaScript can improve the accessibility of online maps. In addition, we have proposed the application of the WCAG 2.0 guidelines, developed by W3C specifically for accessibility.

The tests performed by blindfolded participants produced a majority of positive results, in that most of the answers obtained assessed the application as 'easy'. In other words, the test subjects found it easy to interact with and navigate the application. In addition, the tests provided feedback of great importance for the improvement of the application.

References

1. Bosch, J.: Design and Use of Software Architectures: Adopting and Evolving a Product-Line Approach. Pearson Education, London (2000)
2. Hofmeister, C., Nord, R., Soni, D.: Applied Software Architecture. Addison-Wesley Professional, Boston (2000)
3. ISO: International Organization for Standardization's ISO 9241-171 Ergonomics of human-system interaction – Guidance on software accessibility (2012). https://www.iso.org/obp/ui/#iso:std:iso:9241:-171:ed-1:v1:en
4. World Wide Web Consortium: Accessibility. W3C Recommendation (2005). https://www.w3.org/standards/webdesign/accessibility
5. World Wide Web Consortium: Web Accessibility Initiative (2005). https://www.w3.org/WAI/gettingstarted/
6. World Wide Web Consortium: Web content accessibility guidelines WCAG 2.0. W3C Recommendation (2008). https://www.w3.org/TR/WCAG20/
7. World Wide Web Consortium: Scalable Vector Graphics (SVG) 1.1 (Second Edition) (2003). https://www.w3.org/TR/SVG/
8. World Wide Web Consortium: Accessible Rich Internet Applications (WAIARIA) 1.0. W3C Recommendation (2014). https://www.w3.org/TR/wai-aria/
9. Götzelmann, T., Eichler, L.: Blindweb maps-an interactive web service for the selection and generation of personalized audio-tactile maps. In: 16th International Conference on Computers Helping People with Special Needs, pp. 139–145 (2016)
10. Brulé, E., Bailly, G., Brock, A., Valentin, F., Denis, G., Jouffrais, C.: MapSense: multi-sensory interactive maps for children living with visual impairments. In: 34th Conference on Human Factors in Computing Systems (CHI), pp. 445–457 (2016)
11. Cervenka, P., Brinda, K., Hanousková, M., Hofman, P., Seifert, R.: Blind friendly maps. In: 16th International Conference on Computers Helping People with Special Needs, pp. 131–138 (2016)
12. Calle-Jimenez, T., Luján-Mora, S.: Accessible map visualization prototype. In: 13th Web for All Conference, pp. 15–16 (2016)
13. World Wide Web Consortium: User Agent Accessibility Guidelines (UAAG) (2015). https://www.w3.org/WAI/intro/uaag
14. World Health Organization: Guidelines for drinking-water quality, vol. 1 (2004)
15. World Health Organization: World Report on Disability (2013). http://www.who.int/workforcealliance/members_partners/member_list/iapb/en/
16. Calle-Jimenez, T., Luján-Mora, S.: Accessible online indoor maps for blind and visually impaired users. In: 18th International Conference on Computers and Accessibility (SIGACCESS) (2016)

17. Calle Jiménez, T.: Aportaciones técnicas y pedagógicas a la creación de mapas en línea accesibles (2017). https://rua.ua.es/dspace/bitstream/10045/71469/1/tesis_tania_elizabeth_calle_jimenez.pdf
18. Brisaboa, N.R., Luaces, M.R., Paramá, J.R., Trillo, D., Viqueira, J.R.: Improving accessibility of web-based GIS applications. In: 6th International Workshop on Database and Expert Systems Applications, pp. 490–494 (2005)
19. Zhang, C., Zhao, T., Li, W.: Geospatial data interoperability, geography markup language (GML), scalable vector graphics (SVG), and geospatial web services. In: Geospatial Semantic Web, pp. 1–33 (2015)
20. Bose, R., Jurgensen, H.: Online graphics for the blind: intermediate format generation for graphic categories. In: 16th International Conference on Computers Helping People with Special Needs, pp. 220–223 (2016)
21. Nielsen, J., Clemmensen, T., Yssing, C.: Getting access to what goes on in people's heads?: reflections on the think-aloud technique. In: Second Nordic Conference on Human-Computer Interaction (2002)

Analysis and Improvement of the Web Accessibility of a Tele-rehabilitation Platform for Hip Arthroplasty Patients

Tania Calle-Jimenez[1(✉)], Sandra Sanchez-Gordon[1],
Yves Rybarczyk[2,3], Janio Jadán[4], Santiago Villarreal[2],
Wilmer Esparza[2], Patricia Acosta-Vargas[2], César Guevara[4],
and Isabel L. Nunes[5,6]

[1] Escuela Politécnica Nacional, Quito, Ecuador
{tania.calle,sandra.sanchez}@epn.edu.ec
[2] Intelligent & Interactive Systems Lab, Universidad de Las Américas,
Quito, Ecuador
{yves.rybarczyk,santiago.villarreal,wilmer.esparza,
patricia.acosta}@udla.edu.ec
[3] CTS/UNINOVA, DEE, Nova University of Lisbon,
Monte de Caparica, Portugal
[4] Universidad Tecnológica Indoamérica, Ambato, Ecuador
{janiojadan,cesarguevara}@uti.edu.ec
[5] Faculty of Science and Technology, Universidade NOVA de Lisboa,
2829-516 Caparica, Portugal
{y.rybarczyk,imn}@fct.unl.pt
[6] UNIDEMI, Campus de Caparica, 2829-516 Caparica, Portugal

Abstract. This paper explains some of the challenges that exist to make accessible the web interfaces of a Tele-rehabilitation platform for hip arthroplasty patients and propose an iterative method to improve the level of accessibility using automatic evaluation tools. Web accessibility is not concerned with the specific conditions of people who use the Web, but with the impact that their conditions have on their ability to use it. If the web interfaces of the Tele-rehabilitation platform for hip arthroplasty patients are not accessible enough, the patients will not be able to understand, perceive or operate adequately the platform to benefit completely of the physical therapy. The Web Content Accessibility Guidelines (WCAG 2.0) provides a set of rules and recommendations to help solve the problems of Web accessibility. Additionally, there are evaluation tools that allow identifying main web accessibility problems. These tools are best exploited when used by accessibility experts. The purpose of this research is threefold. First, to present the results of a web accessibility evaluation of the web interfaces of the Tele-rehabilitation Platform for Hip Arthroplasty Patients using three the evaluation tools: WAVE, AChecker and TAW. Second, to analyze the results presented by the tools according to the WCAG 2.0 guidelines to define a list of accessibility improvements. Third, to implement the improvements through the re-factorization of the existing code and re-testing the improved web interfaces to verify that they meet acceptable accessibility levels.

© Springer International Publishing AG, part of Springer Nature 2019
I. L. Nunes (Ed.): AHFE 2018, AISC 781, pp. 233–245, 2019.
https://doi.org/10.1007/978-3-319-94334-3_24

Keywords: Telemedicine · Tele-rehabilitation · Hip arthroplasty
Physical therapy · Web accessibility · Web accessibility evaluation tools
WAVE · TAW · ACHECKER

1 Introduction

Physical rehabilitation, also known as physical therapy or physiotherapy, is the branch of medicine that aims to develop, maintain and restore maximum movement and functional ability through a person's life. Physical rehabilitation is provided in circumstances where movement and function are affected by age, pain, surgery, illness, disorder, injury or environmental factors [1].

That is the case of patients who have hip arthroplasty, also known as hip replacement, who usually have mild or moderate long-term impairments and disabilities postoperatively. The disabilities include contracture of hip joint, muscle weakness of hip abductors, pain, and inability to perform certain activities of daily living, such as putting on and off socks. A home program of physical rehabilitation is very convenient for these patients. The program should be one that patients can do easily and safely in their homes without direct supervision from their therapist [2].

The situations in which hip replacement surgery may be required are many. Among the most common are osteoarthritis and hip fracture. Osteoarthritis is a condition that affects approximately 3.3% of the global population and becomes a more common situation among people between 60 and 80 years of age [3]. Hip fracture affects more to female population; approximately 15% of women suffer this injury throughout their lives [4]. By 2030, the demand for primary total hip arthroplasties is estimated to grow by 174% to 572,000 only in the Unites States. Hence, the demand for hip revision procedures is projected to grow by 137% by the year 2030. These large projected increases in demand provide a quantitative basis for the deployment of appropriate technological alternatives to serve this need [5].

Tele-rehabilitation is the application of telecommunication technologies to provide support, advice and intervention remotely to people in need of physical rehabilitation. Tele-rehabilitation provides new opportunities to implement rehabilitation services in different situations where face-to-face rehabilitation is complicated [6]. Specifically, the group of people that would benefit from a Tele-rehabilitation platform for hip arthroplasty patients are mostly elderly people.

The International Organization for Standardization (ISO) defines accessibility as "the usability of a product, service, environment or facility by people with the widest range of capabilities" [7]. Tim Berners-Lee, father of the Web, stated in 1999 that "accessibility is the art of ensuring that, to as large an extent as possible, facilities (such as, for example, web access) are available to people whether or not they have impairments of one sort or another" [8]. The World Wide Web Consortium (W3C), organization in charge of developing web standards, created the Web Accessibility Initiative (WAI) to develop guidelines for universal access. In 2008, with the help of accessibility experts and disabled users, WAI published WCAG 2.0. In 2012, ISO recognized WCAG 2.0 as the international standard named ISO/IEC 40500:2012 [9]. Many countries around the world, such as Australia, Brazil, Bolivia, Canada,

Colombia, Chile, Ecuador, France, Germany, India, Italy, Japan, Mexico, Norway, Netherlands, Norway, Portugal, Spain, UK, USA, United States, Uruguay, reference WCAG 2.0 in their accessibility laws [10–12].

WCAG 2.0 establishes four principles that give the foundation of web accessibility [13]: perceivable, operable, understandable, and robust, known as POUR. The perceivable principle states that users must be able to perceive with their able senses both the content and the user interface. The operable principle states that users must be able to operate the interface through interaction that users can perform. The understandable principle states that users must be able to understand the content as well as the operation of the user interface. The robust principle states that users must be able to access the content as technologies advance. Under POUR principles, there are 12 guidelines and 61 success criteria. Criteria belonging to level of conformance A must be satisfied to make the content accessible for all users. Criteria from level AA should be satisfied to remove the accessibility barriers. Criteria from level AAA may be satisfied to make the web content more comfortable for different groups of users. In most of the countries listed above, level AA is required.

The rest of this paper is organized as follows. Section 2 presents a literature review. Section 3 explains the research method. Section 4 details the results of the web accessibility analysis of the web pages of the Tele-rehabilitation platform for hip arthroplasty patients. Section 5 shows the implementation and re-testing of the web accessibility improvements. Finally, Sect. 6 presents conclusions and future work.

2 Literature Review

In previous relevant published researches, some Tele-rehabilitation solutions have been presented. Nevertheless, none of them has addressed the specific problem of web accessibility of Tele-rehabilitation of hip surgery patients.

In [14], authors present an application as a solution to problems encountered by users with motor skills impairment. This application uses the Microsoft Kinect Sensor and its Visual Studio SDK to write code that interacts with a device originally intended for gaming but used for learning. Preliminary results from prototype testing show that the system is usable if accessibility requirements are included by design.

In [15], authors present the concept of virtual patients as computer simulations that behave in the same way that an actual patient would in a medical context. Since these characters are simulated, they can provide realistic yet repetitive practice in patient interaction since they can represent a wide range of patients. In this work, authors detail a usability evaluation of the scenario creation tool. They found that nurse educators were able to use the tool to create a virtual patient scenario in less than two hours thanks to the usability level of the solution.

In [16], authors explain that people with motor disabilities caused by stroke experience limitations in performing daily tasks independently, and traditional rehabilitation programs have tediousness and accessibility problems for these patients. These authors propose a Microsoft Kinect Sensor-based virtual rehabilitation system, which allows the user to transmit their gesture information to the virtual environment by performing specified actions. Authors present a six-level recovery assessment criterion through a body parts motion monitoring based on a recovery assessment method. Results suggest that the proposed virtual rehabilitation system is effective to assist patients in conducting home-based rehabilitation without a physician's supervision.

In [17], authors present a comparison of motion tracking performance between the low-cost Microsoft Kinect and the high fidelity OptiTrack optical system for use in a rehabilitation tool. In this research, data on six upper limb motor tasks have been incorporated into a game-based rehabilitation application. The experimental results show that Kinect can achieve competitive motion tracking performance compared with OptiTrack and provide pervasive accessibility that can enable patients to take rehabilitation treatment in their home environment.

3   Research Method

The method used in this research is presented in Fig. 1. The method included three inputs, three processes and one final output. The inputs were the Tele-rehabilitation platform (available at http://telerehabilitation.cedia.org.ec/), a set of accessibility requirements based on the WCAG 2.0 guidelines plus the findings of the WAI-AGE project [18] and a testing procedure designed by the researchers.

These three inputs fed the first process, which was a web accessibility analysis using three tools: WAVE, AChecker and TAW. WAVE Accessibility Tool is an online web service or browser extension available as a free community service by WebAIM that evaluate the accessibility of web content [19]. AChecker is an open source evaluation tool that checks single HTML pages for conformance with accessibility standards [20]. TAW is an automatic on-line tool for analyzing website accessibility that uses as technical reference WCAG 2.0.

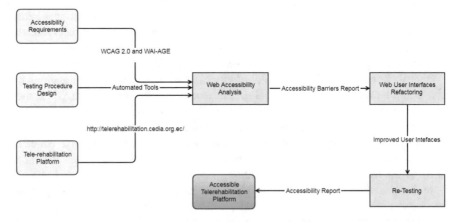

Fig. 1. Research method. The three processes are web accessibility analysis *(blue square)*, web user interfaces refactoring *(blue square)* and re-testing *(blue square)*.

The output of the first process was a report of the accessibility barriers that presented the Tele-rehabilitation platform at that point. This report served as input for the second process, the user interface refactoring performed by the developers using HTML and CSS on the design component of the Django CMS, which is the content management system that host the Tele-rehabilitation platform. The refactorized user interfaces were re-tested using the three tools to verify the accessibility improvements. The final output was a more accessible Tele-rehabilitation platform for elderly hip arthroplasty patients.

4   Results of the Web Accessibility Analysis

In this section, researchers present a set of accessibility requirements for the web interfaces of the Tele-rehabilitation platform for elderly hip arthroplasty patients. The first requirement list is based on WCAG 2.0 whereas the second requirement list is based on the findings of the WAI-AGE project. For better understanding, all the requirements has been categorized into five groups.

4.1   Accessibility Requirements

Taken in account that the patients' average age is between 60 and 80 years, researchers have defined 31 requirements based on WCAG 2.0 and six requirements based on WAI-AGE findings, organized into five categories. Table 1 shows the web accessibility categories.

Table 1. Web accessibility categories.

#	Category	Description
1	Text size	Older users need large text due to declining vision. This includes not only body text but also text included in form fields and other types of interface controls
2	Text style and text layout	Text style and its visual presentation impacts how hard or easy it is for older users to read the text, taken in account their declining vision
3	Color and contrast	Older users' color perception changes, even becomes color blindness. Older user also lose contrast sensitivity
4	Multimedia	Older users need transcripts, captions, and low background sound due to declining vision and hearing
5	Text-to-speech	Older users use text-to-speech software, also known as speech synthesis software, to help them overcome their visual impairments

Table 2 shows the 31 web accessibility requirements for older users identified by this study based on WCAG 2.0 [13] and organized by category. Each requirement has a unique ID, the correspondent success criterion and level of conformance (A, AA or AAA). For example, the first requirement RW1 corresponds to the WCAG 2.0 success criterion 1.4.4, defined under the Guideline 1.4 'Distinguishable: Make it easier for users to see and hear content including separating foreground from background'. It corresponds to level AA. Table 3 shows six additional requirements based on the findings of the WAI-AGE project [22].

Table 2. Accessibility requirements for the Tele-rehabilitation platform for hip arthroplasty patients based on WCAG 2.0.

ID	Requirement definition	Guideline - level
Text size		
RW1	Text can be resized without assistive technology up to 200% without loss of content or functionality	1.4.4-AA
Text style and text layout		
RW2	For the visual presentation of blocks of text, a mechanism is available to do the following: Users can select foreground and background colors. Width is no more than 80 characters. Text is not justified. Line spacing is at least space-and-a-half within paragraphs, and paragraph spacing is at least 1.5 times larger than the line spacing. Text can be resized without assistive technology up to 200% in a way that does not require the user to scroll horizontally to read a line of text	1.4.8-AAA
RW3	Images of text. If the technologies being used can do the visual presentation, text is used to convey information and not images of text	1.4.5-AA
RW4	Images of text (no exception). Images of text are used only for decoration or for a particular presentation of text that is essential	1.4.9-AAA
RW5	Bypass blocks. A mechanism is available to bypass repeated blocks of content	2.4.1-A
RW6	Page titled. Titles that describe the purpose	2.4.2-A
RW7	Link purpose (in context). The purpose of each link can be determined from the link text alone	2.4.4-A
RW8	Headings and labels. Clear description of the purpose	2.4.6-AA
RW9	Focus visible. Any keyboard operable user interface has a mode of operation where the keyboard focus indicator is visible	2.4.7-AA
RW10	Link purpose. A mechanism is available to allow the purpose of each link to be identified from link text alone, except where the purpose of the link would be ambiguous to users in general	2.4.9-AAA
RW11	Section headings are used to organize the content	2.4.10-AAA
Color and contrast		
RW12	Color is not used as the only visual means of conveying information, indicating an action, prompting a response, or distinguishing a visual element	1.4.1-A
RW13	Contrast (minimum). The visual presentation of text and images of text has a contrast ratio of at least 4.5:1	1.4.3-AA

(continued)

Table 2. (*continued*)

ID	Requirement definition	Guideline - level
RW14	Contrast (enhanced). The visual presentation of text and images of text has a contrast ratio of at least 7:1	1.4.6-AAA
Multimedia		
RW15	An alternative for time-based media is provided that presents equivalent information for pre-recorded audio-only content. Either an alternative for time-based media or an audio track is provided that presents equivalent information for pre-recorded video-only content	1.2.1-A
RW16	Captions are provided for all pre-recorded audio content in synchronized media	1.2.2-A
RW17	An alternative for time-based media or audio description of the pre-recorded video content is provided for synchronized media	1.2.3-A
RW18	Captions are provided for all live audio content in synchronized media	1.2.4-AA
RW19	Audio description is provided for all pre-recorded video content in synchronized media	1.2.5-AA
RW20	Where pauses in foreground audio are insufficient to allow audio descriptions to convey the sense of the video, extended audio description is provided for all pre-recorded video content in synchronized media	1.2.7-AAA
RW21	An alternative for time-based media is provided for all pre-recorded synchronized media and for all pre-recorded video-only media	1.2.8-AAA
RW22	An alternative for time-based media that presents equivalent information for live audio-only content is provided	1.2.9-AAA
RW23	Low or no background audio	1.4.7-AAA
RW24	All non-text content that is presented to the user has a text alternative that serves the equivalent purpose	1.1.1-A
RW25	Information, structure, and relationships conveyed through presentation can be programmatically determined or are available in text	1.3.1-A
RW26	Meaningful sequence. When the sequence in which content is presented affects its meaning, a correct reading sequence can be programmatically determined	1.3.2-A
RW27	Sensory characteristics. Instructions provided for understanding and operating content do not rely on sensory characteristics	1.3.3-A
RW28	On focus. The focus does not initiate a change of context	3.1.1-A
RW29	On input. The input does not automatically cause a change of context	3.3.2-A
RW30	Parsing. Elements have complete start and end tags and do not contain duplicate attributes	4.1.1-A
RW31	Name, role, value. For all user interface components, the name and role can be programmatically determined	4.1.2-A

Table 3. Accessibility requirements for the Tele-rehabilitation platform for hip arthroplasty patients based on WAI-AGE.

ID	Requirement definition
Text style and text layout	
RE1	Avoid bold body-text
RE2	Avoid underlined text other than links
RE3	Ensure links change color after visit
RE4	Clearly separate links
RE5	Make search results visible
RE6	Make sure the user notices small page changes/updates

4.2   Testing Procedure Design

This section presents the description of the testing environment to obtain the web accessibility barriers of the Tele-rehabilitation platform, the set of sample pages used in the testing, and the results obtained.

Testing Environment. The test environment used to check the compliance of the web accessibility requirements included three evaluation tools: WAVE, TAW and AChecker. These tools allow to test the accessibility of the web interfaces, to measure compliance with the accessibility guidelines, and if these interfaces are adequate to display information without affecting usability and navigation. The evaluation tools were used on a Chrome web navigator under a Windows operating system.

Set of Sample Web Pages. Table 4 details the three web pages selected for the testing with their URL and description. This set of web pages were selected due to the importance that they have in the context of the operation of the Tele-rehabilitation platform.

Table 4. Sample web pages for testing.

#	Page	URL	Description
1	Login page	http://telerehabilitation. cedia.org.ec/accounts/ login/?next=/patient/	This interface is important because it is main gate to the Tele-rehabilitation platform. It also provides security of the resources assigned to the user and user identification for progress monitoring purposes
2	Questionnaire page	http://telerehabilitation. cedia.org.ec/extra/ questionnaire/	Users must fill this questionnaire before doing the rehabilitation exercise. It defines whether the user is in adequate health conditions to practice
3	Exercise assessment page	http://telerehabilitation. cedia.org.ec/assessment/	This interface is a sample of the set of interfaces that show the exercises to the user and provides instructions

Test Results. The three automatic tools were applied to the set of pages of the sample. Each tool detected different results of non-compliance with WCAG 2.0. As an example, Fig. 2 shows the results of the accessibility evaluations for the Login web page. As can be seen, there are different errors detected by each of the tools. This is because each tool has different criteria for evaluation. In summary, WAVE detected five errors, AChecker detected 14 errors and TAW detected 28 errors.

The detected errors from each evaluation tool were mapped to the accessibility requirements. Not all the requirements from Table 3 had errors detected by the tools. Table 5 presents requirement ID, WCAG principle, WCAG level, and the number of errors detected for the set of sample pages by WAVE, AChecker and TAW respectively. The web page with most errors was Questionnaire with 78 errors, followed by Login page with 61 errors. It is important to notice that some of these errors are triplicate or duplicate among the evaluation tools. In addition, some of the errors correspond to more than one WCAG success criterion.

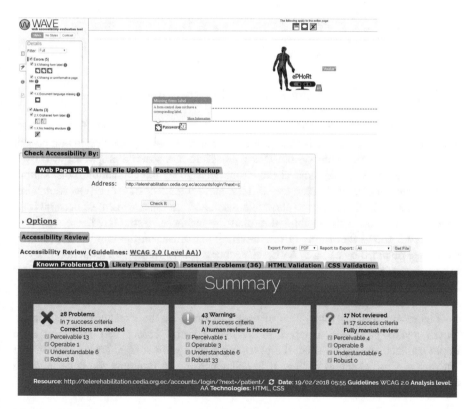

Fig. 2. Test results for login web page. The three tools are WAVE *(up)*, AChecker *(middle)* and TAW *(below)*.

Table 5. Summary of testing results for the login and Questionnaire pages

Id	WCAG 2.0	Login			Questionnaire		
		Wave	ACHECKER	TAW	Wave	ACHECKER	TAW
RW1	1.4.4-AA		4				
RW3	1.4.5-AA			1			
RW5	2.4.1-A	1					
RW6	2.4.2-A		1	1	1		1
RW7	2.4.4-A				1		1
RW8	2.4.6-AA	3			7		
RW12	1.4.1-A			1			
RW13	1.4.3-AA			2			
RW14	1.4.6-AAA						1
RW24	1.1.1-A	3		6	9	2	9
RW25	1.3.1-A	3	6	8	7	7	9
RW27	1.3.3-A			1			
RW28	3.1.1-A	1	2	1			2
RW29	3.3.2-A	3	3	5	7		7
RW31	4.1.2-A			5			7

5   Implementation of the Accessibility Improvements

To implement the accessibility improvements in the web interfaces of the Tele-rehabilitation platform, it was necessary to modify code on the design component of the content management system Django CMS to add specific properties at HTML and CSS level that solved most of the errors detected by the evaluation tools. Due to space restrictions, the web interface with the highest number of errors (Questionnaire) has been used in this section to show the results of the improvement of accessibility. However, the entire Tele-rehabilitation platform has been re-structured.

Figure 3 shows the results of the re-testing performed on the improved Questionnaire page obtained by using WAVE, AChecker y TAW. There are a reduction from 10 to zero errors in WAVE, from six to zero errors in AChecker and from 36 to four errors in TAW. Therefore, there is a reduction of errors of 100% in WAVE, 100% in AChecker and 89.2% in TAW.

Figure 4 shows a comparative of errors reported by the evaluation tools for the Questionnaire page and mapped to WCAG criteria, before and after the code improvements. As can be seen, the re-testing results show a significant improvement with respect to the initial testing. It is still necessary to fine-tune the code to reach conformance with success criteria 1.3.1 and 3.2.2 to comply fully with WCAG.

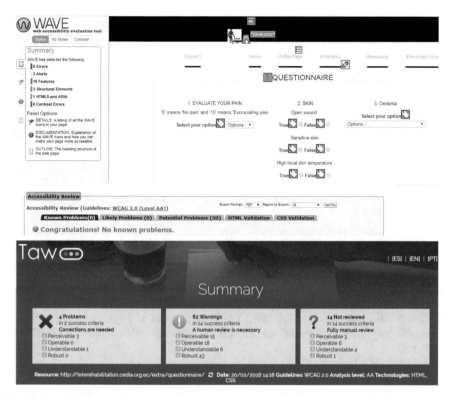

Fig. 3. Re-testing results of improved Questionnaire page. The three tools are WAVE *(up)*, AChecker *(middle)* and TAW *(below)*.

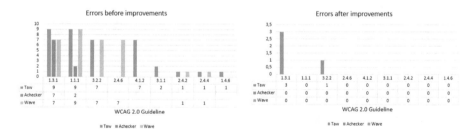

Fig. 4. Comparative results before and after code improvement of Questionnaire page.

6   Conclusion

The iterative method to improve the level of accessibility of web applications using automatic evaluation tools presented in this study allowed the researchers to improve the accessibility of the Tele-rehabilitation platform for hip arthroplasty patients. Testers and developers can use this method to test, debug, and re-test web applications to improve web accessibility. Nevertheless, they must have an understanding of the accessibility requirements and the WCAG 2.0 guidelines in order to take full advantage of the automated tools.

As future work, it is necessary to carry out a third iteration of the method to reduce to zero the errors detected by the automated tools in the re-testing process. In addition, due to the experimental conditions, researchers were not able to carry out clinical experiments with patients with hip replacement surgery. Researchers will try to carry out such experiments in future research. Additional experiments with elderly patients with vision and cognitive disabilities must also be carried out to improve further the accessibility of the Tele-rehabilitation platform.

Acknowledgments. The authors thank *Consorcio Ecuatoriano para el Desarrollo de Internet Avanzado (CEDIA)* for partially funding this study through the project "CEPRA XI-2017-15 Tele-rehabilitación".

References

1. World Confederation for Physical Therapy: Description of Physical Therapy (2017). https://www.wcpt.org/policy/ps-descriptionPT
2. Sashika, H., Matsuba, Y., Watanabe, Y.: Home program of physical therapy. Effect on disabilities of patients with total hip arthroplasty. Arch. Phys. Med. Rehabil. **77**(3), 273–277 (1996)
3. Kassebaum, N.J., Arora, M., Barber, R.M., Bhutta, Z.A., Brown, J., Carter, A., Cornaby, L., et al.: Global burden of disease: global, regional, and national incidence, prevalence, and years lived with disability for 310 diseases and injuries, 1990–2015: a systematic analysis for the global burden of disease study. Lancet J. **388**(10053), 1545–1602 (2016)
4. Ferri, F.: Ferri's Clinical Advisor. Elsevier Health Sciences (2017)
5. U.S. Department of Health and Human Services: Hip Replacement Surgery (2012). https://www.nih.gov/news-events/news-releases/hip-replacement-information-now-nihseniorhealthgov
6. Schmeler, M.R., Schein, R.M., Michael, M.: Telerehabilitation clinical and vocational applications for assistive technology. Research, opportunities, and challenges. Int. J. Telerehabilit. **1**(1), 59–72 (2009)
7. International Organization for Standardization: ISO 9241-171 Ergonomics of human-system interaction – Guidance on software accessibility (2012). https://www.iso.org/obp/ui/#iso:std:iso:9241:-171:ed-1:v1:en
8. W3C: Weaving the Web Berners Lee (1999). http://www.w3.org/People/Berners-Lee/Weaving/glossary.html
9. Sanchez-Gordon, S., Luján-Mora, S.: Accessible blended learning for non-native speakers using MOOCs. In: Proceedings of International Conference on Interactive Collaborative and Blended Learning (ICBL), pp. 19–24. IEEE (2015)
10. Rogers, M.: Government accessibility standards and WCAG 2 (2017). https://www.powermapper.com/blog/government-accessibility-standards/
11. WebAIM: Web Accessibility in Mind. World Laws (2018). https://webaim.org/articles/laws/world/
12. Urgilés, C., Célleri-Pacheco, J., Maza-Córdova, J.: Web accessibility: a challenge for the developers of Latin America. In: Conference Proceedings UTMACH, vol. 1, no. 1 (2017)

13. W3C: Web Content Accessibility Guidelines WCAG 2.0 (2008). http://www.w3.org/TR/WCAG20/

14. Alzubaidi, L., Elhassan, A., Alghazo, J.: Enhancing computer accessibility for disabled users: a kinect-based approach for users with motor skills disorder. In: Proceedings of the International Conference on Advances in Information Technology for the Holy Quran and Its Sciences, pp. 113–117. IEEE (2013)

15. Dukes, L.C., Meehan, N., Hodges, L.F.: Usability evaluation of a pediatric virtual patient creation tool. In: Proceedings of the International Conference on Healthcare Informatics (ICHI), pp. 118–128. IEEE (2016)

16. Zhao, L., Lu, X., Tao, X., Chen, X.: A Kinect-based virtual rehabilitation system through gesture recognition. In: Proceedings of International Conference on Virtual Reality and Visualization (VRVC), pp. 380–384. IEEE (2016)

17. Chang, Ch., Langel, B., Zhang, M., Koenig, S., Requejo, P., Somboon, N., Sawchuk, A., Rizzo, A.: Towards pervasive physical rehabilitation using microsoft kinect. In: Proceedings of Sixth International Conference on Pervasive Computing Technologies for Healthcare (PervasiveHealth), pp. 159–162 (2012)

18. W3C: Web Accessibility Initiative: Ageing Education and Harmonization (WAI-AGE) (2012). https://www.w3.org/WAI/WAI-AGE/

19. Wave: Web Accessibility Evaluation Tool (2018). http://wave.webaim.org

20. AChecker: Web Accessibility Checker (2011). https://achecker.ca/checker/index.php

21. TAW: Web Accessibility Test (2018). http://www.tawdis.net/

22. W3C: Developing Websites for Older People: How Web Content Accessibility Guidelines 2.0 Applies (2010). http://www.w3.org/WAI/older-users/developing.html

23. W3C: Web Accessibility Initiative Guidelines and Older Web Users: Findings from a Literature Review (2009). http://www.w3.org/WAI/WAI-AGE/comparative.html

Interaction with a Tele-Rehabilitation Platform Through a Natural User Interface: A Case Study of Hip Arthroplasty Patients

Yves Rybarczyk[1,2]([✉]), Santiago Villarreal[1], Mario González[1],
Patricia Acosta-Vargas[1], Danilo Esparza[1], Sandra Sanchez-Gordon[3],
Tania Calle-Jimenez[3], Janio Jadán[4], and Isabel L. Nunes[5,6]

[1] Intelligent & Interactive Systems Lab, Universidad de Las Américas,
Quito, Ecuador
{yves.rybarczyk, santiago.villarreal,
mario.gonzalez.rodriguez, patricia.acosta,
wilmer.esparza}@udla.edu.ec
[2] CTS/UNINOVA, DEE, Nova University of Lisbon, Monte de Caparica,
Portugal
[3] Escuela Politécnica Nacional, Quito, Ecuador
{sandra.sanchez, tania.calle}@epn.edu.ec
[4] Universidad Tecnológica Indoamérica, Ambato, Ecuador
janiojadan@uti.edu.ec
[5] Faculty of Science and Technology, Universidade NOVA de Lisboa,
2829-516 Caparica, Portugal
{y.rybarczyk, imn}@fct.unl.pt
[6] UNIDEMI, Campus de Caparica, 2829-516 Caparica, Portugal

Abstract. Using a tele-rehabilitation platform for motor recovery obliges the user to interact at a remote distance from the computer. In such a situation, a natural user interface can be used both to record the therapeutic movements performed by the patient, and to navigate in the web application. Nevertheless, it is necessary to assess the user experience to validate the system usability. The present paper describes an experiment to test the usability of a platform designed to enhance the recovery of patients after hip replacement surgery. The user experience is evaluated through several metrics (completion time and usability questionnaire) and the results are related to a sociodemographic questionnaire. A clustering approach is implemented to identify a relationship between the user's profile and the interaction performance with the platform.

Keywords: Usability assessment · User experience · Kinect
Health computing · Data analysis · Clustering

1 Introduction

During the last decades, tele-rehabilitation has been rapidly growing [1]. This trend is supported by the boom of the high velocity internet and a will to reduce the costs of the medical care. The concept consists of enabling patients to complete part of a

© Springer International Publishing AG, part of Springer Nature 2019
I. L. Nunes (Ed.): AHFE 2018, AISC 781, pp. 246–256, 2019.
https://doi.org/10.1007/978-3-319-94334-3_25

rehabilitation program at home, using a Web application that permits them to share with health professionals the evolution of the recovery process [2]. The current applications are mainly developed for the rehabilitation of cognitive disabilities [3]. Due to a complex and relatively expensive implementation, the technological tools for physical therapy are rarer. Nevertheless, the recent emergence of Natural User Interfaces (NUI), such as the Kinect, allows for an affordable system to capture the human movement. Once the three-dimensional coordinates of the body joints are automatically extracted by the device, motion data can be processed for applications in different fields, such as gaming [4], ergonomics [5], psychology [6], health [7], among other.

This paper presents a Web-based platform for motor tele-rehabilitation applied to patients after hip arthroplasty surgery [8]. This orthopedic procedure is an excellent case study, because it involves people who need a postoperative functional rehabilitation program to recover strength and joint mobility. The development of a tele-rehabilitation system is justified by the condition of these individuals that makes difficult their transportation to and from the physiotherapist's office. Such a physical rehabilitation platform implies two mandatory requirements: (i) being affordable, and (ii) being as fully usable as possible by the patients themselves. Considering that the patient has to perform the exercises standing up and at a certain distance from the computer, the Kinect can be used both to capture the therapeutic movements, and to control remotely the navigation throughout the Web application. The manuscript describes a usability study to assess the navigation by NUI into the platform.

The remainder of the paper is organized into five parts. Section 2 presents related work on usability evaluation of gesture-based interaction with computational applications. Section 3 consists of a description of an experiment to test the user experience. Section 4 is a presentation of the results in terms of completion time, assessment of the usability, and relationship between sociodemographic and performance data. The last part draws conclusions on the usability of a NUI to interact with a tele-rehabilitation platform.

2   Related Work

The usability is defined as the extent to which a product can be used by specified users to achieve specified goals with effectiveness, efficiency and satisfaction in a specified context of use. Assessing usability can be carried out through two possible approaches. The first method is called Cognitive Walkthrough and consists of evaluating the extent to which naïve users can use a system and achieve predefined tasks [9]. The second technique is the Heuristic method, in which experts analyze the compliance of the interface features against the principles of usability defined by the heuristic. The heuristic defined by Nielsen is the most popular method [10]. This paper assesses the usability by using the former technique.

Different metrics can be used to evaluate the user experience through a Cognitive Walkthrough approach. In pretest, participants can be submitted to a sociodemographic and/or health questionnaire to assess previous experience with a system, and cognitive and/or physical disability (e.g., Physical Activity Readiness Questionnaire) [11], respectively. During the test, it is possible to gauge the user activity through multimodal information, such as: raw performance (e.g., speed and accuracy), behavioral data (e.g., cursor tracking, scan path…), physiological activity (e.g., electrodermal response, heart rate…), and images and speech (e.g., think-aloud protocol) [12]. Finally, qualitative (e.g., interview) and quantitative (e.g., User Experience Questionnaire, Technology Acceptance Model, Computer System Usability Questionnaire, System Usability Scale, Physical Activity Enjoyment Scale, Task Load Index, Presence Questionnaire, among others) assessments can be recorded in posttest [13]. For the purpose of this study, several of these metrics are applied, in order to find a relationship between features through a clustering analysis of the data.

3   Usability Evaluation

Thirty subjects (20 males and 10 females; ages 20–55, Mean = 31, SD = 9.32; 28 right-handed and 2 left-handed) participated in the experiment. They were informed about the purpose of the study and they gave us their consent to participate. During the experiment, the participants were in stand up position at approximatively two meters in front of a Kinect camera. This vision-based sensor enabled the users to remotely control a cursor with their right hand.

The study was divided into three stages. The first stage consisted of explaining the task to the subjects and gave them two minutes to get familiar with the use of the system. The second stage was the experiment per se, which was composed with three trials to assess the learning effect. Each participant had to navigate in the platform by executing predetermined steps asked by the experimentalist. A trial was divided in the nineteen steps or tasks as follows:

1. Click on "PRACTICE" (Fig. 1)

MAIN MENU

Fig. 1. Main menu.

1. Click on "GO"
2. Change the "ASSESSMENT SCALE" to 5 (Fig. 2)

Fig. 2. Questionnaire to assess the medical condition.

1. Click on "Verification"
2. Access to "PRACTICE MENU"
3. Go to "STAGE 1" (Fig. 3)

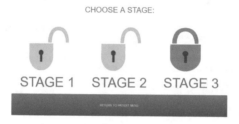

Fig. 3. Stages menu.

1. Click on "ACTIVE EXERCISE"
2. Go to and click on the last exercise "WALK AND CLIMB STAIRS"
3. Run the video (Fig. 4)

Fig. 4. Example of exercise of the stage 1.

1. Pause the video
2. Click on "NEXT SET" until you get 3 blue dots
3. Go back to "EXERCISES MENU"

4. Return to "STAGE 1"
5. Return to "MAIN MENU"
6. Go to "STAGE 2" (Fig. 5)

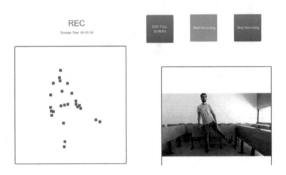

Fig. 5. Example of exercise of the stage 2.

1. Activate "FULL SCREEN"
2. Start "RECORDING"
3. Stop "RECORDING"
4. Go to "HOME MENU"

After the last trial, the subjects were asked to fill in two questionnaires: a sociodemographic, and an adaptation of the System Usability Scale (SUS) [14]. The content of the sociodemographic questionnaire was:

– Q1: What is your identification? Student/Faculty Employee/Other
– Q2: What is your gender? Male/Female
– Q3: What is your age? 20–29/30–39/40–49/ > 50
– Q4: What is your laterality? Right-handed/Left-handed/Both
– Q5: Experience with Kinect? None/Few/A lot
– Q6: Experience with other NUIs? None/Few/A lot
– Q7: Do you play computer games? Never/Occasionally/Regularly

The SUS was chosen as usability test, because it enables us to get a reliable subjective evaluation by the users in few questions (high correlation with other tests of subjective measures). SUS is a questionnaire based on a Likert scale, in which a statement is made and the respondent then indicates the degree of agreement or disagreement with the statement on a 5-point scale, from "strongly disagree" (0) to "strongly agree" (4). The statements, adapted for the purpose of this specific study, were as follows:

– S1: If I was a patient, I think that I would like to use this system frequently.
– S2: I found the control interface too complex to navigate in the website.
– S3: I thought the control interface was easy to use.
– S4: I think that I would need the support of a technical person to be able to use this system.

- S5: I found the visual interface well designed to be used with the Kinect.
- S6: I felt confused by the way to control the cursor.
- S7: I would imagine that most people would learn to use this system very quickly.
- S8: I found the control interface less practical than the traditional one (e.g., mouse).
- S9: I felt very comfortable using the system.
- S10: I will need to train a lot before I could master the control interface.

The odd statements are positive observations and the even are negative. To calculate the SUS score, first sum the score contributions from each item. Each item's score contribution ranges from 0 to 4. For items 1, 3, 5, 7 and 9, the score contribution is the scale position minus 1. For items 2, 4, 6, 8 and 10, the contribution is 5 minus the scale position. Multiply the sum of the scores by 2.5 to obtain the overall value of system usability. SUS score has a range of 0 to 100.

4 Results

4.1 Completion Time

A one-way ANOVA shows that the completion time significantly deceases from trial 1 to 3, $F(2, 87) = 80.22$, $p < 0.000$. This result reveals a training effect (Fig. 6). To note that the standard deviation also diminishes across the sessions (from $SD = 41.6$ to $SD = 13.2$), which suggests that all the participants improve their performance whatever their sociodemographic profile.

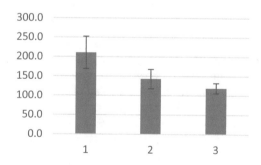

Fig. 6. Completion time against the experimental sessions.

4.2 SUS Score

The overall value of the usability is obtained by averaging the SUS scores of all the participants. The average value is 76.1 out of 100, which can be considered as a good evaluation of the usability of the platform. Figure 7 represents the average assessment for each statement (range from 0 to 4). To note that the highest scores are obtained for the statements 7 (value = 3.5) and 3 (value = 3.3), which are related to the facility to use the system. The lowest scores are obtained for the statements 8 (value = 2.5) and 2 (value = 2.6), which compare the traditional to the natural user interface.

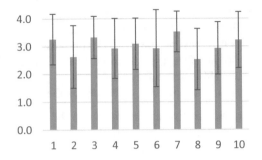

Fig. 7. Average SUS score for the 10 statements.

4.3   Clustering

Algorithm. In order to find a relationship between the attributes of the sociodemographic questionnaire and the dependent variables (SUS score and completion time), we try to group the instances in clusters. The chosen algorithm is KMeans. KMeans does iterative distance-based clustering. It can take different distance metrics, but the most used is the Euclidean distance. First, we specify the desired number of clusters, which is the k value. Second, the algorithm chooses k points at random as cluster centers. Third, all the instances of the dataset are assigned to their closest cluster center. Fourth, the centroid (or mean) of all the instances in each cluster is calculated, which transforms these centroids in new cluster centers. Then, the algorithm goes back to the beginning and carries on until the cluster centers do not change. In other words, this algorithm searches for a minimization of the total squared distance from the instances to their cluster centers.

Model. The model is built from a total of 30 instances and 8 attributes (SUS score, completion time, participant ID, gender, age, experience with Kinect, NUI and games). The function to calculate the distance between the instances and their cluster centers is set to Euclidean distance. And the selected number of clusters is $k = 2$. The resulting model is described in Table 1. The distribution of the instances is 16 (53%) and 14 (47%) for cluster 0 and 1, respectively. The total squared error is 57.8. Overall, the attribute grouping seems coherent, since the highest average value of SUS (77.2) and

Table 1. Cluster centroids.

Attributes	Cluster 0 (16 instances)	Cluster 1 (14 instances)
SUS	77.19	74.82
Time	145.97	170.68
ID	Student	Employee
Gender	Male	Female
Age	20–29	30–39
Kinect usage	Average	None
NUI usage	High	Average
Game usage	High	Average

the lowest mean value of time (146 s) are gathered in the same cluster (cluster 0). In addition, this cluster is represented by participants with a higher experience with the Kinect, the NUI and the computer games than in the other cluster (cluster 1). Another difference between the 2 groups is the fact that cluster 0 is more masculine and younger (mainly students) than cluster 1.

We can visually assess the quality of the clusters' separation in terms of the first three principal components as shown in Fig. 8. The principal components are the orthogonal axis for which the data have the most variability along. This allows us to better depict the data to check in a 3D scatter plot the clusters separability or overlap-ping. Figure 8 shows that the cluster projections are completely separable in the first three principal components denoting a good partition of the data in the suggested two clusters.

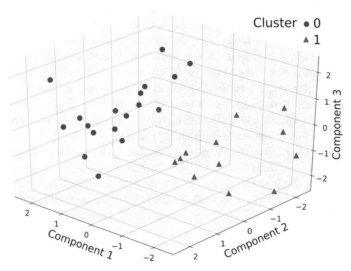

Fig. 8. Representation of the data divided in two groups (cluster 0, in blue circles, and cluster 1, in red triangles) according to the first three principal components (explaining 77.1% of the variance).

Principal component analysis (PCA) is a multivariate non-parametric method for extracting relevant information from data observations that are described by several inter-correlated quantitative dependent variables. This information is then represented as a set of new orthogonal variables called principal components [15]. PCA is used for data reduction, to simplify and uncover hidden structures in the data, and to represent the separability of the clusters. The association coefficients between the principal components and the variables (loadings) are listed in Table 2 for the first three principal components. Such coefficients in linear combination allow for predicting a variable by the (standardized) components.

Table 2. Loadings for the first three principal components (empty cells are approximately zero).

	Comp. 1	Comp. 2	Comp. 3
Time	−0.369	0.407	
ID	−0.457		0.102
Gender	0.315	−0.322	0.406
Age	−0.469		0.118
Kinect	0.350	0.355	0.397
NUI	0.195	0.556	−0.467
Games	0.408	0.250	−0.122
SUS		−0.478	−0.642

Validation The evaluation of the cluster separability through the KMeans model is done by classification via clustering. The attribute chosen as class is SUS, because it is expected to be the most relevant feature to assess the usability of the system. Thus, to perform a classification from the initial dataset the SUS values must be transformed from numeric to nominal. The median of the scores is used as threshold to divide the SUS values in two categories: high (or above the median) and low (or below the median). The algorithm of classification via clustering works by ignoring the classes, clustering the data, and assigning to each cluster its most frequent class. For this validation, the classification is performed by using KMeans with $k = 2$. The evaluation is done by 10-folds cross-validation, which takes 90% to form a clustering and a classification based on that clustering, and then test how well the clusters classify on the held-out 10% of the dataset. The performance of the model is provided by the confusion matrix displayed in Table 3. The clusters to classes mapping is: Cluster 0 = High SUS, and Cluster 1 = Low SUS. The overall precision of the classification is 63.3%, which means that 19 instances are correctly classified and 11 instances are misclassified.

Table 3. Confusion matrix.

		Predicted classes	
		Low SUS	High SUS
Actual	Low SUS	9	6
Classes	High SUS	5	10

5   Conclusions

This study consisted of performing a Cognitive Walkthrough experiment to assess the usability of a Kinect NUI in the context of interaction with a tele-rehabilitation plat-form. Different metrics were used to evaluate the user experience. The first result shows a significant reduction of the time required to perform the task across the sessions. It indicates that the participants can rapidly learn to master the gesture-based interaction.

This positive observation is confirmed by the high score reported on the SUS questionnaire (76.1%). A more refined analysis of the data by clustering enables us to discriminate between two profiles of users. The best performances (highest SUS and lowest Completion Time) tend to be obtained by male students with the youngest age, and the highest experience and practice of new technologies. Nevertheless, the accuracy of this classification is limited (63.3%), which suggests that older people are also able to use the system. A certain overlapping in the discrimination of the clusters can be explained the noisy subjective SUS answers provided by the subjects. Future work will consist of applying other clustering methods (e.g., Expectation Maximization) and considering additional features (e.g., cursor and event recording) to go further in the relationship between user's profile and usability assessment. In addition, the comment made by the participants will be implemented, in order to upgrade the visual and control interfaces.

Acknowledgments. The authors would like to thank CEDIA for partially funding this study through the project "CEPRA XI-2017-15 Telerehabilitación".

References

1. Rybarczyk, Y., Kleine Deters, J., Aladro Gonzalvo, A., Gonzalez, M., Villarreal, S., Esparza, D.: ePHoRt project: a web-based platform for home motor rehabilitation. In: Rocha Á., Correia A., Adeli H., Reis L., Costanzo S. (eds.) WorldCIST 2017. Advances in Intelligent Systems and Computing, vol. 570, pp. 609–618. Springer, Heidelberg (2017)
2. Rybarczyk, Y., Vernay, D.: Educative therapeutic tool to promote the empowerment of disabled people. IEEE Lat. Am. Trans. **14**, 3410–3417 (2016)
3. Rybarczyk, Y., Gonçalves, M.J.: WebLisling: a web-based therapeutic platform for the rehabilitation of aphasic patients. IEEE Lat. Am. Trans. **14**, 3921–3927 (2016)
4. Gameiro, J., Cardoso, T., Rybarczyk, Y.: Kinect-sign: teaching sign language to listeners through a game. In: Rybarczyk, Y., Cardoso, T., Rosas, J., Camarinha-Matos, L.M. (eds.) eNTERFACE 2013. IFIP Advances in Information and Communication Technology, vol. 425, pp. 141–159. Springer, Heidelberg (2014)
5. Plantard, P., Shum, H., Le Pierres, A.S., Multon, F.: Validation of an ergonomic assessment method using kinect data in real workplace conditions. Appl. Ergon. **65**, 562–569 (2016)
6. Coelho, T., De Oliveira, R., Cardoso, T., Rosas, J., Rybarczyk, Y.: Body ownership of virtual avatars: an affordance approach of telepresence. In: Rybarczyk Y., Cardoso T., Rosas J., Camarinha-Matos L.M. (eds.) eNTERFACE 2013. IFIP Advances in Information and Communication Technology, vol. 425, pp. 3–19. Springer, Heidelberg (2014)
7. Brook, G., Barry, G., Jackson, D., Mhiripiri, D., Olivier, P., Rochester, L.: Accuracy of the microsoft kinect sensor for measuring movement in people with Parkinson's disease. Gait Posture **39**, 1062–1068 (2014)
8. ePHoRt project. http://telerehabilitation.cedia.org.ec/
9. Wharton, C., Rieman, J., Lewis, C., Polson, P.: The cognitive walkthrough method: a practitioner's guide. In: Nielsen, J., Mack, R.L. (eds.) Usability Inspection Methods, pp. 105–140. Wiley, New York (1994)
10. Nielsen, J.: Heuristic evaluation. In: Nielsen, J., Mack, R.L. (eds.) Usability Inspection Methods, pp. 25–62. Wiley, New York (1994)

11. Sheu, F., Lee, Y., Yang, S., Chen, N.: User-centered design of interactive gesture-based fitness video game for elderly. In: Chen, G., Kumar, V., Kinshuk, Huang R., Kong, S. (eds.) Emerging Issues in Smart Learning. Lecture Notes in Educational Technology, pp. 393–397. Springer, Heidelberg (2015)
12. Nakai, A., Pyae, A., Luimula, M., Hongo, S., Vuola, H., Smed, J.: Investigating the effects of motion-based kinect game system on user cognition. J. Multimodal User Interfaces **9**, 403–411 (2015)
13. Simor, F., Brum, M., Schmidt, J., Rieder, R., De Marchi, A.: Usability evaluation methods for gesture-based games: a systematic review. JMIR Serious Games **4**, e17 (2016)
14. Brooke, J.: SUS – A quick and dirty usability scale. In: Usability Evaluation in Industry, pp. 189–194. Taylor & Francis, London (1996)
15. Abdi, H., Lynne, J.W.: Principal component analysis. Wiley Interdisc. Rev.: Comput. Stat. **2**, 433–459 (2010)

Comparison of Theory of Mind Tests in Augmented Reality and 2D Environments for Children with Neurodevelopmental Disorders

N. Tugbagul Altan Akin[(⊠)] and Mehmet Gokturk

Department of Computer Engineering, Gebze Technical University, Gebze,
Kocaeli, Turkey
tualtan@gmail.com

Abstract. In this work, we use AR and 2D virtual environments to do TOM tests on the computer with the suitable interface properties for CWND (Children with Neurodevelopmental Disorders). The study is the first to test ToM on 2D and 3D virtual environments and include their comparisons. The study is a vital instance to merge the psychology and the computer science disciplines to provide benefits about the developments of learning skills of and interaction with CWND. We believe that the results will give the new ideas for the researchers who study in this area.

Keywords: Human computer interaction (HCI) · Augmented reality (AR)
2D virtual environment · Children with autism (CWA) · Theory of mind (ToM)

1 Introduction

Human Computer Interaction (HCI) is a field of study on developing a new approach and design for the interaction between humans and computers. This filed encapsulate varieties technologies and types of human senses to improve the interaction between humans and computers. In addition, it includes multiple disciplines such as computer science, cognitive science, universal design approaches on the computer, and human-factors engineering. Augmented Reality (AR) is a sub-title of HCI. AR-Virtual Environment provides the real world with virtual objects. This is the key point for reality without a completely disappearing. AR remembers us the other types of Virtual Environments such as Virtual Reality (VR). The main difference between AR and VR is related to the key point. VR supplies completely virtual environments. In this work, we chose AR because of properties of users. The next section, we explain the reason for our choice in details.

As we know, the world has 3 dimensions. To learn systems, we modelled them on a paper (2 dimensions). Finding digital environments, the paper environment was carried to digital environments and then we recognized losing of the knowledge with using 2D to define it. Before 3D virtual environments, mostly 2D environments have been used. 2D environments are reason for loosing properties. These loosing properties are meaningful and gaps for some of us while it does not matter for others. This is mostly

© Springer International Publishing AG, part of Springer Nature 2019
I. L. Nunes (Ed.): AHFE 2018, AISC 781, pp. 257–264, 2019.
https://doi.org/10.1007/978-3-319-94334-3_26

related to the perception. Perception is the first step for understanding. The goal of the study is to decide the best suitable environments for the students with LDs according to their perception abilities. Especially, the children with neurodevelopmental disorders (CWND) have more characteristic properties about the perception through nearly all of them has concentration problems due to being vulnerable from environmental perceptional warnings. In general, CWN *have:*

- Diffuculties explaining one's own behaviours
- Difficulties understanding people's emotions
- Problems Understanding the perspective of others
- Problems inferring the intentions of other people

These skills are related to Theory of Mind (ToM) and ToM is descripted as ability to imagine the thoughts, beliefs, knowledge, emotion, goals, and desires of others [1]. To measure lack of ToM or some abilities related to Tom, some tests were developed for children such as Theory of Mind Task Battery. It was designed to assess the theory of mind understanding of younger and older children who vary widely in their cognitive and linguistic profiles (Hutchins et al. 2008). The test is appropriate for non-verbal individuals as respondents can indicate responses either verbally or through pointing. The test is implemented on the paper. The Theory of Mind Task has internal consistency and it correlates strongly with other measures of theory of mind [2]. Test–retest reliability of theory of mind tasks representing a range of content and complexity and adapted to facilitate the performance of children with ASD. Focus on Autism and Other Developmental Disabilities. Merve Altıntas implemented this test for Turkish Children with/without autism age range 4 to 5. She tested the reality of the Test and her participants were tested on the paper. As a result, the test was reliable yet, task 2 in the test was decided not to suitable for Turkish children after statically analyzing methods [3]. The question is related to a perspective of the statue and ToM. In this work, we changed test environment. TST (Test System Tool) was developed. TST encapsulates 2D and 3D (AR) questions for the participants to do the questions including measure their perspectives and ToM abilities. Using computer environment provide individual knowledge from participant directly and more precisely [4].

Low- concentration of participants with neuro-developmental disorders may occur with their cognitive process. 2D environments need quite more cognitive process than 3D environments according to abstractions. In literatures about Theory of Mind approach, the general judgment is that their abstraction abilities, imaginations of CWND are lower than the children without neurodevelopmental disorders. Therefore, in this work we compared Theory of Mind tests in AR and 2D environments for CWND. We adopted Theory of Mind Test into 2D and AR environment. Therefore, using virtual environments' properties gave us vital opportunities to adjust some environmental effects. Evaluating ToM tests on the virtual environment is a fundamental way. This paper encapsulates 2D and 3D version of ToM test and their comparable results on Turkish CWND. The paper is comprised of the following sections: Sect. 1 introduces the topic, Sect. 2 includes the related work, Sect. 3 describes the model and implementation, the experiment environment and the experiment are talked in Sect. 4, Sect. 4.1 displays experimental results and discussion about them and then Sect. 5 concludes the paper with general analysis about the research.

2 Related Work

In contrast to VR (synthetic environments), AR allows the user to see the real world in other words, AR supplements reality, rather than completely replacing it [5]. AR combines real world and virtual objects. The property- 'supplementing reality' of AR is the reason to choose it rather than VR. From our point of view, this property emphasizes for CWA being in the similar environment rather than completely replacing it. For example, to do their research test for CWA, the authors preferred to do retest in participants' home [2]. In general, when CWA change their environment completely, they may feel more unconfident than being in the similar environment.

According to our researches and literatures about related studies, children with autism and children with mental retardation have the enormous interaction with the computer. Especially, touch-screen has the significantly important role for their interaction [4]. The application in [6] is developed for CWND to take order. Touch-screen propability of the device provides them to take order without literacy. In this study, we use 2D touch screen monitor to supply CWND the environment in which do test ToM by themselves.

ToM is used in Daily life-every where to predict some's thoughts. There are many studies about evaluating ToM abilities for children. Furthermore, ToM is adapted to the system to recognize emotions.

Today, technology predicts emotions from voice, facial expressions, body language [7]. On the other words, the model based on computer science gains ToM abilities to do specialized tasks. Today, ToM is used in everyday in our daily life to predict and produce answers or feedback to some's behaviors.

The primary emotions can be grouped into anger, fear, happiness and sadness. Human emotions are related to the brain [8]. The brain processes cognitive steps to extract meaning from someone's feels. The body behaviors, gestures, face expressions and sounds give us the key about the mood situations. They all need cognitive processes that may affect the different part of the brains although extractions of the meanings are related with ToM.

The autism is assumed that one type of the LDs. Children with mental disorders have learning difficulties but they can be taught [9]. In most dyslexics, metacognitive (Wilder and Williams 2001) – the process of thinking about thinking deficiency is appeared [10, 11] is talked about that dyslexic children who is unable to display their latent mathematical abilities yet, they often exhibit superior mathematical skills after training. This interesting result encourages us to study on CWND in the different platforms such as computer-based systems 2D or 3D. In this work, we changed test environment. We developed 2D and 3D (AR) applications for participants to do questions about perspectives and ToM abilities. Touch-screen tablet is preferred.

For 3D environment. Virtual Reality and Augmented Reality may be used. In terms of CWA, generally, they do not like and feel uncomfortable when their environment is changed completely. In this project, AR is preferred. The applications related to AR are developed very frequently and AR for CWA is the novel approach to contribute to their abilities. One of the studies is about that having lack of imagination by CWA motives Bai et al. to design playing scenarios based on AR. Their hypothesized that AR may encourage the mental representation of pretense by presenting a reflection of the world in which a simple play objects (a wooden block) is replaced by an imaginary alternative (a car). As a result, participants were highly engaged with the AR system and produced a diverse range of play ideas [12].

Another virtual environment for people with Autism Spectrum Disorder (ASD) is designed to teach social skills. The research provides two scenarios: a virtual café and a virtual bus to find a seat [13].

3   Model Description and Implementation

3.1   Model Description

The number one principle of universal design for all users, provides the same means [14]. Using image is very old historical method to explain somethings. In ancient Egyptian, they used to present their languages using Egyptian Hieroglyphs. This method challenges throughout history. We can understand without the old literacy. Using images makes learning easy for all users, universally. For CWND, using images not only for CWND but also for researchers makes understanding easy. Children with mental disabilities has long, working- term memories problems. Palmer suggests that dyslexic learners use phonological codes less efficiently and having bad effects on working- memory capacity [15].

Interface model for them should take customized interface properties into account. In the work, we have 2D and 3D designs with some customized properties. We compared them in terms of these properties. 2D models are created with C#. In this research, 3D models are constructed on Unity. Unity is the ultimate game development platform. Use Unity to build high-quality 3D and 2D games, deploy them across mobile, desktop, VR/AR, consoles or the Web, etc. [16]. Unity is supported by C#. In this model, to create screens based on animations and to connect them with each other, C# is used.

3.2   Model Implementation

In this study, the examples of the developed 2D and 3D applications' screens are shown in Figs. 1 and 2, respectively.

4   Experiment Environment and Experiment

To collect data from CWND, Test System Tool (TST) was developed in C# with Unity. TST recorded CWND answers, their test duration and demographic informations of CWND such as age, genders, the school information, kinds of disabilities, etc.

TST encapsulates 2D and 3D (AR) questions for the participants to do the questions including measure their perspectives and ToM abilities. The general test scenarios include a girl and a horse. General test question: "Which side of the horse does the girl see?". In each question, the position of the horse and the girl are changed. Each

Fig. 1. One of the examples of 2D TST

Fig. 2. One of the examples of 3D TST

position combination of the girl and the horse become each test question separately. In shortly, all perspectives of the horse are asked according to the girl in TST. The scenarios are related to their ToM abilities which are measuared. Touch-screen tablet is preferred for the test platform because they can use it easily.

Figure 3 shows some CWND when they do TST.

Fig. 3. CWND doing 2D Test.

4.1 Sample Size and Demographic Information of Participants

Sample size is very important to supply reliability test results. According to [7], example of how confidence intervals change as a function of sample size are shown in Table 1.

Table 1. Example of how confidence intervals change as a function of sample size [17].

Number successful	Number of participants	Lower 95% confidence	Upper 95% confidence
4	5	36%	98%
8	10	48%	95%
16	20	58%	95%
24	30	62%	91%
40	50	67%	89%
80	100	71%	86%

In terms of Table 1, 13 participants were tested in this work. In terms of Table 1 [17], the results are confident.

ToM test is implemented on 4–5 years children because ToM abilities are developed on the children about 4–6. On the other hand, developments of the children with learning disabilities and their abilities is lower than the typical development children. Thus, ToM test can be implemented to higher than 4–6 age CWND.

TST has two parts: 2D and 3D environments. The age range of the male and female participants is 7 to 13. They did 2D and 3D tests. 15 participants and each test were done 2-3 times for each participant. They interested the tests. Some of them were surprised while the girl appeared in 3D test environments. They controlled the back of the tablet.

According to [3], the perspective question is the most difficult question in the ToM test which is in [3]. In this test, just only perspective questions are asked. In this work, we changed test environment. We developed 2D and 3D (AR) applications for participants to do questions about just only perspectives and ToM abilities. Each question in both of the environments are answered.

In the 2D environment, the question which answer is "the near side of the horse" is found easier than the other questions while the question which answer is "the back of the horse" is found easier than the other in 3D environment.

5 Conclusion

In this research, AR and touch-screen tablet made contributions to ToM test more efficiently. This is the novel approach doing ToM test with AR (3D) and 2D touch screen environment. All CWND finished the 2D and 3D TSTs. Moreover, they paid afford to do although they have low-consantrations.

The results show that 2D and 3D ToM is suitable for CWND to do ToM test.

The next work, the test cases will be developed and the number of the participants will be increased.

Acknowledgments. Thank you for Tuzla Kardesler Special School and Gebze Special School.

References

1. Baron-Cohen, S.: Mindblindness: An Essay on Autism and Theory of Mind. MIT Press, Cambridge (1995)
2. Hutchins, T.L., Bonazinga, L.A., Prelock, P.A., Taylor, R.S.: Beyond false beliefs: The development and psychometric evaluation of the Perceptions of Children's Theory of Mind Measure – Experimental Version (PCToMM-E) (2008)
3. Altintas, M.: Cocuklar icin Zihin Kurami Test Bataryasi'nin Gecerlik Güvenirlik Calismasi. Thesis, Istanbul (2014)
4. Akin, N.T.A., Gokturk, M.: Providing individual knowledge from students with autism and mild mental disability using computer interface. pp. 686–697 (2018). https://doi.org/10.1007/978-3-319-60492-3_65

5. Azuma, R.: A survey of augmented reality. In: Presence: Teleoperators and Virtual Environments. vol. 6, no. 4, pp. 355–385, August 1997

6. Akin, N.T.A., Gokturk, M.: Order interface model for individuals with down sydrome and emotion analysis. In: SMC 2016, Hungary (2016)

7. Fukuda, S.: Extracting emotion from voice. In: 1999 IEEE International Conference on Systems, Man, and Cybernetics, 1999. IEEE SMC 1999 Conference Proceedings (1999). https://doi.org/10.1109/icsmc.1999.812417

8. Akin, N.T.A: Theory of relativity, energy and the digitization of human emotions. In: ICCESEN 2015, processing, December 2015

9. Zervas, P., Kardaras, V., Sampson, D.G.: An online educational portal for supporting open access to teaching and learning of people with disabilities. In: 2014 IEEE 14th International Conference on Advanced Learning Technologies, Athens, pp. 564–565 (2014). https://doi.org/10.1109/icalt.2014.165

10. Flavell, J.: Metacognition and cognitive monitoring: a new area of cognitive- developmental inquiry. Am. Psychol. **34**(10), 906–911 (1979)

11. Martens, W.L.: Designing the "Spatial Math Tutor": a non-symbolic, on-line environment for teaching mathematical skills to dyslexic children. In: Second International Conference on Cognitive Technology. Humanizing the Information Age. Proceedings, Aizu-Wakamatsu City, pp. 262–270 (1997). https://doi.org/10.1109/ct.1997.617706

12. Bai, Z., Blackwell, A.F., Coulouris, G.: Using augmented reality to elicit pretend play for children with autism issue. IEEE Trans. Vis. Comput. Graph. **21**(05), 598–610 (2015)

13. Kerr, S.J., Neale, H.R., Cobb, S.V.G.: Virtual environments for social skills training: the importance of scaffolding in practice. Paper presented at the Assets, 8–10 July 2002, Edinburgh, Scotland (2002)

14. Jacko, A.: The Human- Computer Interaction Handbook: Fundamentals, Evolving Technologies, and Emerging Applications, Third Edition, p. 926 (2012)

15. Palmer, S.: Phonological recoding deficit in working memory of dyslexic teenagers. J. Res. Read. **23**(1), 28–40 (2000)

16. Unity programming. https://unity3d.com. Accessed 24 Nov 2017

17. Tullis, T., Albert, B.: Measuring the User Experience. Elsevier, New York City (2008)

A Real-Time Algorithm for Movement Assessment Using Fuzzy Logic of Hip Arthroplasty Patients

César Guevara[1], Janio Jadán-Guerrero[1(✉)], Yves Rybarczyk[2,3],
Patricia Acosta-Vargas[2], Wilmer Esparza[2], Mario González[2],
Santiago Villarreal[2], Sandra Sanchez-Gordon[4], Tania Calle-Jimenez[4],
and Isabel L. Nunes[5,6]

[1] Universidad Tecnológica Indoamérica, Ambato, Ecuador
{cesarguevara, janiojadan}@uti.edu.ec
[2] Intelligent & Interactive Systems Lab, Universidad de Las Américas,
Quito, Ecuador
{yves.rybarczyk, patricia.acosta, wilmer.esparza,
mario.gonzalez.rodriguez,
santiago.villarreal}@udla.edu.ec
[3] CTS/UNINOVA, DEE, Nova University of Lisbon, Monte de Caparica,
Portugal
[4] Escuela Politécnica Nacional, Quito, Ecuador
{sandra.sanchez, tania.calle}@epn.edu.ec
[5] Faculty of Science and Technology, Universidade NOVA de Lisboa,
2829-516 Caparica, Portugal
{y.rybarczyk, imn}@fct.unl.pt
[6] UNIDEMI, Campus de Caparica, 2829-516 Caparica, Portugal

Abstract. The present work proposes a model of detection of movements of patients in rehabilitation of hip surgery in real time. The model applies the Fuzzy Logic technique to identify correct and incorrect movements in the execution of rehabilitation exercises using the motion capture device called XBOXONE Microsoft's Kinect. An algorithm generates a multivalent logical model that allows the simultaneous modeling of deductive and decision-making processes. This model identifies the correct and incorrect movements of the patient during the execution of rehabilitation exercises. This model uses all the information collected from 24 points of the body with their respective axes of coordinates (X, Y, Z). Using data mining analysis, the algorithm selects the most remarkable attributes, eliminating non-relevant information. The main contributions of this work are: creation of a patient profile based on the movement of the human being in a generic way for the detection and prediction of execution of physical exercises in rehabilitation. On the other hand, an avatar was developed in 3D, which copies and evidences graphically the exercises performed by patients in real time.

Keywords: Hip arthroplasty patients · Fuzzy logic
Tele-rehabilitation platform · Movement assessment · Kinect

© Springer International Publishing AG, part of Springer Nature 2019
I. L. Nunes (Ed.): AHFE 2018, AISC 781, pp. 265–273, 2019.
https://doi.org/10.1007/978-3-319-94334-3_27

1 Introduction

Currently, the application of artificial intelligence (AI) in multiple areas of knowledge is crucial to solve problems efficiently. One of the fields of study is Medicine, where its application aids significantly the improvement results of patients handling different ailments. An example of this is the application of AI for the rehabilitation process of people that have had hip surgery. For these cases, body movement detection cameras are used in three dimensions (Kinect Xbox One). These cameras identify each one of the body points of human beings which allows to identify a person´s movements almost to perfection. This method detects dynamically when a patient is executing an exercise in a correct or incorrect manner, allowing to correct the patient's movements to improve the results of their rehabilitation exercises.

The main contribution of this study is to apply equipment learning techniques for the selection of attributes and the application of fuzzy logic to identify the exercises performed correctly or incorrectly by patients recovering from hip surgery. The detection frame established within an automatic learning focus uses the angles of the extremities of the body, speed and relationship between body points, as it is presented by Torres [1], applying a time series with optimal results.

The document is organized as follows; Sect. 2 describes related studies used as base for this research. Section 3 briefly describes the data used and the algorithms that were applied. Section 4 describes the proposed model for the detection of correct movements in real time for rehabilitation patients. Section 5 presents and analyzes the results. The article finishes with conclusions and work to follow.

2 Related Studies

The paper developed by Hebesberger [2], presents the implementation of an autonomous robot for a long term as the assistant of a walk group in an elder attention center facility. The robot accompanied two walk groups during a month offering visual and acoustic stimulation. The experience of therapists, the influence of the robot in the group dynamics and the estimation of therapists on the utility of the robot was evaluated thru a design of mixed methods that consisted on the observations, interviews and grading scales. The obtained results have proved to be reliable turning the robot into a useful assistant to overcome the extra workload and exhaustion for physical therapists.

The study performed by Klamroth-Marganska [3], describes a robotic training of an arm with Arm hemiparesis. The main object was to use an exoskeleton that allows the specific training of movement of the extremity in three dimensions, reducing motor deterioration more effectively than with conventional therapy. The study analyzed the evolution of patients that had faced motor deterioration for more than 6 months and moderated to severe arm paresis after a cerebrovascular accident that met with our eligibility criteria from four centers in Switzerland. Therapy was scheduled for at least 45 min, three times a week during 8 weeks (a total of 24 sessions). The main result was the change in score on the Fugl-Meyer (FMA-UE) evaluation for the arm section (high extremity). Results were pretty optimal with respect to the improvement of patients and the mobility of their extremities, offering an efficient solution for this medical condition.

Going up and down the stairs is a big part of daily mobility; it allows patients to gain independence from wheel chairs. The work developed by Hesse [4], has developed an innovative walking robot, G-EO-Systems (EO, Lat: I walk). It is based on the end effector principle. This robot uses plates in its feet to create walking trajectories which are designed freely. This allows not only to practice walking on a simulated floor but also practice going up and down the stairs. The study compares the activation patterns of the lower extremity muscles for hemiparesis subjects during real walks on the floor and while going up the stairs. This paper has demonstrated an improvement while executing a walk after the application of the machine on the patient.

Portable exoskeletons allow patients with walking dysfunctions to perform training walks on-site, even right after an acute event. The work performed by Molteni [5], investigates the viability and the clinical effects of a walking training on the ground with a portable exoskeleton for patients who suffered subacute and chronic strokes. Patients submitted themselves to 12 sessions of ambulatory rehabilitation training using Ekso (a bionic portable suit), that allows people with disabilities on the lower extremities to perform tasks such as standing up, sitting down and walking on a hard surface floor with a walking aid that stands all the weight. The results obtained using this robot have been very promising, they have shown a pretty good motility index and statistically significant improvements.

3   Materials and Methods

This section presents materials and techniques used to develop the algorithm of detection of the exercise execution of hip rehabilitation.

3.1   Materials

The database used for this study includes 7 patients who were subjected to hip surgery. There are two types of exercises suited for the rehabilitation process and recovery from this surgery, functional and complementary exercises. Functional exercises are those that allow to perform common tasks such as raising knees, move articulations (hip, shoulders, etc.) and extremities (arms and legs). Complementary exercises are the combination of several functional exercises in sequence, which eases the recovery of regular movement for patients in their daily lives. This study focuses on two types of functional exercises which are "side step" and "front step", as shown on Fig. 1.

The rehabilitation process took 4 weeks in average, where patients have permitted the capture of their correct and incorrect movements with the help of the Kinect 2 equipment (Xbox ONE). This equipment captures 25 body points three-dimensionally (X, Y, Z), which means, 75 attributes, as shown on Fig. 2. The database includes entries of patients with well executed exercises as well as badly executed ones.

Attribute Selection for the Detection Model. Selecting the most relevant attributes for the study was done by applying artificial intelligence techniques such as the one of

Fig. 1. Image of "side step" and "front step" exercise.

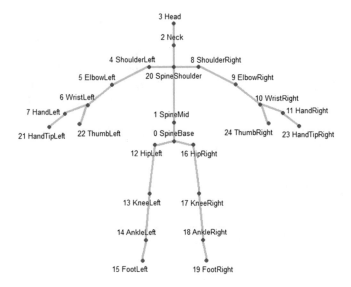

Fig. 2. Body points detected by Xbox One Kinect 2 (MSDN Microsoft).

GreedyStepwise. This technique permits an exhaustive research towards the front or the back thru the space of subsets of attributes [6]. It can start with none or all the attributes, or from an arbitrary point in the space. The algorithm stops when the addition or the elimination of any remaining attribute results in a lower score on the evaluation (1). It can also produce a classified list of attributes while crossing the space from one side to the other and register the order in which the attributes are selected [7].

$$Goodness\ of\ an\ attribute\ subset = \frac{\sum_{all\ attributes\ x} C(x, class)}{\sqrt{\sum_{all\ attributes\ x} \sum_{all\ attributes\ y} C(x, y)}} \quad (1)$$

Attribute selection thru the application of this algorithm results in obtaining 18 relevant characteristics, since in our case we are only dealing with exercises focused on the hip and the lower extremities (arms and legs). Specifically, these attributes are:

0y, 6x, 6y, 10x, 10y, 12x, 12y, 13x, 13y, 14x, 14y, 15x, 15y, 16x, 16y, 17x, 17y and 19x. As you can see, these attributes identify the patients movements of lower extremities in coordinates (x, y) which will allow to detect movement corrections.

Furthermore, this result has been validated with a physical therapist to assess efficiently the involvement of these points in the performance of the patient during the execution of the exercise. The result has been a set of points for the "side step" and "front step" exercise as shown on Figs. 3 and 4.

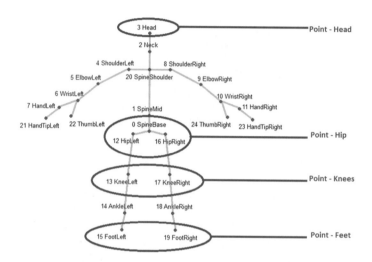

Fig. 3. Selected Points for "Front Step" exercise.

3.2 Methods

This section describes in detail an automatic learning technique called Fuzzy Logic which has been used for the development of the movement detection model for the rehabilitation exercises.

Fuzzy Logic is a multivalued logic that permits to represent uncertainty and vagueness mathematically. This technique provides formal tools to handle information and better adjusts to the real world we live in. This logic was formulated in 1965 by engineer and mathematician Lotfi A. Zadeh.

The main purpose of Fuzzy Logic is that any problem in the world can be solved with a set of input variables (entry space), obtaining an appropriate value of output variables (exit space). Fuzzy logic allows to establish this mapping adequately following criterion of meaning (and not precision).

The term fuzzy logic was first used in 1974. Currently it is used in a broader sense, grouping the fuzzy set theory, yes-then rules, fuzzy arithmetic, quantifiers, etc. [8].

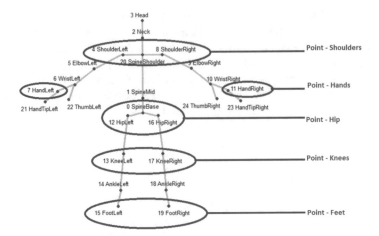

Fig. 4. Selected Points for "Lateral Step" exercise

4 Proposed Model

The proposed model uses the points described on Sect. 3 to further calculate angles and maximum/minimum speeds for lower extremities so the patient can perform a rehabilitation exercise correctly. Additionally, it determines movement optimal distances for all complementary points such as arms, shoulders and head. To calculate the appropriate angels for each of the exercises it is necessary to identify the initial pointB (X_i, Y_i), ending point $C(X\_j, Y\_j)$ and the value of the α angle of the triangle formed by the trajectory, as shown on Fig. 5.

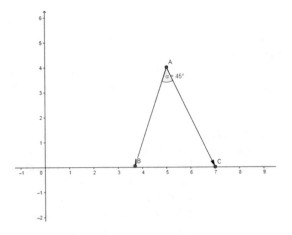

Fig. 5. Extremity opening angle.

Table 1 shows the referential angles and distances for the exercises performed correctly and incorrectly. Data was obtained from the patients that collaborated with the study.

Table 1. Rules to determine correct or incorrect movement of rehabilitation exercises in terms of opening angles of extremities, speed and 6 complementary movements

	Low		Good		High	
	Min	Max	Min	Max	Min	Max
Angle of legs (degrees) α	0	24	25	45	46	90
Hip movement (cm) μ	–	–	0	10	11	20
Shoulder movement (cm) β	–	–	0	10	11	20
Head movement (cm) δ	–	–	0	10	11	20
Execution speed (cm/s) ρ	20	30	10	19	5	9

Based on these rules, the development of a fuzzy model has been proposed with the input variables described in Table 1; obtaining a movement detection scheme for both exercises as shown on Fig. 6.

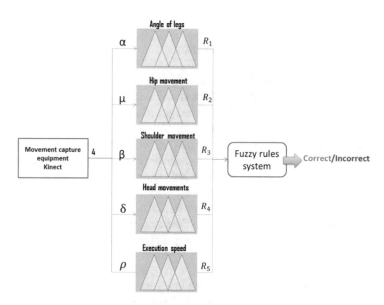

Fig. 6. Implementation of Simulink of the detector of exercises of rehabilitation with Kinect.

During the application of this fuzzy control, several tests have been performed with different patients who have obtained optimal results, which we are presenting in the following section.

5   Results and Discussion

The simulation results indicate that according to this detection scheme: 18% of the times patients perform an exercise correctly with the corresponding angle and only 27% does it at an acceptable speed. This means that developing an efficient strategy for a patient rehabilitation is to start at a lower angle and with a medium speed. On the other hand, during the testing stage, the optimal exercise detection rate found was 97.42% with a low false positive percentage of 2.58% as shown in Table 2. Tests were made with 500 entries for each user that included 200 correct movements and 300 incorrect movements for each patient.

Table 2. Matrix of confusion for the fuzzy model of rehabilitation exercise detection.

	Correctly classified	Incorrectly classified
Well executed exercise	2100	0
Badly executed exercise	1310	90
Total	**3410**	**90**

Detection time is close to 0.002 s in tests with 7 patients and an average of 0.0045 s with random information generated by the Matlab simulator.

6   Conclusions and Further Studies

The development of this model has evidenced that the application of fuzzy logic for human movement problems provides optimal and certain results. In addition of being an adjustable, easy and very dynamic technique to synthetize information of blurry data. On the other hand, the information identified with Kinect provides precise data for the detection of any kind of movement of human beings in three dimensions. Bear in mind that the selection of the appropriate attributes is needed for the performance of each task. This is why selection algorithms need to be implemented such as GreedyStepwise, which has excellent results.

Finally, as further studies derived from this research, genetic algorithms will be applied to improve in a dynamic way the detection of body movements. Moreover, pattern detection based on multi-layer neural networks have proven to have good results in other published papers.

Acknowledgments. The authors thank CEDIA for partially funding this study through the project "CEPRA XI-2017-15 Telerehabilitación". They also thank Esteban Farías, student of Faculty of Engineering & Design for his collaboration in the design of the avatar.

References

1. Torres, R., Huerta, M., Clotet, R., González, R., Sagbay, G., Erazo, M., Pirrone, J.: Diagnosis of the corporal movement in Parkinson's disease using Kinect sensors. In: World Congress on Medical Physics and Biomedical Engineering, 7–12 June 2015, Toronto, Canada, pp. 1445–1448. Springer, Cham (2015)
2. Hebesberger, D., Dondrup, C., Koertner, T., Gisinger, C., Pripfl, J.: Lessons learned from the deployment of a long-term autonomous robot as companion in physical therapy for older adults with dementia: a mixed methods study. In: The Eleventh ACM/IEEE International Conference on Human Robot Interaction, pp. 27–34. IEEE Press (2016)
3. Klamroth-Marganska, V., Blanco, J., Campen, K., Curt, A., Dietz, V., Ettlin, T., Luft, A.: Three-dimensional, task-specific robot therapy of the arm after stroke: a multicentre, parallel-group randomised trial. Lancet Neurol. 13(2), 159–166 (2014)
4. Hesse, S., Waldner, A., Tomelleri, C.: Innovative gait robot for the repetitive practice of floor walking and stair climbing up and down in stroke patients. J. Neuroeng. Rehabil. 7(1), 30 (2010)
5. Molteni, F., Gasperini, G., Gaffuri, M., Colombo, M., Giovanzana, C., Lorenzon, C., Farina, N., Cannaviello, G., Scarano, S., Proserpio, D., Liberali, D.: Wearable robotic exoskeleton for over-ground gait training in sub-acute and chronic hemiparetic stroke patients: preliminary results. Eur. J. Phys. Rehabil. Med. (2017)
6. Gevrey, M., Dimopoulos, L., Lek, S.: Review and comparison of methods to study the contribution of variables in artificial neural network models. Ecol. Model. 160, 249–264 (2003)
7. Xue, B., Zhang, M., Browne, W.N.: Particle swarm optimization for feature selection in classification: a multi-objective approach. IEEE Trans. Cybern. 43(6), 1656–1671 (2013)
8. Rastelli, J.P., Peñas, M.S.: Fuzzy logic steering control of autonomous vehicles inside roundabouts. Appl. Soft Comput. 35, 662–669 (2015)

An Integrated System Combining Virtual Reality with a Glove with Biosensors for Neuropathic Pain: A Concept Validation

Claudia Quaresma[1,2(✉)], Madalena Gomes[1], Heitor Cardoso[3],
Nuno Ferreira[3], Ricardo Vigário[1,2], Carla Quintão[1,2],
and Micaela Fonseca[1,2,3,4]

[1] Departamento de Física, Faculdade de Ciências e Tecnologia,
Universidade Nova de Lisboa, 2892-516 Monte da Caparica, Portugal
{q.claudia,r.vigario,cmquintao,
micaelafonseca}@fct.unl.pt,
ml.gomes@campus.fct.unl.pt
[2] Laboratório de Instrumentação, Engenharia Biomédica e Física da Radiação
(LIBPhys-UNL), Departamento de Física, Faculdade de Ciências e Tecnologia,
Universidade Nova de Lisboa, 2892-516 Monte da Caparica, Portugal
[3] VR4NeuroPain, Rua Capitão Renato Baptista nº15, 1150-085 Lisbon, Portugal
cardoso.heitor@gmail.com, nmmfer@gmail.com
[4] Universidade Europeia, Laureate International Universities, Estrada da Correia,
nº53, 1500-210 Lisbon, Portugal

Abstract. Spinal cord injuries are among the most traumatic situations, having relevant repercussions on an individual's occupation performance. Although loss of functionality is considered to be the most significant consequence, neuropathic pain can determine an individual's inability to return to daily activities. Therefore, it is imperative to develop new technologies with significant impact on the rehabilitation process of the spinal cord injuries. VR4NeuroPain combines virtual reality with a glove "GNeuroPathy", covered with a variety of biosensors, that allows for the collection of physiological parameters and motor stimulation. The main purpose of this paper is to describe the system VR4NeuroPain, and to validate the "GNeuroPathy" concept. With that in mind, and after calibrating the VR4NeuroPain system using with a group of 16 individuals, the validation results showed that "GNeuroPathy" was comfort, accessibility in place and the collection of physiological parameters was performed as expected.

Keywords: Rehabilitation · Virtual reality · Biosignals · Spine injury

1 Introduction

The Neuromotor disorders are motor disturbances that follow an injury located at the central nervous system. Motor deficits arise as the main consequence, usually accompanied by cognitive problems [1]. These injuries can be acquired or congenital, that is,

© Springer International Publishing AG, part of Springer Nature 2019
I. L. Nunes (Ed.): AHFE 2018, AISC 781, pp. 274–284, 2019.
https://doi.org/10.1007/978-3-319-94334-3_28

they can be the result of a pre-, peri- or post-natal injury. For these cases, rehabilitation plays a crucial role in the restoration and/or maintenance of motor skills [2].

The spinal cord is the main communication route between the brain and the rest of the body, so injuries at this level are devastating to the patient, both physically and psychologically. A global incidence study of traumatic vertebro-medullary injuries [3] carried out in 2011 estimated that the overall incidence rate of these cases is 23 cases per million, what represents 179 312 new cases per year. Several incidence rates from Western European countries were reported in that study, with a median value of 16 cases per million.

In another study [4], an incidence rate of traumatic vertebro-medullary injuries was reported in Portugal in the order of 57.8 individuals per million.

These disabilities affect patients' ability to accomplish real-life activities of daily living (ADLs), such as drinking and eating [5]. These ADLs often involve critical sub movements, including reaching and/or grasping [6]. Given the level of severity that the injury can cause, a variety of resources are needed, from rehabilitation therapy to psychological support, so that the patient can recover as effectively as possible [2].

Despite the loss of functionality is considered the most significant consequence, the neuropathic pain can be determinant for the patient's inability to return to ADLs, production and entertainment.

Therefore, it is imperative and urgent to develop new technologies that have a significant impact on the rehabilitation process of the vertebro-medullary lesion patient. With technological advances virtual reality has become increasingly popular and available being integrated into intervention programs, such as for pain and stress reduction, skills training and rehabilitation [7]. Thus, we developed the "VR4Neur-oPain", which consist on a system that associates virtual reality with sensory and motor stimulation. The system is an integrative technology solution that enhances sensory rehabilitation from the creation and development of the "GNeuroPathy" glove, which allows real-time monitoring of physiological parameters in patients with neuropathic pain.

Aiming to provide patients with an immersive and innovative environment for the rehabilitation process., "VR4NeuroPain" system promotes:

- motivation in the rehabilitation process;
- active role in the rehabilitation process;
- quality of life and well-being;
- production and movement control;
- perception and distinction of tactile sensorial stimuli used throughout the rehabilitation process;
- decrease in time spent in the rehabilitation process;
- greater economic sustainability of health sector units;
- use of interactive technologies by health professionals, promoting greater technological literacy.

According to Tatla et al. [8], when a technology is too complicated or time-consuming to set up, the clinical practitioners will not use it [9]. That is why, during technology development it is essential to establish partnerships with clinical practitioners, and focus on usability and user experience, through a validation process that includes and assesses all different types of users.

This validation of the concept allows us to ascertain the user's interest in adopting the technology, to verify if it is user-friendly and easy to apply [10].

During the validation process, the following usability parameters must be analyzed:

- Efficacy - the users' ability to complete tasks in interaction with the device;
- Efficiency - the amount of resources (cognitive, motor and time) that each user needs to perform a task in order to obtain a positive result;
- Satisfaction- the satisfaction degree demonstrated by users during the interaction with the device.

So, during the usability test, we intend not only to assess the feasibility of the device, the user's preferences, but also the performance of the users during the tasks, in terms of efficiency, effectiveness and satisfaction.

In order to perform these tests, it is necessary that both health's professionals and patients, use the device without any help and without previous training, in order to identify the type of challenges, times for completion of tasks, as well as comments and suggestions for improvement [11]. According to Nielsen [12], the number of participants required to guarantee the validity of the usability tests are at least 5 people and Satisfaction tests should also be performed using questionnaires or interviews if they can perceive the users' opinions.

In short, the main purpose of this paper is to describe the "GNeuroPathy" glove and to validate its concept.

2   Methods

"VR4NeuroPain" combines virtual reality with a glove, covered with a variety of biosensors, that allows for the collection of physiological parameters. The complete solution includes also a suite of software applications responsible for stimulating the subjects, as well as analysing the signals recorded by the sensors. The system allows:

- Use of biosensors - EDA and EMG - simultaneously, for monitoring muscular activity and endodermic activity. A Bitalino* acquisition module were, also, used;
- Use of different textures that in contact with patient's hand allow the stimulation of superficial and deep sensitivities ("pin", "sandpaper, "cotton", "hot" and "cold" stimuli);
- Identification of temperature through virtual objects;
- Manipulation of virtual objects promoting the hold.

2.1   Glove "GNeuroPathy" System Requirements

The design of our glove "GNeuroPathy" was centered on a set of essential requirements that were carefully analyzed during the design process [13]. The core requirements are outlined as follows:

1. Biosensors - The glove should allow the integration of EMG and EDA sensors and not interfere with the performance of the therapy. It must also be portable.
2. Comfort and durability - The glove should be lightweight, comfortable and not interfere with the tasks performed by the subject. It should also be adaptable to various hand sizes.
3. Function -The glove should enable a more real interaction with the subject and the object that is being manipulated.
4. Cost - The glove should be inexpensive for use by clinicians and patients.
5. Haptic feedback - By applying vibration patterns along mini vibrator-motors, the glove should recreate different sense of touch.

To meet these requirements, the developed Glove system consists in two main parts: a glove with EDA sensors and EMG, integrated with an haptic feedback technology. In the present paper we will only study and analyse a glove that only incorporates biosensors.

2.2   Design of "GNeuroPathy"

Figure 1 shows the glove "GNeuroPathy" prototype. After several glove prototypes using various material combinations, we determined that lycra and silk fabrics provided maximum flexibility in addition to being lightweight. We incorporated open-ended fingertips to allow haptic feedback and assist in grasping manipulation and hand shaping. It was incorporated open-ended fingertips to allow the inclusion of mini vibrator-motors (in two proximal phalanges of the 1st and 2nd fingers). This part (haptic feedback) will be validate in the future.

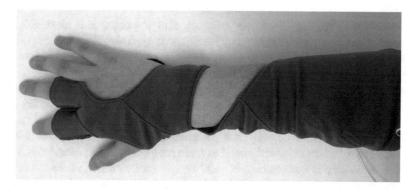

Fig. 1. The glove "GNeuroPathy"

The use of velcro solutions allows a practical handling and easily user´s application even if the user has restricted hand movements. The biosensors (EDA and EMG) are incorporated in the glove (see Fig. 2). For recording the EMG and EDA signals, a Bitalino* acquisition module, 2 EMG and 2 EDA sensors were used.

For haptic feedback, the mini vibrator-motors embedded in the glove are wired connected to a mini-controller hardware with bluetooth connection paired with the

Fig. 2. The figure shows the inside of the glove and the position of the electrodes

computer to synchronize the user's virtual touch (in the virtual environments objects) with the different vibrations patterns during the tasks, aiming the hand's sensory stimulation.

BITalino is an all-in-one board that was designed as a set of modular blocks to allow maximum versatility. It integrates multiple measurement sensors on-board for bioelectrical and biomechanical data acquisition [14, 15]. The Bitalino is capable of recording 4 biosignal simultaneously with a resolution of 16 bits and sampling frequencies up to 1000 Hz. As with the force platform, the data is streamed via Bluetooth. To connect the sensors to the subject, 2 Ag/AgCL with solid adhesive pregelled electrodes were used per sensor (TIGA-MED Gold 01-7500, TIGA-MED GMBH, Germany).

Electrodermal activity (EDA), or galvanic skin response, is a biosignal that is associated with the sympathetic nervous system activity. This signal indicates that the skin's electrical features variated, as a result of sweat glands' activity caused by various internal or external stimuli, which is translated into conductivity [16].

Low amplitude bioelectrical signals sent from motor control neurons trigger muscle activation, which can be used as a reliable measure of the activation level of the muscle [17]. EMG allows these electrical signals to be translated into numerical values. With BITalino, a surface EMG (sEMG) was performed, this type of EMG uses a bipolar differential front end for enhanced signal to noise ratio (SNR). This means that, while a three lead accessory is used to obtain EMG sensor data, two leads correspond to the common positive and negative voltage and the third lead is a reference lead. So, measurements from each of the negative/positive poles are subtracted, providing a 1-D time series [17].

For recording the data streamed from the Bitalino, the software used was OpenSignals from Plux.

2.3   Glove "GNeuroPathy" Data Collection

The data collection procedure is shown in Fig. 3. In the hand (at the level of the fingers) will be applied five different stimuli: 1- pin; Sandpaper; 3- cotton; 4- hot; 5- cold. Each stimulus will be applied for 10 s. During the application of each stimulus, data will be collected with the EMG and the EDA (the individual looking in another direction). Three collections are carried out with the same stimulus. By doing this we can verify the difference associated to each of the stimuli.

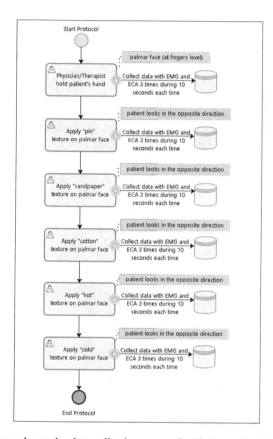

Fig. 3. The figure shows the data collection protocol with the application of the stimuli

2.4   Validation Process

The validation process was conducted at Faculdade de Ciências e Tecnologias da Universidade Nova de Lisboa. The experimental protocol was approved by the Centro de Reabilitação de Alcoitão ethical committee and all participants gave their written informed consent to participate in the study.

A total of 16 participated in the validation concept process. Of the total sample, 15 subjects were healthy and 1 has a spinal cord injury.

The participant with spinal cord injury is 37 years old, with a weight of 105 kg and a height of 1.86 m, while the group of healthy subjects consisted in 10 females and 5 males, with an average age of 29 ± 12 years old, an average weight of 65 ± 13 kg, and an average height of 1.68 ± 0.10 m.

We validated the "GNeuroPathy" in a two-part process:

1. Usability validation that examines subject's degree of satisfaction when using the glove;
2. Data collection procedure validation that involves assessing the performance of the data collection protocol.

2.5   Usability Tests

These tests were conducted to evaluate the parameters of Visual aspect, Accessibility in place, Comfort, Fixation and Texture in a scale where each of them was considered as Unsuitable, Partially adequate or Suitable.

During the usability tests it was found that the parameters considered as "suitable" by the healthy subjects were the visual aspect refereed by 80% of the participants in the study, accessibility in place, comfort and fixation with 87% of responses, and texture considered as suitable by the all participants of this study (Fig. 4).

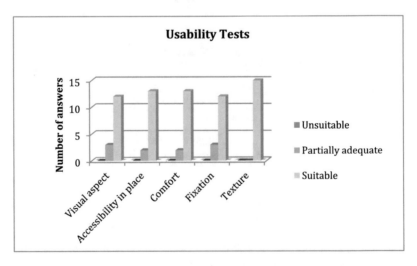

Fig. 4. The figure shows the healthy subject's degree of satisfaction

2.6   Validation of Data Collection Procedure

This procedure involves assessing the performance of data collection protocol. Therefore, the following protocol was defined and applied to all subjects:

1. Explanation of the study, as well as its objectives, and obtainment of participants' informed consent;

2. Filling out a questionnaire regarding the sample characterization and subject's degree of satisfaction;
3. Placement of the electrodes (for collection of biosignals) on the anterior face of the right hand and flexor muscles of the wrist and hand. A ground electrode is also placed in the pisiform bone;
4. Glove placement;
5. Photographic registration of upper limbs (hand and forearm) for validation of collection points as well as glove placement;
6. Application of data collection protocol (detailed in Fig. 3);
7. Collection of data;
8. Removal of electrodes and the glove.

We have verified that no adjustments to the protocol are required. It is easy to apply, and the application of stimuli and data collection takes more than 10 min. However, it is required a comprehensive process design to understand sources of variability and achieve process understanding.

Tables 1 and 2 present some illustrative examples of the type of results that can be gathered during stimulation of different textures. While Table 1 presents results of data collection in healthy subjects, Table 2 have the results obtained in the subject with spinal injury.

Regarding the values of healthy subjects, it was found that EMG values do not present significant differences throughout data collection. However, it was observed that, in average, both the amplitude of EDA and the standard deviation are always higher in pin and hot stimuli.

Table 1. The figure shows the healthy subject's

Stimulus	Measure	Channel	Average (μS)	Standard deviation (μS)
Basal	1	EDA	7.9	3.3
		EMG	12.5	0.0
Pin	1	EDA	12.6	5.2
		EMG	12.5	0.0
	2	EDA	12.6	5.1
		EMG	12.5	0.0
	3	EDA	12.6	5.3
		EMG	12.5	0.0
Sandpaper	1	EDA	12.1	4.7
		EMG	12.5	0.0
	2	EDA	11.7	4.7
		EMG	12.5	0.0
	3	EDA	11.6	4.7
		EMG	12.5	0.0

(continued)

Table 1. (*continued*)

Stimulus	Measure	Channel	Average (μS)	Standard deviation (μS)
Cotton	1	EDA	11.7	4.6
		EMG	12.5	0.0
	2	EDA	11.7	4.7
		EMG	12.5	0.0
	3	EDA	12.0	5.0
		EMG	12.5	0.0
Hot	1	EDA	14.1	5.7
		EMG	12.5	0.0
	2	EDA	14.3	6.3
		EMG	12.5	0.0
	3	EDA	14.2	6.4
		EMG	12.5	0.0
Cold	1	EDA	11.4	5.1
		EMG	12.5	0.0
	2	EDA	10.7	4.9
		EMG	12.5	0.0
	3	EDA	10.4	5.1
		EMG	12.5	0.0

Table 2. The figure shows the EDA and EMG values of the subject with spinal cord injury

Stimulus	Measure	EDA A (μS)	EDA M (μS)	EDA Ma (μS)	EMG A (μS)	EMG M (μS)	EMG Ma (μS)
Basal	1	0.9	0.7	1.2	12.5	11.7	13.1
Pin	1	1.0	0.7	1.4	12.5	12.1	13.5
	2	0.9	0.6	1.2	12.5	12.3	13.0
	3	1.0	0.7	1.2	12.5	12.3	13.0
Sandpaper	1	1.0	0.8	1.5	12.5	12.2	13.1
	2	1.0	0.8	1.2	12.5	12.3	13.0
	3	1.0	0.8	1.3	12.5	12.2	13.0
Cotton	1	1.0	0.8	1.3	12.5	12.2	13.2
	2	1.0	0.6	1.3	12.5	12.3	13.0
	3	1.0	0.8	1.2	12.5	11.8	13.0
Hot	1	1.0	0.8	1.3	12.5	12.4	12.9
	2	1.1	0.9	1.3	12.5	12.4	12.8
	3	1.0	0.9	1.3	12.5	12.3	12.9
Cold	1	1.1	0.9	1.2	12.5	12.3	12.8
	2	1.2	0.9	1.4	12.5	12.3	12.8
	3	1.2	0.8	1.4	12.5	12.3	12.8

A - Average; M - Maximum; Ma - Minimum

Regarding the values of the subject with spinal injury, it was observed that in EMG there are few differences throughout the data collection. However, it was observed that, in average, both the values of the EDA and the standard deviation are always low when compared to healthy subjects.

3 Conclusions

The work here presented is part of an ongoing study to the design and develop a system named "VR4NeuroPain". In this paper we focused on testing a prototype of the glove "GNeuroPathy", which consists on a physical part of the more complete "VR4Neur-oPain" system.

This study helped to better define and obtain clear feedback on prototype design and feasibility as well as on the procedure of data collection.

The future of this project will comprise the implementation of software, biosignals processing algorithms. Furthermore, the glove - "GNeuroPathy" must also be validates with the other parts of the system and in a virtual reality environment. Tests in subjects with neuropathic pain will be performed with the "VR4NeuroPain" and compared with the conventional procedure in order to prove that this is reliable system.

The system can be used by physicians, occupational therapists and physiotherapists and will allow us to apply innovative and interactive intervention methodologies favoring intergenerational and family relations, and promoting the integration of the rehabilitation process.

Acknowledgments. The authors would like to thank Collide for the help and support provided in this investigation.

References

1. Hagen, E.M., Rekand, T.: Management of neuropathic pain associated with spinal cord injury. Pain Ther. **4**(1), 51–65 (2015)
2. Klein, B.: Mental health problems in children with neuromotor disabilities. Paediatr. Child Health **21**, 1 (2016)
3. Lee, B.B., Cripps, R.A., Fitzharris, M., Wing, P.C.: The global map for traumatic spinal cord injury epidemiology: update 2011, global incidence rate. Spinal Cord **52**, 110–116 (2014)
4. Van den Berg, M.E.L., Castellote, J.M., Mahillo-Fernandez, I., de Pedro-Cuesta, J.: Incidence of spinal cord injury worldwide: a systematic review. Neuroepidemiology **34**, 184–192 (2010)
5. Luk, K.H.K., Souter, M.J.: Spinal cord injury. In: Khan, Z. (ed.) Challenging Topics in Neuroanesthesia and Neurocritical Care. Springer, Cham (2017)
6. Nathan, D., Johnson, M., McGuire, J.R.: Design and validation of low-cost assistive glove for hand assessment and therapy during activity of daily living-focused robotic stroke therapy. J. Rehabil. Res. Dev. **46**(5), 587–602 (2009)
7. Chen, C., Jeng, M., Fung, C., Doong, J., Chuang, T.-Y.: Psychological benefits of virtual reality for patients in rehabilitation therapy. J. Sport Rehabil. **18**, 258–268 (2009)

8. Tatla, S., Shirzad, N., Lohse, K., et al.: Therapists' perceptions of social media and video game technologies in upper limb rehabilitation. JMIR Serious Games **3**, e2 (2015)
9. Proeça, J., Quaresma, C., Vieira, J.: Serious games for upper limb rehabilitation: a systematic review. Disabil. Rehabil.: Assist. Technol. **13**(1), 95–100 (2018)
10. Kushniruk, A., Patel, V.: Cognitive and usability engineering methods for the evaluation of clinical information systems. J. Biomed. Inform. **37**(1), 56–76 (2004)
11. Wiklund, M.: Usability Testing: Validating User Interface Design. Americas (2007). https://www.mddionline.com/usability-testing-validating-user-interface-design. Accessed 28 Feb 2018
12. Nielsen, J.: Why you only need to test with 5 users. Nielsen Norman Group (2000). https://www.nngroup.com/topic/user-testing/. Accessed 28 Feb 2018
13. Ulrich, K., Eppinger, S.: Product Design and Development, 3rd edn. McGraw-Hill, Philadelphia (2003)
14. Guerreiro, J., Martins, R., Silva, H., Lourenço, A., Fred, A.: BITalino-A multimodal platform for physiological computing. In: Proceedings of the International Conference on Informatics in Control, Automation and Robotics (ICINCO), pp. 500–506 (2013)
15. Guerreiro, J., Lourenço, A., Silva, H., Fred, A.: Performance comparison of low-cost hardware platforms targeting physiological computing applications. In: Conference on Electronics, Telecommunications and Computers – CETC 2013 Procedia Technology, vol. 17, pp. 399–406 (2014)
16. Kim, K.H., Bang, S.W., Kim, S.R.: Emotion recognition system using short-term monitoring of physiological signals. Med. Biol. Eng. Comput. **42**(3), 419–427 (2004)
17. Krantz, G., Forsman, M., Lundberg, U.: Consistency in physiological stress responses and electromyographic activity during induced stress exposure in women and men. Integr. Physiol. Behav. Sci. **39**(2), 105–118 (2004)

Development of a Low-Cost Eye Tracker – A Proof of Concept

Ricardo Vigário[1,2(✉)], Filipa Gamas[1,2], Pedro Morais[1,2], and Carla Quintão[1,2]

[1] Departamento de Física da Faculdade de Ciências e Tecnologia da Universidade Nova de Lisboa (FCT-UNL), Monte da Caparica, 2892-516 Caparica, Portugal
{r.vigario, cmquintao}@fct.unl.pt,
{f.gamas, jpe.morais}@campus.fct.unl.pt
[2] Laboratório de Instrumentação, Engenharia Biomédica e Física da Radiação (LIBPhys-UNL), Monte da Caparica, 2892-516 Caparica, Portugal

Abstract. Humans live in, and interact with, a very complex and multi-sensory environment. Yet, one parcels efficiently the incoming information through attention mechanisms. Even if restricted to the visual sensory system alone, full understanding of one's perception often implies a suitable identification of that person's direction of gaze. There is a considerable amount of new and commercially available eye tracking devices. We investigate the use of a mobile phone to acquire precise information on the direction of gaze, in a controlled visual stimulation environment. The main advantages of the proposed new approach are its price, ease of use and ubiquitous availability. The results attained in this proof-of-concept study display fairly high accuracy and precision for the estimation of the direction of gaze.

Keywords: Low-cost eye tracker · Visual perception · Attention
Portability

1 Introduction

Our everyday life is filled with a cacophony of multi-sensory stimuli (mostly meaningless noise), speckled with congruent and meaningful information. What mediates between noise and information is the amount of attention one assigns to decode a particular stimulus. Effortlessly, and without even noticing it, we constantly perform some sort of (cognitive) beamforming, to reduce the plethora of stimuli into the desired information, which can then be dealt with efficiently. By doing so, we shut down the arrival of unwanted information, and focus solely onto what really matters to us then.

When restricting to visual stimuli alone, beamforming can be related to the increase in density of rods and cones observed around the fovea [1]. This means that only the direct view of an object captures its full texture and colour characteristics. Peripheral viewing hold crucial contextual information, but lack in detail. Hence, knowing where one directs one's gaze gives immediate information onto what we are really attending to. This is the basic principle underlying the construction of most eye trackers [2].

I. L. Nunes (Ed.): AHFE 2018, AISC 781, pp. 285–296, 2019.
https://doi.org/10.1007/978-3-319-94334-3_29

Since the introduction of the first eye trackers [3], proposed around an electrooculogram, mounted on a contact lens, there have been a number of different, and rather successful, ways to track one's direction of gaze.

It is out of the scope of this manuscript to go in detail over the practical use of eye trackers. Nevertheless, and to give an idea of the breadth of application areas they are in, we may still highlight those related to Human-Machine Interaction [4], for example for gaming [5]; to clinical evaluation of strabismus [6] and several types of neurodegenerative diseases [7]; and as a perfect tool for the analysis of customer impact of publicity, in neuromarketing [8].

The current study is based on Gamas' M.Sc. thesis [9], and provides a proof-of-concept for the use of simple mobile phone technology, together with an inexpensive cardboard virtual reality phone support, as a low cost eye-tracking device. In that way one should be able to stay close to 10% of the costs of the least expensive such devices.

2   Materials and Methods

In this study, the stimulus management and delivery, as well as the capture and storage of all eye images were performed with the phone, through a dedicated app, was developed using Android Studio 2.3. All processing of the acquired images and videos, as well as their relations to the stimuli, was performed offline, using Matlab[1].

The proposed device was tested with 7 subjects, with varying colour of their irises. Of those, 6 produced results in line with the illustrations presented herein. The outlier subject failed the calibration procedure, with errors propagating to the use of the device.

2.1   The Experimental Setup

The Phone. All image recording, as well as stimulus delivery was done using an *Alcatel Idol 4*, mobile phone, with Android 6.0 operating system. It has a frontal camera with 8 M pixel resolution, and a flash directed also to the front. The latter particular feature allowed for specific conditions of illumination during the collection of images from the eyes. The front camera insures a wider angular objective, since it is mostly used for short range image capturing. The phone has a 5.2" *Full HD* display, with a resolution of 1920 × 1080 p.

The Support. We used Virtoba's V2 mobile phone support ([10], shown in Fig. 1), designed following the specifications in the original Google Cardboard [11], for 3D visualization and VR applications.

Unlike some other eye tracking devices, the one used is firmly connected to the head, through a dedicated elastic strap. Hence, the relative position between the display/camera and the eyes is kept constant throughout the study, even when it requires some degree of head movement. The total weight of the experimental device

---

[1] MATLAB Release 2012b The MathWorks, Inc., Natick, Massachusetts, United States.

a)

b)

Fig. 1. Eye tracker's cardboard mobile phone support. Front view, showing the lenses for close eye focus on the stimuli (a). Top view (b) of the cardboard. The strap insures a stable relative position between the phone and the eyes. Note the 1.5 cm offset between the phone and the edge of the support.

shown in Fig. 1, cardboard plus phone is 223 g (88 g + 135 g), which makes it easy to hold on for experiments that last for a rather long periods of time.

The distance between the pupil and the back edge of the lens (see Fig. 1a) is around 18 mm, and between the front edge and the display is 39 mm. The lens has a focal length of 35 mm, rendering the apparent distance from the lens to the object displayed of 667 mm.

Monocular Excentricity. If we keep the traditional binocular stimulus layout of standard VR systems, the camera will invariably collect artefacts from the reflections of the stimulus on the lens (clearly visible in Fig. 2a). One solution found to remove such artefacts was to deliver the stimulus to one eye alone, and measure the direction of gaze from the other (Fig. 2b). The assumption is that both eyes, moving together, will convey redundant information. In addition, we noticed that the camera is positioned close to one edge of the phone, which makes it excentrical, both horizontally and vertically, when compared to the corresponding eye. Also, since it is located higher than the eye line, any stimulus presented at the lowest possible location of the screen often resulted in significant occlusion of the pupil by the eyelid (Fig. 2c). To reduce part of the horizontal excentricity, we chose to place the phone with a small offset of 1.5 cm (as shown in Fig. 1b). To handle the last mentioned artefact, we reduced the available display area, as shown in Fig. 2d, still retaining a stimulus with a resolution of 1280×960 p.

This approach is not without drawbacks. In fact, the rather strong binocular redundancy assumption fails for some subjects. In those cases, a different strategy needs to be employed.

2.2 Stimuli

Throughout the study we used two types of visual stimuli. To extract information regarding the direction of gaze, either for calibration (Fig. 3a), or to assess the regions of particular cognitive interest (Fig. 3b), we used still images as stimuli. We were also interested in knowing how the system fared with recording the subject's attention to a moving stimulus. Hence, we designed a very simple "track the ball" video.

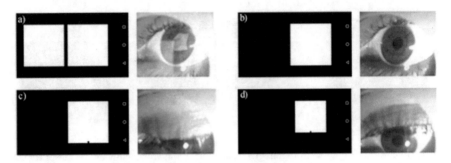

Fig. 2. Reduction of reflection artefacts cause by the stimuli (a–b); and avoidance of occlusions caused by extremely low display stimuli (c–d).

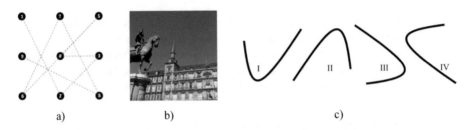

Fig. 3. (a) Illustration of the complete set of possible locations used for calibration, together with a superposition of the nine dots, and their order of appearance. (b) Still image utilized for the proof-of-concept study. (c) Four examples of lines used for the eye tracking accuracy study.

Images. The calibration procedure consisted of attending a dot, successively located in one of nine possible positions of a 3 × 3 grid (Fig. 3a). In order to prevent the subject from moving her/his eyes pre-emptively, with a prediction strategy rather than a tracking one, the order by which each position was displayed was randomised. Figure 3a shows a superposition of the first nine displayed dots, together with one possible order of their appearance. This calibration allows the determination of the transformation between the coordinates found for the centres of the pupil and those in the image stimuli.

The picture in Fig. 3b was used for most of the realistic scenario attention tests. The main characteristics required were that there should be, at least, one region with reduced visual information content – represented by the sky –; and another of sufficient content and texture, to drag the subject's attention – the horseman and the heavy textured background buildings.

The last set of images displayed in Fig. 3c corresponds to four different lines, used separately to study how accurately the eye tracker evaluates the following of an object. The subject, when presented with this type of stimulus should follow the line, from its leftmost extreme to the rightmost, at her/his own rhythm.

Videos. In addition to following the direction of gaze, when a subject inspects a complex still image, it is important to assess as well whether the device is capable also to track accurately particular patterns of evolving stimuli. For that purpose we employed as well a "follow the ball" video, where a white circle moves randomly, yet smoothly, across the display, while asking the subject to keep the attention on the object.

2.3   Pre-processing

Capture the Eye. We are interested only in the area surrounding the pupil. There are a number of possible solutions to reduce the image size to the desired dimensions. Yet, since the cardboard support enforces a constant relative position between the camera, the lens and the eye, throughout the duration of the experiment, one may simply determine the optimal cropped region of the image, which will be analysed, based on the cardboard structures (Fig. 4). These images have a resolution of 1280 × 960 p.

From Colour to Red. Mobile phone images are often acquired in RGB (*Red, Green* and *Blue)* format. For our processing, though, we do not require so much information. Instead, we prefer to handle simple, one-dimensional intensity data. After evaluating each channel data separately, we concluded that the red one presented, for most eye colours, the highest contrast between the pupil and the surrounding iris (Fig. 5). In addition, and to reduce some outlier-type noise, a 3 × 3 median filter is applied to the image. The result is an eight bit grey-scale image, where 0 corresponds to pure black and 255 to white.

Binarization and Pupil Identification. The next processing step is to isolate the pupil. As visible in Fig. 5, R, that structure holds most of the darkest pixels in the image. Hence the pupil should stand out in a simple threshold operation, with its value defined as a percentage of the lowest luminance in the picture. To make sure that dark pixels, visually indistinguishable from black, are also included in the definition of the pupil, the threshold is defined by the lowest intensity value in the image, to which we add the value 30 (Fig. 6a). This value was found, empirically, to be reasonable, yet its exact determination did not appear to be too critical.

In addition, we performed a series of geometric transformations, including erosion and dilation, to remove small edges in the estimated pupil; holes were filled; and too small objects removed. The final step was to reverse the intensity map, so that other functions could be used more easily. The result is displayed in Fig. 6b.

Location and Eccentricity. After identifying all remaining objects in the image, and if there is more than one such object left, one needs to define additional heuristics to reduce them to one. By construction of the image cropping, the pupil will never touch the leftmost border of the cropped image. Hence, any object in those conditions, such as the portion of the *plica semilunaris,* highlighted in Figs. 7a–c with a red rectangle, should be disregarded.

In addition, the shape of the pupil should always be an oval/ellipse. Since the pupil has an approximate circular morphology, and because of the existence of a lens and an offset of the camera's location, when compared to the eye, one may assume the ellipse as the best geometric model for that eye structure. Furthermore, an excessive

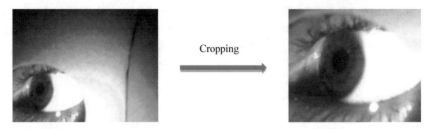

Fig. 4. Image capture of the eye, and the effect of the cropping operation.

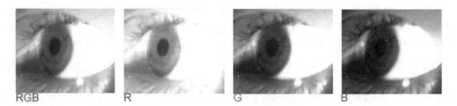

Fig. 5. Leftmost frame shows the grey-scale direct conversion of an RGB image of the eye. The subsequent frames hold, from left to right, a separate colour channel – red, green, and blue, respectively.

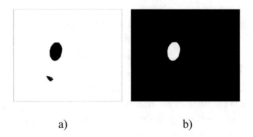

a)   b)

Fig. 6. (a) Image binarization. (b) Result of erosion/dilation transformations, as well as the removal of small patches in the image, and its intensity reversal.

eccentricity in the estimated ellipse should suggest the existence of an artefact. This can be seen as the shaded lower-left part of the iris, in Fig. 7d, which should also be discarded. The result should, once more, be a sole object, corresponding to the pupil.

Cardboard Occlusion. It is not possible to guarantee that the pupil will not be partially occluded by the cardboard, while the subject attends particular locations in the display. The result is a poor estimation of the pupil (Fig. 8a). Fortunately, the light reflected by the cardboard is brighter than that of the iris. Hence, we will consider, as frontier between the pupil and the iris, only those points that are not in too high intensity transitions. The result of this procedure is visible in Fig. 8b.

Pupil Dimensions and Continuity. Another important feature, which can be used in pupil identification, is the fact that such a structure, for a rather uniform light exposure,

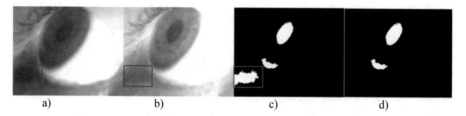

a)   b)   c)   d)

Fig. 7. Another example of image captured from the eye (a), its red channel (b) and the binarization procedure described earlier in the text (c). The darker portion of the *plica semilunaris* is highlighted with a red rectangle, and touches the leftmost border of the image limits, (c). The lower portion of the iris, visible as the lower-left white object, can be identified through its eccentricity.

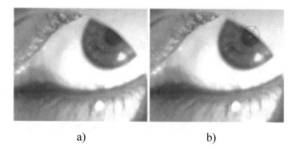

a)   b)

Fig. 8. (a) Partially occluded pupil. (b) Estimation of the pupil using only frontier points that do not have very bright neighbours.

has a fairly stable shape. Several artefacts result in clear changes to that shape. Figure 9a shows an artefact resulting on an object combining the pupil and some eyelids. Although the middle frame suggests that a single object is already found, its shape, as well as the size of the object, clearly deviates from the expected. For the experiments we made, we established empirically that the pupil should not have more than 8000, nor less than 4000 pixels. If the object found exceeds the upper threshold, we decreased the binarization threshold, rendering it more restrictive. After going through all processing steps as described earlier, the pupil was successfully identified, as shown in Fig. 9b.

In the opposite direction, Fig. 9c gives us an example of underestimation of the pupil extension. After increasing the binarization threshold, a new set of objects is detected. As before, a series of criteria are then applied to reduce it to a single object. The remaining estimate of the pupil, although not perfect, is much closer to the targeted shape.

Removal of Flash Artefacts. The last artefact we need to handle is the reflection of the flash, used to improve the visualization of the structures of the eye (clearly visible in Fig. 10a). Invariably, it will appear as the brightest elements in the image, typically with a circular format. The first step to solve this, is to identify the location of those

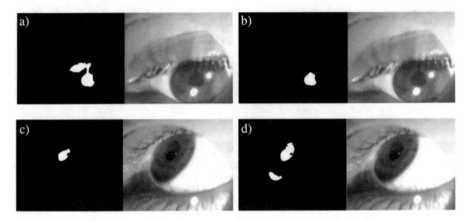

Fig. 9. Correcting artefacts identifiable through the estimated size of the pupil, and corrected by readjusting the binarization threshold. Area greater than expected (a), corrected by taking a more restrictive threshold (b). Too small pupil (c), corrected by allowing a more permissive threshold (d).

brightest spots, around which we can define a small region (b). We can then consider those regions as missing data, and reproduce the image that should be there, by taking into account their existing, surrounding pixels. We can then estimate the pupil, following all the steps described earlier (c).

2.4   Establishing the Direction of Gaze

The processing performed up to the last section resulted in one shape that should be almost elliptic. To determine which ellipse fits best the shape above, we used *fitellipse,* a public domain function designed for Matlab [12]. It is a Least-Squares estimate of 2D ellipses, from a given set of frontier data points (x, y). The least amount of required such points are five. Figure 11 displays a series of successfully fitted ellipses, together with their main axes. The centre of those lines is assumed as the centre of the pupil.

From Pupil Centre to Attended Location. Once we have established the location of the pupil, for any given direction of gaze, it is important to find how those relate to each position in the stimulus image. We find this 2D, 2nd order polynomial transformation function during calibration, utilising the 3×3 calibration grid, as mentioned earlier, and reproduced in Fig. 12a – note that the low order of the polynomial is justified by the smooth transformations caused by the optical setup. With the 9 pairs of coordinates, the fitted transformation can be illustrated by the dotted lines in Fig. 12b. In certain conditions, the lower right stimulus is incorrectly estimated. Then the calibration should avoid said position (Fig. 12c). Since one only needs 6 distinct points to fit the 2nd order polynomial, the estimate should still be effective.

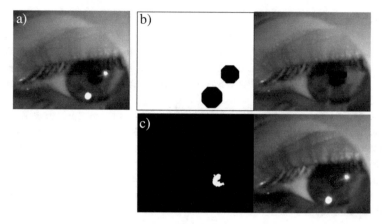

Fig. 10. Handling artefacts caused by the reflections of the flash on the cardboard lens (a). Exclude the regions of brightest intensity, and replace them by extrapolations from surrounding pixels (b). The subsequent pupil estimate can then follow the standard procedure (c).

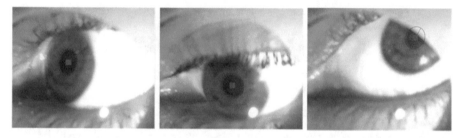

Fig. 11. Three examples of successfully fitted ellipses, and the estimated centre of the pupil.

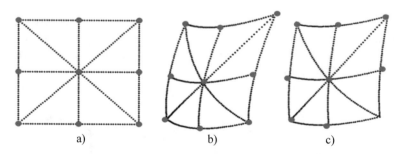

Fig. 12. Calibration locations (a); transformed by the optics, and evaluated by the pupil estimation (b); transformation calculated from the most reliable eight, of the nine calibration locations.

3 Results

The most important measures of usability of an eye tracker are its precision, and accuracy. The former assesses how the reproducible are the results suggested by the device. The latter deals with how close, in average, the proposed solutions are to the correct ones.

As stated earlier, this manuscript describes a proof of concept. More than a thorough evaluation of the precision and accuracy of the proposed device, we want to show evidence of suitability, leaving a more detailed quantification study for future work.

In Fig. 13 we show the results of the calibration procedure for 6 passes over the 9 calibration points. It can be seen as a visualization of how precise the eye tracker is. Note that the extreme left-right and top-right corners are the areas of the image that contain highest degree of deviance.

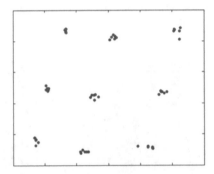

Fig. 13. Assessment of the precision of the proposed eye tracker. The stimulus consisted of the calibration image shown in Fig. 3a. Each red spot corresponds to one of six passes over the respective location.

Figure 14 shows how accurately the eye tracker allows us to follow the direction of gaze during the very specific task of following, with the eyes, a given line. There is a fairly good accuracy throughout. At time, a certain degree of pre-emptive eye movement is easily detected as well.

Fig. 14. Accuracy testing, in four different "follow-the-line" tasks.

As an illustration of how one may use the eye tracker to identify areas of interest in one image, Fig. 15 recovers the stimulus in Fig. 3b. Notice how the eyes fixated more often the statue and the building structures, rather than the little informative sky.

Fig. 15. Example of the information collected by the eye tracker when one of the subjects observes a complex visual stimulus.

4   Conclusions

As we started this study, our goal was to show that it is possible to collect, with a fairly good degree of precision and accuracy, direction of gaze information, using a rather low-cost apparatus, mostly based on highly pervasive mobile phone technology. What we concluded was that it is not only possible, but that the results surpassed our initial expectations.

Naturally, the study is not without some limitations. One is that we utilized a mobile phone with frontal flash, which is not yet in wide application. A continuation of the present work should include other alternatives. Although we fully fulfilled our proof-of-concept goals, one could argue that the biggest drawback of our study is a certain lack in statistics and formal evaluation of the required "accuracy" and "precision" factors. A follow-up to it will have to address as well those issues.

In the future, we plan to move all the post-processing to the Android platform, which will render the device truly portable and closer to a real-time tool for the evaluation of one's visual attention.

References

1. Breedlove, S.M., Neil V. Watson, N.V.: Biological Psychology: An Introduction to Behavioral, Cognitive, and Clinical Neuroscience. 7th edn. Sinauer Associates (2013)
2. Duchowski, A.T.: Eye Tracking Methodology, 2nd edn. Springer, Heidelberg (2007)
3. Huey, E.B.: The Psychology and Pedagogy of Reading. The Macmillan Company, pp. 15–50 (1908)
4. Nehete, M., Lokhande, M., Ahire, K.: Design an eye tracking mouse. Int. J. Adv. Res. Comput. Commun. Eng. 2(2), 1118–1121 (2013)
5. Isokoski, P., Martin, B., Eye tracker input in first person shooter games. In: 2nd Conference on Communication by Gaze Interaction: Communication by Gaze Interaction-COGAIN 2006: Gazing into the Future, pp. 4–6 (2006)
6. Chen, Z., Fu, H., Lo, W.L., Chi, Z.: Eye-tracking aided digital system for strabismus diagnosis. In: 2015 IEEE International Conference on Systems, Man, and Cybernetics, SMC 2015, pp. 2305–2309 (2016)
7. MacAskill, M.R., Anderson, T.J.: Eye movements in neurodegenerative diseases. Curr. Opin. Neurol. 29(1), 61–68 (2016)
8. dos Santos, R.O.J., de Oliveira, J.H.C., Rocha, J.B., Giraldi, J.D.: Eye tracking in neuromarketing: a research agenda for marketing studies. Int. J. Psychol. Stud. 7(1), 32–42 (2015)
9. Gamas, F.: Desenvolvimento de um Eye Tracker de Baixo Custo. M.Sc. thesis in Biomedical Engineering, Faculty of Sciences and Technology, University NOVA of Lisbon, September 2017
10. Virtual Reality Viewer V2. http://www.virtoba.com. Accessed 14 Feb 2018
11. Google Cardboard for Manufacturers. http://static.googleusercontent.com/media/vr.google.com/pt-PT//cardboard/downloads/manufacturing-guidelines.pdf. Accessed 14 Feb 2018
12. Gal, O.: fitellipse. https://www.mathworks.com/matlabcentral/fileexchange/3215-fit-ellipse. Accessed 10 Feb 2018

Management of Productivity in Smart and Sustainable Manufacturing - Industry 4.0

Socio-Technical Capability Assessment to Support Implementation of Cyber-Physical Production Systems in Line with People and Organization

Fabian Noehring[1(✉)], René Woestmann[1], Tobias Wienzek[2], and Jochen Deuse[1]

[1] Department for Production Systems,
RIF Institute for Research and Transfer e.V., Joseph-von-Fraunhofer-Str. 20,
44227 Dortmund, Germany
{Fabian.Noehring, Rene.Woestmann,
Jochen.Deuse}@rif-ev.de
[2] Research Area Industry and Labour Research, TU Dortmund University,
Otto-Hahn-Straße 4, 44227 Dortmund, Germany
Tobias.Wienzek@tu-dortmund.de

Abstract. Cyber-Physical Production Systems (CPPS) enable the intelligent, horizontal and vertical interconnection of people, machines and objects throughout the enterprise in real-time by information and communication technologies providing a basis for increasing transparency and productivity of production processes. However, especially small and medium-sized enterprises with limited resources and personal competencies need support in planning and evaluation of CPPS. Former developments, as the CIM-era, showed that changes in production systems focusing only on technology failed. Due to the interconnection of CPPS, a holistic approach, taking likewise humans, technology and organization into account is necessary. This paper presents requirements as well as an evaluation of existing approaches. Furthermore, this paper presents the approach of a socio-technical capability assessment, enabling companies to evaluate effects of CPPS as well as deriving implementation measures. It concludes with a validation based on a use case of a worker information system.

Keywords: Cyber-Physical Production Systems · Industrial Internet
Industry 4.0 · Small and medium-sized enterprises
Socio-technical capability assessment

1 Cyber-Physical Production Systems (CPPS)

1.1 Benefits and Barriers of CPPS

Increasing digitalization leads to an industrial revolution called Industry 4.0 or Industrial Internet and enhances the implementation of Cyber-Physical Production Systems (CPPS) in enterprises. CPPS enable the intelligent horizontal and vertical

© Springer International Publishing AG, part of Springer Nature 2019
I. L. Nunes (Ed.): AHFE 2018, AISC 781, pp. 299–311, 2019.
https://doi.org/10.1007/978-3-319-94334-3_30

interconnection of people, machines and objects throughout the enterprise in real-time via information and communication technologies [1]. This allows enterprises, among others, to increase transparency and productivity of production processes [2]. E.g., current machine and order statuses can be visualized in an operator information system to increase process transparency and optimize order allocation. Furthermore, machine data can be processed and used for prediction of machine failures [3].

Since CPPS are complex systems, they are, so far, mostly limited to demonstrators, research in universities or implementations in large-scale enterprises [4]. Thus, especially small and medium-sized enterprises (SME) with limited resources and/or personal competencies are still restrained and observing, needing support in planning and evaluating CPPS [5, 6]. Former developments as the Computer-Integrated-Manufacturing (CIM)-era showed, that changes in production systems focusing only on technology, failed, due to the missing consideration of effects on or adaption to the organization and especially to the staff [7]. Therefore, all aspects as organizational structures (e.g. hierarchy in communication), technological structures (e.g. machines and their connectivity) as well as personnel (e.g. competences and customs procedures) have to be taken into consideration. As these factors also interact, considering inter-relations is important for a successful implementation of CPPS [8]. Already existing socio-technical approaches (e.g. [9]) require high effort of both personnel and time and do not support recent technological advances. While other evaluation approaches for CPPS do not allow the comprehensive view required (e.g. [10]).

Thus a holistic systematic for socio-technical capability assessment is developed to support planning and implementation of CPPS step by step, which is part of the joint research project STEPS[1] (http://www.steps-projekt.de) [4, 11]. In this paper the importance of a socio-technical view on CPPS is emphasized, existing approaches and its evaluation are presented deriving the main requirements for a capability assessment. Based on these results, a new method for a socio-technical capability assessment has been designed and is presented in this paper.

1.2   Socio-Technical View on CPPS

The socio-technical view on CPPS is based on the concept of the "socio-technical system" as an analysis and design approach [12]. In a socio-technical system concept view, the issue is not one of either technology or humans, but one of a total socio-technical system designed holistically. This concept takes into account how the structural and economic requirements of the respective field of application and the various knowledge domains of organizations are included. The central focus is the total context of a production process with its subsystems human, technology and organization. Therefore, it focusses on subsystems interdependencies instead of single operations or conversion processes of single subsystems. Specifically, it concerns the interpretation of the functional relationships and interfaces between the three aspects

[1] This research and development project is/ was funded by the German Federal Ministry of Education and Research (BMBF) within the Program "Innovations for Tomorrow's Production, Services, and Work" (funding number 02P14B101) and managed by the Project Management Agency Karlsruhe (PTKA). The author is responsible for the contents of this publication.

human, technology and organization (see Fig. 1), since they affect each other. In the following, probable effects and design possibilities in the context of Industry 4.0 are presented [13, 14].

The first relevant interface is the relationship between "technology" and "human". That means especially the distribution of behavioral responsibility between technology and human action. The ways functions are delineated between these two systems is one of the fundamental questions for work design in Industry 4.0. The aim should be an interface design where humans obtain or retain control over production processes and intelligent assistance systems support humans physically and by the evaluation and presentation of information.

The second relevant interface is the relationship between "organization" and "human". The structure of an organization determines job profiles and qualifications required. The freedoms given by organizational design allow a basic revaluation of all jobs and skills. Work situations may emerge which are characterized by specific qualification requirements, possibly a high degree of operational freedom and poly-valent utilization of human actors, and various possibilities to "learn on the job".

Fig. 1. Socio-technical view of CPPS and fields of design [14]

The third relevant interface is the interface between "technology" and "organization". Here, challenges emerge for the design of work in different ways: First, the level of automation. The subsystem decides what techniques have to be chosen on for organizational design, e.g. flexible allocation of work content. Secondly, because digital technologies allow a wide temporal and functional decoupling of technology and work, these systems permit the realization of different organization forms, e.g. remote maintenance. Third, with networked systems, organizational design needs to include a horizontal and vertical dimension of organization, as well as supply chain. New forms of communication enable companies to connect direct areas, e.g. production and logistics, as well as indirect areas, e.g. planning, management and engineering functions. In terms of a human-oriented organizational design, this could indicate a turn toward far-reaching decentralization and delayering of hierarchies.

Design aspects and possible benefits described above are not inherent in the implementation of CPPS. The design of a CPPS, which is beneficial for the company and its employees, requires effort and consideration of the given fields of design.

2   Definition of Requirements

In order to support SME in the introduction of CPPS, it is necessary to provide a systematic assessment of existing and required abilities to introduce and operate a CPPS solution. Based on the statements in Sect. 1 this part aims at deriving requirements for a socio-technical capability assessment. Seven requirements are identified and defined below.

The demand for *multi-perspectivity* results from the failure of the CIM concept. Main reason for this circumstance is the focus on technical aspects only while ignoring the effects on social ones. Therefore, multi-perspectivity implies a holistic perspective on a work system and all related intersections. In this context, the term "holistic" means the consideration of a socio-technical system in the range of human, technology and organization.

The implementation of Industry 4.0 exposes problems for small and medium sized enterprises due to a lack of approaches covering this topic. An orientation towards *implementation* ensures rating of the actual and target state on the one hand and development of appropriate recommendations for actions on the other hand, leading to a purposive strategy for the implementation.

Developing a specific procedure model for different sectors is not expedient due to the effort resulting. This requires a *universal* model, which is suitable for industrial enterprises regardless of their branch. Thus, categories considered in the assessment must be sufficiently broad to address different kinds of industrial companies, but also allow a view on company-specific details.

The concept of Industry 4.0 affects a number of different business divisions. When considering the integration of CPPS, this *interdisciplinary* perspective poses a main challenge. Hence, the inclusion of different organizational units, specializations and hierarchy levels means a benefit due to the expertise of the participants. Furthermore, it aims at granting a neutral and objective valuation. A strategical and employee oriented point of view as well as the technical production and business administration are key objectives of this requirement.

Furthermore, *validity* of the model is required, since the implementation of CPPS is mostly still theory-based. Thus, it is important to integrate practical oriented enterprises because of their experience. An iterative process of development and testing the model entails benefit.

In order to prevent subjective assessments and ensure comparability when valuating different divisions, *objectivity* is another important requirement. It characterizes whether the criteria used in the valuated maturity model are distinct or not. The use of explanation passages or examples allow inter- and intrapersonal utilization of the models. Additionally it improves transparency.

Granularity of a model enables consideration of differing detail levels as well as hierarchy levels of the enterprise. That is a benefit because it allows detailed analysis of the effects to the changes made. The range includes a work system as well as, for example a whole production sector.

3   Assessment of Existing Approaches

In the last years, different approaches as instruments, checklists or maturity models were developed, aiming to support enterprises in the transformation towards digital companies. The benefit, among others, of evaluation is the digital maturity index to increase the transparency of Industry 4.0 within the company and by deriving specific recommendations for action. Kese and Terstegen [15] provide an overview and short summaries of existing maturity models including their main characteristics according to two dimensions: subject areas and application areas. Furthermore, Pokorni et al. [16] categorize models considering different application areas with a technical focus. In contrast to this approach, Bauernhansl and Dick [6] focus on enterprises in general considering the strategic position or risks and benefits that can be effected by Industry 4.0. An overview of approaches according to the requirements defined in this paper, or similar to that, does not exist. Eighteen approaches were identified and analyzed regarding the requirements defined (see Table 1). The characteristics of the requirements were assessed in five levels from "not fulfilled" until "fulfilled".

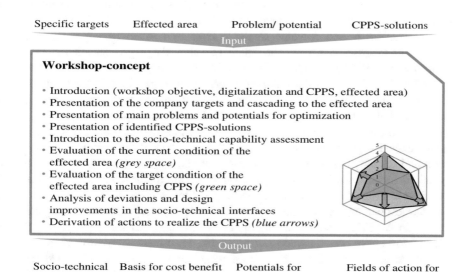

| Specific targets | Effected area | Problem/ potential | CPPS-solutions |

Input

Workshop-concept

- Introduction (workshop objective, digitalization and CPPS, effected area)
- Presentation of the company targets and cascading to the effected area
- Presentation of main problems and potentials for optimization
- Presentation of identified CPPS-solutions
- Introduction to the socio-technical capability assessment
- Evaluation of the current condition of the
  effected area *(grey space)*
- Evaluation of the target condition of the
  effected area including CPPS *(green space)*
- Analysis of deviations and design
  improvements in the socio-technical interfaces
- Derivation of actions to realize the CPPS *(blue arrows)*

Output

| Socio-technical evaluation | Basis for cost benefit evaluation | Potentials for CPPS optimization | Fields of action for implementation |

Fig. 2. Overview of the workshop-concept

In conclusion, the existing approaches especially neglect the requirements multiperspecitvity, interdisciplinarity, granularity and orientation towards implementation. Even though some approaches fulfill certain requirements, they still lack in other categories. Furthermore, online quick checks, e.g. No. 7, 9 and 10, as well as short descriptions with few explanations, e.g. No. 12, do not allow companies to conduct a thorough self-assessment. However, parts of existing approaches are integrated in the developed socio-technical capability assessment, e.g. technical assessment from No. 16 and cultural aspects from No. 3.

Overall, the approach developed aims at supporting the implementation of CPPS considering a socio-technological perspective. The application type of the model is an interdisciplinary workshop. In a first step, the workshop team conducts a socio-technical evaluation, which is the base to derive actions for implementation in a second step. Defined criteria guide this process to ensure an objective result.

4 Design of the Socio-Technical Capability Assessment

4.1 Workshop Concept

Based on the requirements presented in Sect. 3, an approach for a socio-technical capability assessment was developed in the research project STEPS [11]. For this purpose, the approaches presented in Sect. 4 were evaluated and taken into account as well as the requirements defined.

To meet the requirement of *multi perspectivity* based on the socio-technical view, in total 30 criteria were defined concerning human, technology and organization. For each category up to five defined characteristics enable an *objective* and structured evaluation. The criteria are presented more detailed in the following chapter. The focus can be of different *granularity* and contains single workplaces as well as production areas depending on the assessed CPPS. The criteria as well as the procedures are designed to be *universally* applicable in industrial companies by defining universal criteria, able to be transferred on different industrial companies, but also specific to address important aspects in an industrial context. In addition, textual supplements to explain or detail a rating are included. The assessment is conducted in an *interdisciplinary* workshop with decision-makers such as management level as well as process-oriented employees and cross-divisional departments, such as the IT department. The Identification of the current and the target state as well as a gap analysis, including derivation of recommendations for action, supports the *implementation*. Development and validation based on CPPS-use-cases in an industrial context support the *validity* of the concept.

Before explaining the assessment procedure, further prerequisites from the general approach in the STEPS-project are shortly described. Basis for the assessment is that company goals have been derived and cascaded to departments or even work systems. Based on that, possible CPPS have been identified and must be examined more detailed. These steps can be supported and conducted using other modules of the STEPS-project but it is not mandatory.

The capability assessment enables a guided, moderated assessment in an interdisciplinary workshop and leads to a concerted, transparent assessment of company- and sector-specific characteristics of required skills. It supports to identify gaps between current and target state including CPPS and to derive recommendations for action. The standardized, given set of characteristics reduce the effort of the capability assessment, allow a profound discussion, and enable to outline fields of action necessary to implement a chosen CPPS successfully.

The evaluation of the CPPS with regard to implementation effort and potential benefits must be in accordance with company-specific requirements. The assessment is conducted in an interdisciplinary workshop to consider different views and interests in

Table 1. Assessment of existing approaches

No.	Maturity model	Author/ provider	Multi-perspectivity	Implementation	Universality	Interdisciplinarity	Validity	Objectivity	Granularity
1	Production Assessment 4.0	[16]	●	◕	●	◕	●	◔	◕
2	Digitalization in SME	[6]	◔	◔	◔	◕	●	◔	◕
3	Industry 4.0 Maturity Index	[17]	●	◔	●	◕	●	◕	◔
4	Working Environment Industry 4.0	[18]	●	◕	●	◕	●	◔	◔
5	TBS Basic Check Industry 4.0	[19]	●	◕	●	◕	◔	◔	◔
6	Maturity Model Industry 4.0	[20]	◕	●	●	◕	◕	●	◔
7	Digital Maturity Analysis Tool	[21]	○	◕	●	○	○	◔	◕
8	„4i"-Maturity Model	[22]	◕	◔	●	◔	○	◔	◕
9	Digitalization Index	[23]	○	○	◔	○	◔	◔	◕
10	Industry 4.0 Maturity Assessment	[24]	◔	○	●	○	●	◔	◕
11	Guideline Industry 4.0	[25]	○	○	◔	○	◕	◔	◕
12	Industry 4.0 Readiness	[26]	◔	◔	◔	◕	●	●	●
13	Digital Acceleration Index	[27]	◔	○	●	○	◕	◔	◕
14	Quick Check Industry 4.0 Maturity	[28]	○	○	●	○	◕	◕	◕
15	Industry 4.0 Readiness Model	[29]	◔	◔	◔	○	◔	●	◕
16	Toolbox Industry 4.0	[30]	◕	◔	◔	●	●	●	◕
17	Industry 4.0 Self Assessment	[31]	○	○	●	○	◕	◔	◕
18	Company Overview: Industry 4.0 and	[32]	◔	◕	●	◕	●	◕	◔
	Socio-technical capability assessment		●	●	●	●	●	●	●

Legend: ○ not fulfilled   ◔ slightly fulfilled   ◑ partially fulfilled   ◕ mostly fulfilled   ● fulfilled

CPPS planning and evaluation, allowing a profound and objective discussion. Thus, workers and their interests can be considered as well as strategic matters to increase enterprise goals like productivity. Embedded in existing company structures and requirements this discussion builds the ground for an economic calculation, based on identified actions to be taken. The level analyzed can vary according to company requirements (e.g. single workspaces or departments).

4.2 Morphology

Part of the workshop concept is the evaluation of the current and target state of a production system using the morphology, developed based on literature review (see Table 1) and complemented with additional criteria and categories to allow a comprehensive view on production systems in context of CPPS. Criteria are assigned to one of three categories: technology, human and organization, basing on their main effect on the three categories to structure the morphology. This does not imply that these criteria do not influence the other categories but is necessary due to transparency. During analysis and derivation of actions after the assessment the effect on each category, respectively the interface must be considered (see Sect. 1). Each category consists of different criteria, divided in up to five specifications. The higher the specification, the more oriented is a work system towards Industry 4.0. An exemplary excerpt from the assessment of competences is shown in Fig. 3.

In the category *human*, it is assessed which requirements are placed on the employees by introducing new CPPS and which skills already exist in the affected work system. This includes, for example, the evaluation of competencies, the way employees deal with information and the need to carry out qualification measures. In addition, ergonomic aspects such as physical and mental exposure are assessed. In the field of *technology*, it is predominantly assessed which technical framework conditions exist in the selected work system and which are compared to the requirements of a specific solution. This includes subcategories such as data collection, for example, through sensors, required interfaces and protocols of machine communication, data management up to data analytics. In addition, the human-machine interface, such as human-robot collaboration, is assessed at the interface between human and technology. At the interface between human and organizations, it also assesses which options or requirements of cross-company networking are to be taken into account in the supply chain. In the area of *organization*, the organizational framework of the company is evaluated. For example, at a higher level, the Industry 4.0 strategy as well as the current operational and organizational structure can be taken into account. On the other hand, different organizational criteria such as decision-making autonomy, the activity structure and required cooperation of the employees are evaluated at the interface between human and organization.

	Criteria	Specification 1	Specification 2	Specification 3	Specification 4	Specification 5
Organization	Activity structure of the employee	Clearly defined and limited to a few tasks (e. g. assembly steps)	Alternating activities with similar demands on competencies and knowledge (e.g. Job Enlargement)	Various defined tasks with different demands on competencies and knowledge (e.g. Job Enrichment)	Tasks and coordination take place in a group, autonomous group work	
	Cooperation	Little communication and collaboration	Regular communication and cooperation	Regular communication and cooperation, occasional coordination	Regular communication, cooperation and coordination	High degree of communication, cooperation and coordination
	...	...	...	...	...	...
Human	Adaptability	Innovations do not occur, no adaptability of the employees	Willingness of the employees to deal with innovations	After acclimatisation, adaption to innovations and new processes	Actively implement innovations and participate in the design process	Innovative spirit, curiosity and positive acceptance of innovations
	Support of the employee within the process	Only knowledge and experience of the employee to carry out the job	Oral instructions or experience to carry out the activity	Written documents for the execution of the activity	Information in electronic form prior to the activity	Individually adapted information in real time during the activity
	...	...	...	...	...	...
Technology	Data acquisition	Information is collected manually and documented in written form	Information is stored in electronic databases and updated manually	Uniform IT systems and inde-pendent informa-tion management	Collection and automated evaluation of information	Real-time control and updating of information possible
	Design of the human-machine interface	No exchange of information between human and machine	Electronic input devices (keyboard, panels, touch screen)	Centralized / decentralized Production supervision / -control	Mobile devices with intuitive software design and easy operability	Gesture and voice control, extended reality
	...	...	...	...	...	...

Legend (for use case in chapter 5):	Current state	Target state

Fig. 3. Excerpt of the socio-technical capability assessment

5   Validation in the Context of STEPS

For validation, the socio-technical capability assessment was applied within the STEPS project for various use cases. In the following, an example of the application in the logistics sector in an SME is presented.

The pilot area is a packing work system where automotive components are packed for shipment. The initial process envisages that all packaging orders are sorted manually by the area manager based on a given prioritization list according to urgency. The forklift driver receives a package with several orders, whereupon all materials, including packaging, are picked and stored in a preparation zone. The worker takes pre-picked materials from the preparation zone and assigns them to a free packing station, after which the actual packaging process starts. After completing the job, the forklift driver removes the packing table and moves the material to the shipping area.

The company intends to introduce a forklift control and operator assistance system that digitally supports the storage and retrieval process as well as the packaging process in order to be able to visualize work instructions of the large product range accurately and to provide status reports (Fig. 4). The forklift driver receives paperless information on the next order and the corresponding storage locations on a smart device, also including an order prioritization. Furthermore, information on free packing tables is visualized so that the forklift driver can supply the next packing orders with material in a targeted manner. Furthermore, a visualization of packing station status in traffic light colors was implemented at the packaging workstations. This allows direct synchronization of material supply and packaging, work instructions can be visualized digitally and there is an automated feedback of the order status.

To sensitize the employees for implications resulting from the CPPS and to evaluate as well as optimize the planned CPPS and derive required actions for implementation, an interdisciplinary workshop with management, IT department and workers was held, based on the workshop concept shown in Fig. 2. In the following, selected excerpts from the socio-technical capability assessment will be used to show which requirements the new work system places on human, technology and organization and which measures have been derived in a targeted manner for a successful implementation of the CPPS. Excerpts of the evaluation are shown in Fig. 3.

Human: Changes in the work process results in new demands for the workers. *"Adaptability"* is required to adapt to technical changes in daily work and to participate in the innovation process. In addition, the *"Support of the employee within the process"* takes place on a more sophisticated level, because current information about the individual packing orders is visualized and reported back via smart devices in real time. Relevant data of the order and packaging guidelines are available to the worker online and can be retrieved on the tablet if required. In addition, an illustrated work instruction is planned, which can be deposited individually for each job. Thus, a significant simplification of work for employee and supervisor is realized.

Technology: Significant changes can also be observed in the category of technology. At the interface between human and technology, the *"Design of the human-machine interface"* changes fundamentally. The order is placed automatically after a prioritization list by smart devices on the forklift. By using tablets at the packing station, employees communicate their current status and the forklift driver can precisely identify the different

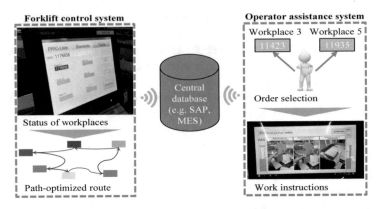

Fig. 4. Concept of the forklift control and operator assistance system

occupation stages of a packing table by means of a simple visualization (traffic light colors). This eliminates the need for ongoing cooperation and duplicate handling on call. The communication among the smart devices occurs via Wi-Fi in the company network. Appropriate protocols and interfaces were designed and the solution integrated into the production planning and control system in self-programming. In addition, the system leads to new possibilities of *"Data acquisition"* as relevant information is collected automated in real-time and can be used both for controlling the orders and for supporting the employee.

Organization: There are also a few changes in the organization. With focus on the direct packaging jobs, there is no change in the *"Activity structure of the employee"* as well as in the evaluation of *"Cooperation"*. The CPPS provides information required and allows operating more efficient and error-free. However, the supervisor requires less *"Cooperation"* and communication to fulfill the work task, but the *"Activity structure of the employee"* increases to the actual management and leadership activities. This illustrates that granularity of evaluation plays a major role and that transparent assessment requires the assessment of impact at different levels and roles. Some requirements will decrease (e.g. direct assembly processes), whereas others will increase (e.g. supervisor) and even entirely new areas of expertise can be necessary (e.g. IT department) to implement and sustain CPPS.

    The results of the workshop regarding the current and the target state were documented in the morphology. All three categories are affected significantly by the CPPS requiring changes throughout the work system. Most of the criteria indicate a development of the work system towards Industry 4.0. Based on the socio-technical capability assessment, further actions required were discussed and work packages defined. This includes trainings for employees of different work systems as well as a discussion of the technical implementation. On this basis, implementation costs and efforts as well as benefits can be estimated. It has been shown that the solution solves existing problems and can help in an economic way to support the company's goals. The socio-technical capability assessment therefore resulted successfully in a decision of implementing the CPPS and a concrete action plan that leads the way to it.

6   Conclusion and Outlook

Digitalization in regards of Industry 4.0 can lead to a competitive advantage for enterprises. Nevertheless, solely technological advances must not drive the implementation of CPPS. CPPS must support company targets and have to be adapted to company specific requirements leading to a holistic design aligning technology, human and organization. This paper defines requirements and presents an approach for a holistic socio-technical capability assessment to support companies within the implementation of CPPS. It consists of an interdisciplinary workshop concept to discuss and adapt possible solutions based on a morphology addressing technology, human and organization from a socio-technical viewpoint. The approach has been validated based on use cases from the STEPS research project.

Further enhancements of the approach are planned. Besides the project-specific evaluation of CPPS, the design of a software tool for a structured analysis and documentation of the current state and planned CPPS throughout the company is in progress. This allows assessing the current state of Industry 4.0 in production-related areas and departments throughout the company. In addition, the socio-technical capability assessment serves as the basis for a work-oriented impact assessment. After evaluation of the status of a certain production area including CPPS, a deeper analysis of the effect on work with specific regard on the human resources is conducted in a next step. The development as well as the evaluation will be carried out with the Industrial Union of Metalworkers in Germany (IG Metall).

References

1. Bauer, W., Schlund, S., Marrenbach, D., Ganschar, O.: Industrie 4.0: Volkswirtschaftliches Potenzial für Deutschland. BITKOM/Fraunhofer IAO, Berlin/ Stuttgart (2014)
2. Bauernhansl, T.: Die Vierte Industrielle Revolution – Der Weg in ein wertschaffendes Produktionsparadigma. In: Bauernhansl, T. et al. (eds.): Industrie 4.0 in Produktion, Automatisierung und Logistik, pp. 5–35. Springer-Vieweg, Wiesbaden (2014)
3. Wöstmann, R., Strauß, P., Deuse, J.: Predictive Maintenance in der Produktion: Anwendungsfälle und Einführungsvoraussetzungen zur Erschließung ungenutzter Potentiale. Werkstatttechnik Online 107(7/8), 524–529 (2017)
4. Wöstmann, R., Nöhring, F., Deuse, J.: Katalog zur zielgerichteten Auswahl von Industrie 4.0-Lösungen. Betriebspraxis Arbeitsforschung 228(11), 38–40 (2016)
5. Jeske, T., Lennings, F., Stowasser, S.: Industrie 4.0-Umsetzung in der deutschen Metall- und Elektroindustrie. Zeitschrift für Arbeitswiss 70(2), 115–125 (2016)
6. Bauernhansl, T., Dick, P.M. (eds.): Digitalisierung im Mittelstand – Entscheidungsgrundlagen und Handlungsempfehlungen. Fraunhofer IPA/Südwestmetall, Stuttgart (2017)
7. Jacobi, H.F.: Computer integrated manufacturing (CIM). In: Westkämper, E., et al. (eds.): Digitale Produktion, pp. 51–92. Springer, Berlin (2013)
8. Deuse, J., Weisner, K., Hengstebeck, A., Busch, F.: Gestaltung von Produktionssystemen im Kontext von Industrie 4.0. In: Botthof, A., Hartmann, E. (eds.) Zukunft der Arbeit in Industrie 4.0, pp. 43–49. Springer, Berlin (2014)

310   F. Noehring et al.

9. Strohm, O., Pardo Escher, O.: Unternehmen arbeitspsychologisch bewerten: Ein Mehrebenen-Ansatz unter besonderer Berücksichtigung von Mensch, Technik, Organisation. In: Strohm, O., Eberhard, U. (eds.) Mensch, Technik, Organisation, vol. 10, pp. 21–37. vdf Hochschulverlag, Zürich (1997)
10. Anderl, R., Fleischer, J.: Leitfaden Industrie 4.0: Orientierungshilfe zur Einführung in den Mittelstand. VDMA, Frankfurt (2016)
11. Nöhring, F., Wienzek, T., Wöstmann, R., Deuse, J.: Industrie 4.0 in nicht F&E-intensiven Unternehmen. Entwicklung einer sozio-technischen Gestaltungs- und Einführungs systematik. ZWF 111(6), 376–379 (2016)
12. Trist, E.L., Bamforth, K.W.: Some social and psychological consequences of the long wall method of coal-getting. Human Relat. 4(1), 3–38 (1951)
13. Hirsch-Kreinsen, H., ten Hompel, M.: Social Manufacturing and Logistics – Arbeit in der digitalisierten Produktion. In: BMWi/BMAS (Hrsg.): Arbeiten in der digitalen Welt. Mensch – Organisation – Technik, Berlin, pp. 6–9 (2016)
14. Ittermann, P., Niehaus, J., Hirsch-Kreinsen, H., Dregger, J., ten Hompel, M.: Social manufacturing and logistics. Gestaltung von Arbeit in der digitalen Produktion und Logistik. Soziologisches Arbeitspapier Nr. 47, Dortmund (2016)
15. Kese, D., Terstegen, S.: Industrie 4.0-Reifegradmodelle, ifaa, Düsseldorf (2017). https://www.arbeits-wissenschaft.net/uploads/tx_news/Tool_I40_Reifegradmodelle.pdf
16. Pokorni, B., Schlund, S., Findeisen, S., Tomm, A., Euper, D., Mehl, D., Brehm, N., Ahmad, D., Ohlhausen, P., Palm, D.: Produktionsassessment 4.0: Entwicklung eines Reifegradmodells zur Bewertung der Lean Management und Industrie 4.0-Reife von produzierenden Unternehmen. ZWF 112(1–2), 21–24 (2017)
17. Schuh, G., Anderl, R., Gausemeier, J., Ten Hompel, M., Wahlster, W.: Industrie 4.0 Maturity Index: Die digitale Transformation von Unternehmen gestalten, acatech - Deutsche Akademie der Technikwissenschaften, Berlin (2017)
18. Bauer, W., Braunreuther, S., Berger, C. et al.: Statusreport: Arbeitswelt Industrie 4.0, VDI/VDE-Gesellschaft, Düsseldorf (2016)
19. Göcking, J., Kleinhempel, K., Satzer, A.: TBS Basis Check Industrie 4.0, Technologieberatungsstelle beim DGB NRW e.V., Dortmund (2016)
20. Jodlbauer, H., Schagerl, M.: Reifegradmodell Industrie 4.0: Ein Vorgehensmodell zur Identifikation von Industrie 4.0 Potentialen. In: Mayr, H.C., Pinzger, M. (eds.) LNI, pp. 1473–1487, Gesellschaft für Informatik, Bonn (2016)
21. Hochschule für angewandte Wissenschaften Neu-Ulm: Digitaler Reifegrad Analysetool. http://reifegradanalyse.hs-neu-ulm.de/quest-ions.php
22. Reuter, T., Gartzen, T., Prote, J.P., Fränken, B.: Industrie 4.0 Audit, VDI-Z Integrierte Produktion, Springer-VDI, Düsseldorf (2016)
23. Deutsche Telekom, A.G.: Digitalisierungsindex Mittelstand: Der digitale Status Quo des deutschen Mittelstands. Telekom/techconsult GmbH, Kassel (2017)
24. Becker, P.: Industrie 4.0 Reifegrad-Test, Vision Lasertechnik GmbH/ bluebiz OHG/ UNIORG Gruppe, Barsinghausen (2016)
25. Vogler, H., Berger, U.: Digitaler Leitfaden für Industrie 4.0. https://www.ihk-industrie40.de
26. H&D International Group: Industrie 4.0 readiness. https://www.hud.de/industrie-4-0/
27. Boston Consulting Group: Digital AccelerationIndex. https://www.bcg.com/de-de/capabilities/technology-digital/digital-acceleration-index.aspx
28. Mittelstand 4.0-Kompetenzzentrum Dortmund: Quickcheck Industrie 4.0 Reifegrad. https://indivsurvey.de/umfrage/53106/uHW7XM
29. Lichtblau, K., Stich, V., Bertenrath, R., Blum, M., Bleider, M., Millack, A., Schmitt, K., Schmitz, E., Schröter, M.: Industrie 4.0-Readiness. IW Consult/ FIR, Köln/ Aachen (2015)

30. Anderl, R., Fleischer, J.: Leitfaden Industrie 4.0: Orientierungshilfe zur Einführung in den Mittelstand, VDMA, Frankfurt (2016)
31. PricewaterhouseCoopers GmbH: Industrie 4.0 Self-Assessment. https://i40-self-assessment.pwc.de/i40/interview/
32. Schilling, G., Nettelstroh, W., Denecke, V., Schormann, D., Vanselow, A., Korflür, I., Weddige, F.: Arbeit 4.0 fair gestalten: Die Betriebslandkarte im Rahmen des Projekts "Arbeit 2020 in NRW", IG Metall, Düsseldorf (2017)

Competences and Competence Development in a Digitalized World of Work

Walter Ganz, Bernd Dworschak, and Kathrin Schnalzer[(✉)]

Fraunhofer Institute for Industrial Engineering IAO,
Nobelstrasse 12, 70569 Stuttgart, Germany
{Walter.Ganz, Bernd.Dworschak,
Kathrin.Schnalzer}@iao.fraunhofer.de

Abstract. Ongoing digitalization opens up major opportunities for companies. Yet this process poses more than just a technical challenge. It is also reshaping the way work and decision-making are divided up between humans and machines, thereby requiring new competences from employees. Various scenarios sketch out future paths of development for technology, work organization, and the requisite competences. This paper elucidates the current state of research on digitalization and its impact on the world of work and competences requirements in Germany. It illustrates – both in general terms and on the basis of specific examples – the kind of (digital) competences that will be increasingly required in tomorrow's workplace.

Keywords: Human factors · Competences · Future of work

1 Introduction

The field of human factors and industrial engineering is tasked with analyzing the changes to working activities brought about by digitalization and shaping those changes in such a way as to create a productive and healthy working environment. The advent of digital networks, technological support systems, and new tools in the workplace has brought about a host of changes in the world of work.

Bauer [1] attributes the drivers of this change to three areas. The first of these drivers is the changing nature of people and the society in which they work. Key topics here are an increased individualization stemming from new values and lifestyles; the aging of society in the wake of demographic change; the emergence of a consciousness with respect to physical and mental health; and growing social and cultural diversity in the workforce, along with the opportunities and potential for conflict that this entails. The second driver of change is new technology, which leads to radical upheavals in an increasingly digital and networked world. The third driver stems from changing business models, which build new value-creation networks founded upon this digital infrastructure. Some of the leading trends here are greater customer focus, increasing supplier diversity, and the development of customized service packages tailored to each context, across the entire product lifecycle, anywhere and anytime. According to a study of Industrie 4.0 [2], digitalization changes work tasks in terms of, for example, a greater individualization of products and services, and their increasing variety. And

I. L. Nunes (Ed.): AHFE 2018, AISC 781, pp. 312–320, 2019.
https://doi.org/10.1007/978-3-319-94334-3_31

once customized solutions have to be developed together with the customer, employees are expected to show greater flexibility with respect to their hours of work [2].

2   Digitalization and Its Impact on the World of Work

The objective of the TransWork[1] project is to analyze how digitalization changes the world of work and to shape that process of change. This research is based on the following definition: digitalization signifies the diffusion of information and communications technology (ICT) throughout all areas of life and work, along with the associated processes of socioeconomic change. In addition, digitalization signifies an intensified and qualitatively different process of computerization. By this is meant the generation, reproduction, and refinement of information and information systems, for which raw data, as such, no longer suffice. Instead, these data must first be processed and interpreted [3]. This is an intensified and qualitatively different process because of the virtually unlimited availability (anytime, anywhere) and analyzability of large volumes of data, which are, in turn, the product of an enormous increase in ICT performance and of the Internet and its continuing development. Furthermore, digitalization is characterized by a permanently available – although, as a rule, used merely as required – digital network of sociotechnical systems involving, where appropriate, the use of mobile devices. Finally, digitalization signifies the use of work systems featuring increasingly intelligent or self-learning system components in which decisions made by either humans or machines exercise a reciprocal influence.

The focus of these investigations is therefore to analyze and shape the changes to the world of work that occur as a result of intensified ICT diffusion, (mobile) digital networking, computerization, and intelligent or self-learning components of work systems. According to our understanding of the impact of digitalization on the world of work, it is a process that has encroached upon different areas of the workplace to varying degrees. We shall therefore consider not only tasks in working environments and value-creation systems that already demonstrate a high level of digitalization but also those in environments and systems that so far, at least, show a lesser degree of digitalization, or are less immediately impacted by this process.

The descriptive model illustrated in the figuere below can serve as a common frame of reference for the process of analyzing and shaping the changes to the world of work through digitalization. It shows interrelated dimensions of change and thereby indicates possible dimensions where this change process can be shaped (Fig. 1).

This model is compatible with the hierarchy and level model [5] from human factors and industrial engineering research; it also served as a descriptive model for the "Statusreport Arbeitswelt Industrie 4.0" [4]. To enable a comprehensive consideration of the topic of work, the model contains variables that are related to the regulation of work by government, associations, trade unions, and other intermediary agents –

[1] This research and development project is funded by the German Federal Ministry of Education and Research (BMBF) within the program "Innovations for Tomorrow's Production, Services, and Work" (02L15A160) and is managed by the Project Management Agency Karlsruhe (PTKA). The authors are responsible for the contents of this publication.

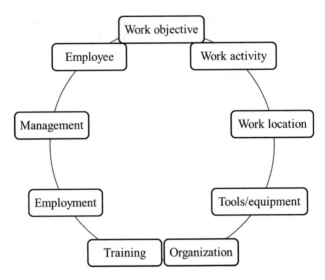

Fig. 1. Dimensions of change in the workplace in the context of digital transformation on the road to "Industrie 4.0" [4] (authors' graphic)

variables, in other words, that are connected not only to the place of employment but also to general factors of employment (e.g., the form of employment, qualifications, and competences). These dimensions also comprise, for example, the changing sources of stress and strain in the workplace. These are described by Hacker in the following terms:

"Greater flexibility in working time, in all its various manifestations, leads to a fragmentation and blurring of the working week and working life. This is a source of indirect stress that can disrupt the rhythm of people's lives, with negative consequences not only for their partnerships and family but also for the social recognition from an involvement in community life, clubs, associations, and local politics, which they would lose due to a lack of regular free time" [6].

Job profiles can also change as a result of digitalization. This is what happens when traditional aspects of a job are eliminated by technology, and when others, involving perhaps an even greater focus on customer communication and development, are added. At the same time, widespread Internet use with mobile devices is changing the tools of the trade in areas such as remote maintenance. This development also includes an increasing degree of network connectivity, new forms of data analysis, and the use of smart machines [7] as either a work tool or work interface. Further changes to the working environment result from a demand for greater flexibility and mobility with regard to the workplace location. According to Hacker [6], strain and stress are caused by longer commutes to work or by problems in maintaining personal relationships as a result of a lack of shared time in one place. There are also potential changes within the dimensions of competences and leadership, where the use of assistance systems in the workplace – such as augmented-reality technology – not only provides employees with real-time support in the work process but also introduces new ways of learning. These changes necessitate a healthy portion of self-management and self-motivation in the workplace. As such, employees are expected to act in a more entrepreneurial way, which can mean, in turn, that they experience a conflict between, for example, what an employer demands in terms of performance and the actual autonomy they are granted: "autonomy in the sense of being entrusted to get on with the job – or, more precisely, being left in the lurch" by the employer [6]. This poses serious challenges in the dimensions of competences and management.

Changes and additions to work tasks and job profiles pose a further challenge. For example, the growing automation of tasks such as remote diagnosis, remote repairs, and remote training will result in certain tasks disappearing completely or being reduced to a merely supervisory capacity. Nevertheless, investigations into workplace innovation have shown how employee involvement has a positive impact on the productivity and quality of work [8]. In other words, employee competences are a key factor here, as is the precise content of the job along with working hours and the working environment. All these aspects must therefore be taken into consideration when defining the job profile.

New approaches to work organization are also urgently needed in complex, knowledge-intensive areas of work. Here, research might well focus on the effect of varying the work load between primary, knowledge-intensive tasks and secondary, "regenerative" tasks. An additional factor in this regard is that employees are required to display a high level of self-organization and therefore have to deal with sometimes conflicting objectives. This, in turn, entails an enhanced sense of autonomy on an organizational level, with employees genuinely experiencing an individual "negotiating autonomy" [9]. Here, the authors were able to demonstrate that "alongside latitude at work, the ability to exercise an organizational dimension of control above and beyond the actual workplace is a relevant factor in the prediction of well-being and innovation in the area of highly qualified knowledge work" [9]. In terms of actual work tasks, this means also being able to influence the framework conditions of employment. This includes having greater freedom with respect to timekeeping and to the disposition of resources and setting of deadlines.

In particular, the kind of jobs to be increasingly found in open, non-demarcated work systems will pose major challenges for the management of employees and teams. Research here should therefore focus also on the different stresses and strains caused by the requirements of job mobility and by the need to reconcile professional and private demands over an entire lifetime. This requires the development of adequate career models (e.g., for specialists and experts) that take account of employees' changing goals and needs over their professional life. Examples here include the sales representative who starts out in their career as an independent, highly mobile young professional. How does their understanding of the job change when they have a home, a family, and many years of professional experience? At the same time, the emergence of new value-creation networks can also accelerate transformation in the workplace. They can lead to a new division of labor, with certain aspects of a job being outsourced, while others are delegated to the customer as a result of closer customer involvement. Network control of processes via mobile devices is already a feature of a host of applications. Similarly, the successive penetration of work systems by smart technology and the increasing use of intelligent control software in production processes have an immediate influence on the division of labor between humans and machines [10–12]. In this respect, the promotion of job quality and healthy working conditions promises to have the greatest impact in those areas where the focus falls on the various gradations of the progressive division of labor between humans and smart technology. On the one hand, smart technology advances incrementally through the introduction of new control functions in, for example, software, production technology, and intelligent objects. On the other, it also progresses in the wake of successive stages of implementation and migration, or it follows a progressive increase in the use, or activation, of optional functions for existing systems. Here, companies continue to choose different configurations of technology and work organization depending on, for example, the precise market and manufacturing requirements they face [13].

With regard to the scope for change in the workplace, there is a particular need for action in the following dimensions: work tasks, work activity, work tools, work organization, competences, and management. Also subject to drastic change as a result of digitalization are the work environment and form of employment. Both of these aspects require further research.

3   Digitalization and Competences Development

The relevant literature contains various scenarios sketching out the future development of technology, work organization, and associated competences. Windelband and Spöttl [14], for example, describe two extreme, diametrically opposed scenarios. In the first scenario – the automation scenario – an ever greater proportion of decisions is made by technology, thereby progressively limiting the scope for autonomous human decision-making and alternative forms of action. The corollary here is the emergence of a competences gap: in an increasingly automated system, humans are required to take action only in the event of a malfunction.

In the second scenario – the specialization scenario – technology serves to support human decision-making and problem-solving at all competences levels. Compared with the automation scenario, employees with at least an intermediate competences level retain responsibility for a significantly greater proportion of decisions, covering more-varied – if not necessarily more-demanding – tasks such as process optimization, intervention in the event of malfunctions, and problem-solving.

Hirsch-Kreinsen [15] likewise describes a broad spectrum of diverging perspectives on the future of work. These range from an upgrading of competences to a polarization of competences.

The competences - upgrading scenario describes a future in which the digitalization of the workplace entails an enhancement and/or increase in competences. This implies that employees will have to learn to operate and master new technology, and that the learning process will be integrated in the working process.

In the study "Produktionsarbeit der Zukunft – Industrie 4.0", Spath et al. [2] present the results of a survey in which companies were asked to rate the importance of future competences in manufacturing industry. The following achieved the highest ratings: a readiness for lifelong learning, strong interdisciplinary thinking and acting, advanced IT competences, the ability to constantly interact with machines and networked systems, and a more-active participation in problem-solving and process enhancement. Distributed work environments require employees to be highly flexible in terms of where and when they work. This, in turn, demands a higher level of initiative and self-organization. At the same time, greater customer involvement in business processes places a premium on communication and coordination competences. With appropriate training, employees can utilize techniques such as enhanced communication, knowledge integration, and affect management in order to influence customer behavior [16]. Similarly, when employees are deployed on site at the customer's location, then problems in managing virtual and distributed teams can certainly arise. The kind of changes involved here require a healthy portion of self-management and self-motivation in the workplace. As such, employees are expected to act in a more entrepreneurial way, which can mean, in turn, that they experience a conflict between, for example, what an employer demands in terms of performance and the actual autonomy they are granted: "autonomy in the sense of being entrusted to get on with the job – or, more precisely, being left in the lurch" [6]. Appropriate measures here include, in particular, a design of work systems and training measures that are conducive to learning, and a use of strategic competences management on the part of the company.

At the other extreme, Hirsch-Kreinsen [15] describes an alternative perspective for the future of work, namely an increasing polarization of work and competences. In this scenario, increasing automation at the intermediate competences level will eliminate tasks in this sector, resulting in a polarization between simple, nonautomated tasks and demanding, highly qualified ones. According to Hirsch-Kreinsen [17], research in the fields of industrial engineering, human factors and industrial psychology shows that advanced automation and an associated increase in system complexity are often accompanied by a limited mastery of the related technology. Hirsch-Kreinsen argues that there is no opportunity in the work process for employees to gain sufficient practical knowledge of the relevant system functions, and that this can lead to a permanent loss of practical competences, practical knowledge, and, most importantly of all, the ability to solve problems, which is indispensable in the unexpected event of a malfunction [18]. In other words, the scenario of an increasing polarization of competences likewise indicates the urgent need for a redesign of work systems and the assistance systems used in the workplace. The use of assistance systems at the human-machine interface, as a remote means of directing employees, cannot be the answer here; this would lead merely to a loss of competences among the workforce.

However, an "early identification" of competences and qualification requirements can serve to determine or estimate the needs of employees in those areas with an especially open and indeterminate path of development, such as those impacted by digitalization. In this case, "identification" means determining which new or modified competences and qualifications will, with a certain degree of probability, be required over the next two to five years, and "early" means identifying those requirements at the stage when they are only just emerging. This is a process that requires the use of qualitative methods.

The qualifications required for some areas of work are pretty clear and comprise a mixture of new and existing competences [19]. For those, however, where the path of development is still open and indeterminate – such as "Industrie 4.0", as a major part of the process of digitalization – it is not yet possible to make a sufficiently clear assessment of requirements. One way of negotiating this difficulty is to employ scenarios of the kind described above. In fact, the extreme scenarios described by Windelband and Spöttl [14] were developed as part of the FreQueNz project, which focused on new competences required in logistics as a result of the Internet of Things [20], and can certainly be applied to "Industrie 4.0" at the very least.

There is a reason why this path of development remains open and indeterminate, and why, as we mentioned at the start of this paper, there is scope for reorganizing work and competences in this area: the technology that has been developed and refined in connection with, specifically, the Internet of Things in industry (Industrie 4.0) and, more generally, the process of digital transformation does not specify any determinate model of work organization. And, as we also mentioned, companies continue to choose different combinations of technology and work organization depending on the market and manufacturing requirements they face [21–23]. In other words, even at the "early identification" stage, we should be paying greater attention than previously to work organization and work design as factors that can mediate between the use of technology and the development of competences requirements. The proposed research initiative is designed to respond to this need.

If the aim is to develop competences and therefore facilitate learning within the work process, and to extend the scope of cognitive models from individuals to networked systems, then research must concentrate on devising new ways of learning. Promising approaches here include the creation of work systems that are conducive to learning, the adaptation of training systems to the needs of individual learners, and the utilization of practical knowledge. As such, the principal dimension where reorganization can have an impact is that of job profile together with training methods and the way in which these are technically implemented and connected with existing systems [24]. In order to create activities that are productive and conducive to learning, it is important here to combine the use of technology with didactically meaningful learning concepts [25]. In addition, companies must establish an effective system to manage the organization and design of in-house training and the provision of individual training measures. This must cover not only an analysis of competences and training requirements but also implementation and monitoring of the outcome and costs [26, 27]. In this respect, research should also focus on the question of how companies can provide the requisite resources for these newly devised learning concepts in the most efficient way. Recent research from the fields of cognitive science und didactics has produced some highly relevant results in connection with human-machine interfaces. However, these insights have yet to be applied in a holistic way. For example, the use of adaptive learning and intelligent tutorial systems enables the development of personalized training measures on the basis of data collected on learning behavior. With the help of learning analytics – e.g., tracking, measuring completion times, and asking learners targeted questions – it is possible to evaluate how much learning assistance is actually provided by the training measure. Similarly, the use of embodiment techniques, along with, for example, spatial representation and gesture control, turns the learning experience into a process of physical interaction that creates the awareness required to learn successfully and efficiently [28]. The transformation of the way in which learning is organized, and the design of digital tools and assistance systems that determine not only the job profile but also the demands that are made of employees – both of these are important fields of investigation, and both are addressed by these research activities.

References

1. Bauer, W.: Working smarter. Menschen. Räume. Technologien. Vortrag Zukunftsforum 2015, Stuttgart (2015)
2. Spath, D. (Hg.), Ganschar, O., Gerlach, S., Hämmerle, M., Krause, T., Schlund, S.: Produktionsarbeit der Zukunft – Industrie 4.0, Stuttgart (2013)
3. Boes, A., Kämpf, T., Langes, B., Lühr, T.: Informatisierung und neue Entwick-lungstendenzen von Arbeit. In: Arbeits- und Industriesoziologische Studien. Jahrgang 7, Heft 1, Mai 2014, S. 5–23 (2014). http://www.ais-studien.de/uploads/tx_nfextarbsoznetzeitung/AIS-14-01-2Boes-u-afinal.pdf
4. VDI, VDE 2016: Statusreport Arbeitswelt Industrie 4.0. https://www.vdi.de/fileadmin/vdi_de/redakteur_dateien/gma_dateien/GMA_Statusreport_-_Arbeitswelt_Industrie_4_0_Internet.pdf
5. Schlick, C., Bruder, R., Luczak, H.: Arbeitswissenschaft, 3. Auflage. Heidelberg u. a. (2010)
6. Hacker, W.: Arbeitsgegenstand Mensch: Psychologie dialogisch-interaktiver Erwerbsarbeit. Ein Lehrbuch, Lengerich (2009)
7. Gartner: The Top 10 Stratetic Technology Trends for 2015 (2014)
8. European Commission: Workplace Innovation. Concepts and indicators, Brussels (2014)
9. Hüttges, A., Moldaschl, M.: Innovation und Gesundheit bei flexibilisierter Wissensarbeit – unüberwindbarer Widerspruch oder eine Frage der Verhandlungsautonomie? In: Wirtschaftspsychologie 4/2009 (2009)
10. Spath, D., Weisbecker, A. (Hrsg.): Potenziale der Mensch-Technik Interaktion für die effiziente und vernetzte Produktion von morgen. Fraunhofer Verlag, Stuttgart (2013)
11. Gratton, L.: Job Future - Future Jobs. Wie wir von der neuen Arbeitswelt profitieren, München (2012)
12. Gombolay, M.C., Gutierrez, R.A., Sturla, G.F., Shah, J.A.: Decision-making authority, team efficiency and human worker satisfaction in mixed human-robot Teams. In: Proceedings of the Robots: Science and Systems (RSS) (2014)
13. Schnalzer, K., Ganz, W.: Herausforderungen der Arbeit industrienaher Dienstleistungen. In: Hirsch-Kreinsen, H., Ittermann, J., Niehaus, J. (Hrsg.) Digitalisierung industrieller Arbeit, S. 121–141, Nomos (2017)
14. Windelband, L., Spöttl, G.: Konsequenzen der Umsetzung des "Internet der Dinge" für Facharbeit und Mensch-Maschine-Schnittstelle. In: FreQueNz-Newsletter 2011, S. 11–12 (2011). http://www.frequenz.net/uploads/tx_freqprojerg/frequenz_newsletter2011_web_final.pdf
15. Hirsch-Kreinsen, H.: Einleitung: Digitalisierung industrieller Arbeit. In: Hirsch-Kreinsen, H., Ittermann, J., Niehaus, J. (Hrsg.) Digitalisierung industrieller Arbeit, S. 9–30, Nomos (2015a)
16. Ganz, W., Tombeil, A.-S., Bornewasser, M., Theis, P.: Produktivität von Dienstleistungsarbeit, Stuttgart (2013)
17. Hirsch-Kreinsen, H.: Digitalisierung von Arbeit: Folgen, Grenzen und Perspektiven. Soziologisches Arbeitspapier Nr. 43/2015 (2015b)
18. Kuhlmann, M., Schumann, M.: Digitalisierung fordert Demokratisierung der Arbeitswelt heraus. In: Hoffmann, R., Bogedan, C. (Hrsg.) Arbeit der Zukunft: Möglichkeiten nutzen – Grenzen setzen. Frankfurt/Main, S. 122–140 (2015)

320 W. Ganz et al.

19. FreQueNz: Zukünftige Qualifikationserfordernisse durch das Internet der Dinge im Bereich Smart House. Zusammenfassung der Studienergebnisse (2011). http://www.frequenz.net/ uploads/tx_freqprojerg/Summary_SmartHouse_final_01.pdf

20. Abicht, L., Spöttl, G. (Hrsg) Qualifikationsentwicklungen durch das Internet der Dinge. Trends in Logistik, Industrie und "Smart House". FreQueNz-Buchreihe Qualifikationen erkennen – Berufe gestalten, Bd. 15, Bielefeld: wbv (2012)

21. Dietzen, A.: Organisation und Kompetenz. In: Dietzen, A., Latniak, E., Selle, B. (Hrsg.) Beraterwissen und Qualifikationsentwicklung: zur Konstitution von Kompetenzanforderungen und Qualifikationen in Betrieben, Schriftenreihe des Bundesinstituts für Berufsbildung, Bielefeld: wbv, S. 23–64. Dietzen 2005 (2005)

22. Dworschak, B., Zaiser, H.: Technologische Innovation und Wissensmanagement. In: Düll, N. (Hrsg.) Arbeitsmarkt 2030 – Fachexpertisen und Szenarien. Trendanalyse und qualitative Vorausschau, Bielefeld, S. 158–190 (2013). https://www.wbv.de/artikel/6004384w

23. TAB (Büro für Technikfolgen-Abschätzung beim Deutschen Bundestag): Zukunftsreport: Arbeiten in der Zukunft – Strukturen und Trends der Industriearbeit. Bericht des Ausschusses für Bildung, Forschung und Technikfolgenabschätzung (18. Ausschuss) gemäß § 56a der Geschäftsordnung. Deutscher Bundestag Drucksache 16/7959 (2008). http://dipbt. bundestag.de/dip21/btd/16/079/1607959.pdf

24. Stich, V., Gudergan, G., Senderek, R.: Arbeiten und Lernen in der digitalisierten Welt. In: Hirsch-Kreinsen, H., Ittermann, J., Niehaus, J. (Hrsg.) Digitalisierung industrieller Arbeit. Nomos. S. 109–130 (2015)

25. Abele, E.: Kompetenzaufbau und Bildung - ein Zukunftsthema für den Maschinenbau. In: Werkstatts-technik online: wt, Springer VDI Verlag, Düsseldorf, 105 (1/2) (2015)

26. Dehnbostel, P.: Betriebliches Bildungsmanagement als Rahmung betrieblicher Bildungsarbeit. In: Weiterbildung. Zeitschrift für Grundlagen, Praxis und Trends, Heft 1/2012, S. 8–11 (2012)

27. Denyer, D., Neely, D.: Introduction to special issue: innovation and productivity performance in the UK. Int. J. Manag. Rev. 5/6(3–4), 131–135 (2004)

28. Gallagher, S.: How the Body Shapes the Mind. Oxford University Press, New York (2005)

Opportunities of Digitalization for Productivity Management

Tim Jeske$^{(\boxtimes)}$, Marc-André Weber, Marlene Würfels, Frank Lennings,
and Sascha Stowasser

Institute of Applied Industrial Engineering and Ergonomics,
Uerdinger Straße 56, 40474 Düsseldorf, Germany
{t.jeske,m.weber,m.wuerfels,f.lennings,
s.stowasser}@ifaa-mail.de

Abstract. The increasing use of digital information and communication technologies in all areas of life – also referred to as digitalization – opens up new opportunities for handling and using information. Thus, also for productivity management in producing companies a wide range of opportunities is expected. For getting further and more detailed information on these expectations, 74 experts of the German metal and electrical industry have been queried in an online survey in 2017. Due to digitalization, the experts estimate an average increase of productivity until 2025 of 32%. This increase of productivity is enabled by a facilitated collection, distribution, analysis and usage of data. Furthermore, the impacts on human factors and the requirements for realizing the expected benefits were queried and are described in this contribution.

Keywords: Productivity management · Digitalization · Industry 4.0
Smart manufacturing · Survey · Metal and electrical industry

1 Digitalization, Industry 4.0 and Productivity Management

Digitalization describes the process of proceeding integration of information and communication technologies in all areas of life [1]. It allows a very efficient handling of information and information flows and leads to a high quality of data – regarding currentness (real-time or close to real-time) as well as level of detail [2].

For producing companies, this opens up a wide range of opportunities for improving and optimizing their production processes [3]. The usage of these opportunities leads to Industry 4.0 and is also referred to as Smart or Advanced Manufacturing [2]. The combination of all elements and components of this digitalized environment is described in a service-oriented architecture which also takes into account a life cycle stream as well as different layers and hierarchical levels: The Reference Architectural Model Industry 4.0 (RAMI 4.0 [4]). The actual and further development of the industry's digitalization as well as the impact on standardization, human factors and legal aspects are described in the German Standardization Roadmap Industry 4.0 [5] which is available to download for free.

A systematical identification and focused implementation of the digitalization's opportunities for each company can be supported by using productivity as a key figure.

© Springer International Publishing AG, part of Springer Nature 2019
I. L. Nunes (Ed.): AHFE 2018, AISC 781, pp. 321–331, 2019.
https://doi.org/10.1007/978-3-319-94334-3_32

Productivity is defined as the ratio between a process' output and input [6]. It is subject of productivity management which aims ideally on decreasing input while increasing output [7]. To do so, control loops for productivity are built and adjusted to specific requirements which are derived from the company's strategic approach.

2   Research Approach

For getting further and more detailed information on the actual practice of productivity management as well as on the expected opportunities and development enabled by the increasing digitalization, a questionnaire was developed (including a pre-study for evaluating comprehensibility and clarity of each question and item).

For reaching many industrial experts as well as for economical purposes regarding distribution and analyses, the questionnaire was designed as an online survey.

2.1   Participants

A total of 74 experts and executives of the German metal and electrical industry participated in the survey. They were invited for participation by the employers' association of the German Metal an Electrical Industry as well as by the help of press releases.

Most participants (69%) are employed in company locations larger than 250 employees which are defined as large companies (according to the European Commision [8]). One third of the participants is working in the executive board; another 44% of them is employed in the personnel department, production and industrial engineering (about 15% each).

The participants answered the questionnaire within 20 weeks between January and June 2017.

2.2   Questionnaire

The questionnaire was structured thematically into eight parts. The first part addressed general information on the participant and the company he/she is working for. The second part started with brief definitions of productivity, productivity management and productivity strategy (one sentence each); queried were the handling of productivity data and expectations on the regarding potential of digitalization.

Subsequently there were questions on the company's actual process of productivity management (third part) as well as on influences of leadership, employees etc. on productivity management (fourth part).

The relation between productivity management and holistic production systems respectively industrial engineering was focused in the fifth part. Afterwards, the strategic orientation of productivity management was stressed (sixth part).

The general importance and usage of digitalization was queried in the seventh part and replicated the same question of a previous study [9]. Finally, the participants were asked to indicate their requirements regarding different kinds of support for different themes in the context of digitalization.

2.3   Statistical Analysis

The statistical analyses were calculated using the statistical software package SPSS Version 23. Correlations have been analyzed with two-tailed tests as rank correlations according to Spearman [10]. The chosen level of significance for each analysis was $\alpha = .05$.

The sorting of alternative answer options rated on Likert scales was based on the summarized proportions of positive and negative assessments.

3   Results and Discussion

Due to the wide range of the questionnaire, only the analyses of selected questions will be presented. These cover the current situation of productivity management, expectations on digitalization, impacts on human factors and requirements for digitalization. The complete results are available to download for free [11].

3.1   Current Situation of Productivity Management

Since key figures are essential for managing productivity, the participants where queried if they measure productivity by using key figures. A majority of 86% uses key figures for this purpose, 4% answered to measure productivity otherwise and 10% do not measure their productivity at all (see Fig. 1).

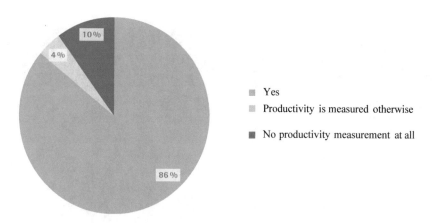

Fig. 1. Is productivity measured in your company by using key figures? (n = 71)

Subsequently participants indicated in which departments of their company key figures are measured. It was found that 47% of key figures are measured in the production area. 14% each are measured to support quality management and assembly (see Fig. 2). The remaining quarter of indicated key figures is nearly equally distributed over the remaining departments logistics, supply chain management, maintenance/service, storage, planning/controlling, administration and development.

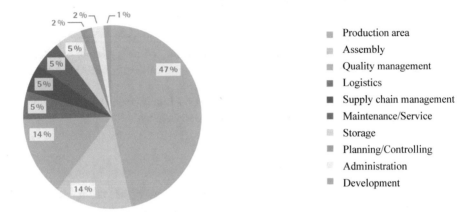

Fig. 2. In which departments do you measure key figures? (n = 62, up to 4 answers allowed, 142 answers in total)

For further differentiation of the key figures, their view focus was indicated by the participants on a scale from a broad view on the company level down to a very focused view on single workplaces. Figure 3 illustrates that key figures are measured on all levels and most key figures are related to single workplaces in production. At the same time the matrix shows large potential for productivity management by further integration of key figures and implementation of related control loops.

Department / Level	Production area	Assembly	Quality management	Logistics	Supply chain management	Maintenance/ Service	Storage	Planning/ Controlling	Administration	Development	Total
Company	9	-	9	1	5	-	-	1	3	1	29
Unit	7	2	3	2	1	1	2	1	-	-	19
Division	13	4	3	1	1	2	2	-	-	1	27
Area	15	7	2	2	-	1	2	1	-	-	30
Workplace	21	7	2	1	-	2	1	-	-	-	34
No information	1	-	1	-	-	1	-	-	-	-	3
Total	66	20	20	7	7	7	7	3	3	2	142

■ many key figures   ■ some key figures   ■ few key figures   ■ no key figures

Fig. 3. Is productivity measured in your company by using key figures? (n = 71)

The collected data respectively key figures are used heterogeneously (see Fig. 4): Most participants (54%) use comparisons between targeted and reached values and derive targeted actions for influencing and improving productivity. Some do at least the comparison (20%), look only on their past development (5%), use them occasionally when needed (11%), view them irregularly (3%) or do not even use them at all (7%). Especially when key figures are measured without using them, their relevance should be evaluated. Also, all those who do not use their data for deriving targeted measures should evaluate their key figures' relevance and check for improvements by closing control loops for supporting productivity management.

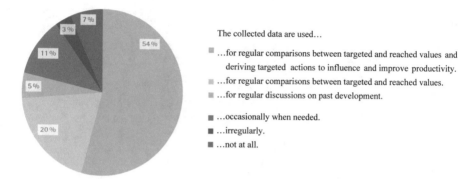

Fig. 4. How are collected data used for productivity management? (n = 61)

3.2 Expectations on Digitalization and Their Causes

Since digitalization as well as related terms as Industry 4.0, Smart or Advanced Manufacturing are linked to great expectations which are usually not quantified or specified, participants were queried for the relevance of productivity management and the digitalization's contribution to productivity for the future.

Asked for the actual relevance of productivity management for their company's success, more than two thirds of the participants (69%) answered on the two highest levels of important (see Fig. 5). Regarding the relevance in five years, this estimation

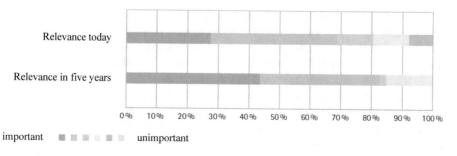

Fig. 5. Which relevance has productivity management for your company's success? (n = 46–51)

increases up to 83%. The relevance of productivity management for the success of the own company is estimated higher in larger companies – actually (n = 51; r_S = ,312; p = ,026) as well as for the future (n = 46; r_S = ,442; p = ,002).

The increasing relevance of productivity management is related to an estimation of an increasing contribution of digitalization to productivity. Participants expect an increase of productivity until the year 2020 of 22% on average (see Fig. 6). According to the answers, this increase will rise up to 32% until the year 2025. Naturally the estimations deviation increases with the forecast horizon.

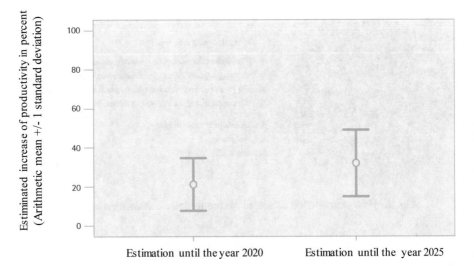

Fig. 6. How much productivity gain do you estimate due to digitalization in your company? (until 2020 respectively until 2025; n = 70–72)

The estimation of an increasing productivity as well as of a rising relevance of productivity management are caused by several expectations: Especially collecting and handling data will be facilitated and enable improvements for all kinds of processes.

This becomes clear when the participants majority answers, more data (60%) can get collected faster (54%) and with less or without manual activities (66%; see Fig. 7). At the same time this describes the appearance of large amounts of data (big data). A comparatively lower agreement on collecting data more precisely (49%) might be due to the fact that data is already very precisely collected nowadays. Only very few participants (4%) expect digitalization to not improve the collection of data.

Based on the expectations regarding the collection of data, the participants expect several advantages of digitalization for productivity management (see Fig. 8). They agree on an improvement of transparency (67%) which allows recognizing relations and causalities more easily. Additionally, they expect more extensive visualizations (59%), fewer manual work on processing data (59%) and an easier use of data due to appropriate software (56%). Only a minority (7%) does not expect any advantages of digitalization for productivity management in their company.

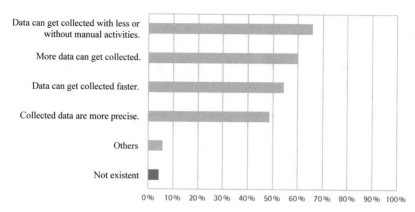

Fig. 7. How do you estimate the potential of digitalization for collecting data in your company? (n = 70, multiple choices possible — excluded for »not existent«)

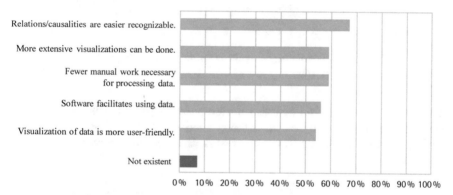

Fig. 8. Which potential has digitalization for your company's productivity management? (n = 61, multiple choices possible — excluded for »not existent«)

3.3 Impact on Human Factors

The impact of productivity management on human factors respectively on employees has been queried in several questions. The results prove mainly positive developments. Since the participants expect many opportunities for productivity management due to digitalization, it can be assumed there are also many opportunities for human factors.

Due to productivity management, the number of reportable accidents at work decreases or decreases strongly (90%) as well as the number of accidental loss hours (87%) and absenteeism of employees (58%; see Fig. 9).

Furthermore, according to 55% of the participants productivity management leads to more or much more motivated and satisfied employees (see Fig. 10).

More influences of productivity management on human factors respectively on employees shows Fig. 11. Central results are: Productivity increased or increased strongly (100%), ergonomic workplace design (98%) and the ability to work increased

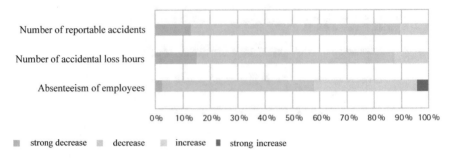

Fig. 9. How changed your company's results in recent years due to productivity management in the following areas? (n = 45–47)

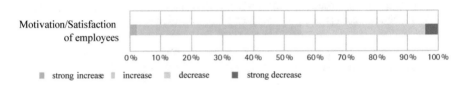

Fig. 10. How changed your company's results in recent years due to productivity management in the following area? (n = 47)

or increased strongly (94%) and also flexibility has improved. The latter applies to content flexibility (range of executable tasks), temporal flexibility (variation of working hours) and spatial flexibility (variation of workplaces and mobile work).

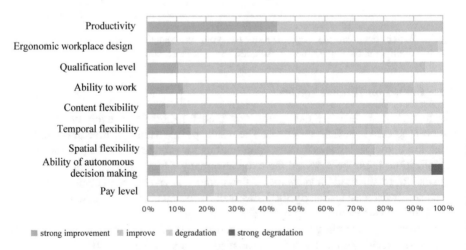

Fig. 11. How are your company's employees influenced by productivity management? (n = 45–49)

3.4   Requirements for Digitalization

The participants' majority considers the opportunities of digitalization during developing processes (75%). In contrary 15% do not consider those opportunities and 10% do not know answering this question (see Fig. 12).

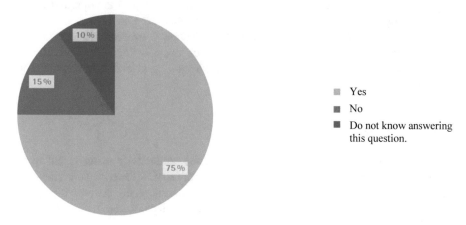

Fig. 12. Are the digitalization's opportunities taken into account during developing processes in your company? (n = 52)

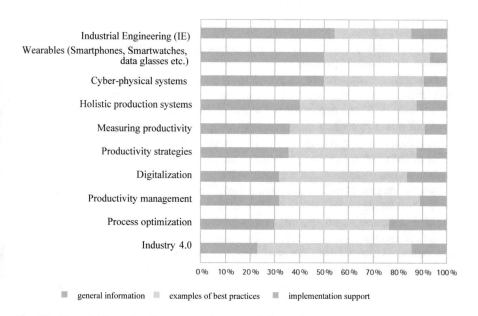

Fig. 13. For which topics do you require more information or support? (n = 40; multiple choices possible, 371 answers in total)

Although the majority considers the opportunities of digitalization during developing processes, there are still different types of support necessary: general information, examples of best practices and implementation support. While there is the largest need for general information on industrial engineering, wearables and cyber-physical-systems, the largest need for examples is indicated for industry 4.0, productivity strategies and productivity management. Implementation support is requested mostly for process optimization, digitalization and Industry 4.0 (see Fig. 13).

For supporting the implementation of digitalization – especially into small and medium sized enterprises –, VDI (Association of German Engineers) published a detailed guideline which contains several examples and is available to download for free [13].

4 Conclusion and Outlook

For exploring the opportunities of digitalization for productivity management in the German metal and electrical industry, the expectations of 74 experts and executives have been queried in an online survey. The applied questionnaire covered the current situation, general expectations, influences on human factors and requirements for implementing digitalization.

The results show that most companies (75%) are using key figures for measuring productivity and most of theses figures are related to the production area (47%). In contrary, there are still many companies that collect necessary data and calculate key figures, but do not derive actions (open loop). Additionally, in many departments – mostly indirect areas – none or only few key figures are measured. Thus, closing control loops and expanding productivity management to indirect areas offers still opportunities for many companies.

Generally, the importance of productivity management is estimated to increase, while improvements in productivity are expected around 22% until 2020 and around 32% until 2025. These expectations are due to the assumption that (1) digitalization allows collecting more data more precisely with less effort and (2) the usage of these data is facilitated by specialized software which leads to easier handling and visualization of data and consequently easier detection of relations and dependencies as a basis for deriving measures for improvements.

Human factors benefit from productivity management since the respondents reported less accidents, less accidental loss hours and less absenteeism as well as higher motivated and more productive employees. Furthermore, ergonomic workplace design and the ability to work are improved. Also, more flexibility regarding the content (e.g. more variance in work tasks), the work time (e.g. flexible shift planning) and the work place (e.g. mobile work) is reported. Since productivity management in general benefits form digitalization, it can be assumed that there will be also further improvements for human factors.

For realizing these expectations, 75% of the respondents are considering the opportunities of digitalization already during process developments. Nevertheless, there is still more information and direct support in implementation requested. The same applies for more examples which can be taken as good-practice.

All in all, the study proved great opportunities of digitalization for productivity management, which do not only affect the essential increase in productivity, but also lead to expected improvements for human factors.

Acknowledgements. The presented results were elaborated in the research project "TransWork – Transformation of Labor by Digitalization", which is publically funded by the German Federal Ministry of Education and Research (BMBF); the grant number is 02L15A164.

Reaching the number of participants was enabled by the help of the employers' association of the German metal and electrical industry.

References

1. Samulat, P.: Die Digitalisierung der Welt. Wie das Industrielle Internet der Dinge aus Produkten Services macht. Springer Gabler, Wiesbaden (2017)
2. Weber, M.A., Jeske, T., Lennings, F., Stowasser, S.: Framework for the systematical design of productivity strategies. In: Trzcielinski, S. (ed.) Advances in Ergonomics of Manufacturing: Managing the Enterprise of the Future, pp. 141–152. Springer, Berlin (2017)
3. Roth, A. (ed.): Einführung und Umsetzung von Industrie 4.0. Grundlagen, Vorgehensmodell und Use Cases aus der Praxis. Springer Gabler, Berlin und Heidelberg (2016)
4. Heidel, R., Hoffmeister, M., Hankel, M., Döbrich, U.: Industrie 4.0 – Basiswissen RAMI 4.0. Beuth, Berlin (2017)
5. German Standardization Roadmap Industry 4.0. Version 3. DIN, Berlin (2018)
6. Wöhe, G.: Einführung in die Allgemeine Betriebswirtschaftslehre, 21st edn. Vahlen, München (2002)
7. Ruch, W.A.: The measurement of white-collar productivity. Glob. Bus. Organ. Excell. **1**(4), 365–475 (1982)
8. European Commission. https://ec.europa.eu/docsroom/documents/15582/attachments/1/translations/en/renditions/native
9. IFAA – Institut für angewandte Arbeitswissenschaft: ifaa-Studie - Industrie 4.0 in der Metall- und Elektroindustrie. Institut für angewandte Arbeitswissenschaft, Düsseldorf (2015)
10. Field, A.: Discovering Statistics Using SPSS. Sage Publications, London (2014)
11. IFAA – Institut für angewandte Arbeitswissenschaft: ifaa-Studie - Produktivitätsmanagement im Wandel – Digitalisierung in der Metall- und Elektroindustrie. Institut für angewandte Arbeitswissenschaft, Düsseldorf (2017)
12. Hammer, W.: Wörterbuch der Arbeitswissenschaft – Begriffe und Definitionen. Carl Hanser, München (1997)
13. VDI Verein Deutscher Ingenieure (Hrsg): VDI/VDE-Gesellschaft Projekt- und Prozessmanagement (Hrsg) Digitaler Transformationsprozess. VDI-Statusreport. VDI, Düsseldorf (2018). www.vdi.de

How Digital Assistance Systems Improve Work Productivity in Assembly

Sven Hinrichsen[✉] and Sven Bendzioch

Ostwestfalen-Lippe University of Applied Sciences,
Liebigstr. 87, 32657 Lemgo, Germany
{Sven.Hinrichsen,Sven.Bendzioch}@hs-owl.de

Abstract. Informational assistance systems contribute decisively to increasing the work productivity in manual assembly processes. However, at this point it is still unclear which problems of information representation actually arise in manual assembly operations and how these problems may be solved through the use of informational assistance systems. In two example cases, this paper identifies problems of information representation by juxtaposing actual and target processes. Using the method of inductive category development, the identified problems of information representation are collected in a total of five categories, for which principles for the design of informational assembly assistance systems are then formulated.

Keywords: Assistance systems · Work productivity
Manual assembly information representation

1 Introduction

Informational assistance systems contribute decisively to increasing the work productivity in manual assembly processes. However, at this point it is still unclear which problems of information representation actually arise in manual assembly operations and how these problems may be solved through the use of informational assistance systems. One goal of this paper is therefore to identify key problems of information representation in manual assembly systems and show their effects on work productivity. Based on these problems, principles for the design of informational assembly assistance systems will be derived.

The approach to achieving these goals is divided into four steps. In a first step, Sects. 2 and 3 of this paper provide theoretical foundations about informational assistance system and productivity management. In a second step potential applications of informational assembly assistance systems are identified (Sect. 4). To this end, work and time studies have been performed in the assembly areas of two companies. Special attention is given to observing the different ways of providing information required for performing the assembly process. In the results, potentials for increasing the work productivity are shown by comparing the existing work process to an optimized process in both companies. To identify the basic approaches for improving the information representation in manual assembly and be able to transfer the results of the two case studies to other manual assembly applications, a categorization of the observed

© Springer International Publishing AG, part of Springer Nature 2019
I. L. Nunes (Ed.): AHFE 2018, AISC 781, pp. 332–342, 2019.
https://doi.org/10.1007/978-3-319-94334-3_33

problems of information representation is developed in the third step. This classification is used to derive design recommendations for informational assembly assistance systems (Sect. 5). In a final fourth step, conclusions are drawn (Sect. 6), along with a brief outlook.

2 Assembly Assistance Systems

2.1 Relevance

Assembly assistance systems are becoming increasingly important. As a result, the dynamism in the development of such systems is high. This trend toward assisted assembly processes can be traced back to three main causes, as outlined in Fig. 1 [1]. First, the basic technologies (e.g. projectors, voice and gesture recognition, image processing) used by assembly assistance systems have attained a high degree of technological maturity for application in manual assembly tasks. In addition, the price of many of these basic technologies has dropped considerably over recent years. Second, the requirements to assembly work processes are changing in a way that tends to increase the complexity of such tasks. This complexity results mainly from high numbers of different parts to be assembled, from a high number of product and parts versions, from small batch sizes, and from a rapid rate of change and development in the products themselves [2]. For example, there is a trend in mechanical engineering towards integrating additional functions. As a result, the question arises of how employees in manual assembly can be supported in a way that addresses these requirements. Third, assembly processes have a high impact on cost, quality, and lead times for many companies in the manufacturing industry [3].

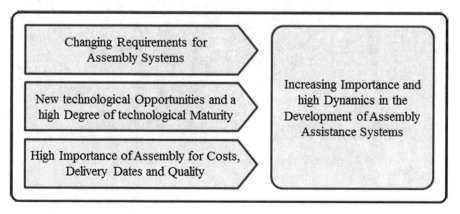

Fig. 1. The increasing importance of assembly assistance systems can be traced back to three main causes.

2.2   Definition and Classification

Manual assembly assistance systems are technical systems that receive and process information via sensors and inputs to assist employees in carrying out their assembly tasks. Based on the type of system support, one can distinguish between energetic (e.g. cooperative robots lifting a load) and informational assistance (e.g. displaying the content of the next work step) [4, 5]. Energetic assistance systems are developed to ensure the feasibility of the task and to reduce the physical strain on employees [4]. Informational assembly assistance systems are designed to provide employees with the right information at the right time according to their requirements in order to reduce uncertainty and mental stress for the employees. These stressors may be additionally increased by pressing delivery deadlines and time pressure in general. In addition, informational assembly assistance systems may contribute decisively to increasing work productivity, e.g. by shortening learning processes for new employees, as well as reducing errors and the time spent searching for information. Figure 2 shows an informational assembly assistance system that projects the information for completing the next process step into the operator's field of view. [7].

Fig. 2. Example of an informational assembly assistance system.

Informational assembly assistance systems can be classified according to various criteria [6]. Systems can be differentiated based on their mobility between stationary assistance systems, mobile assistance systems, hand-held devices, and wearables. Stationary assistance systems are permanently installed at a particular work station (e.g. mounted projection device). Mobile assistance systems on the other hand are solutions that can be moved to the workpiece under assembly. Such solutions can be used, for example, during the assembly and disassembly of injection molding tools [7].

Hand-held devices (e.g. tablet PCs) or wearables (e.g. smartwatch, AR goggles) may either display the information required for the assembly process at the site of assembly in a suitable form (e.g. illustrated assembly instructions) or capture information from the environment (e.g. recognition of gestures or hand positions). In turn, wearables can be classified according to the part of the body where they are worn. Currently, devices worn on the head ("smart glasses"), hand ("smart glove"), and wrist ("smart watch") are the most relevant.

With regard to the scope of process support, a differentiation can be made based on whether the assistance system supports the entire assembly process including setting up the work place, or merely partial processes (e.g., a pick-to-light function supports the process of direct reaching for parts in bin). Another category for the differentiation of assistance systems is the type of process support they offer [8]. For example, a basic assistance system may simply provide step-by-step instructions while more advanced systems provide support specified to the individual user and the concrete situation [2]. Such systems are also referred to as context-sensitive assistance systems [9].

Moreover, it can be distinguished whether the man-machine interface is designed unimodal or multimodal. Unimodal means that a specific channel is available for receiving information, frequently visual, and another for entering information. Multimodal interfaces on the other hand afford various input and output modalities [10]. Another classification can be made according to the type of information output. Optical, acoustic, and haptic output modalities are perceived by humans via their visual, auditory and tactile-kinesthetic sensory modalities, respectively [11]. The term multimedia is used where different forms of coding are used, such as text, voice, and image [11].

Depending on the extent of visual information output, it can be distinguished whether the system is a simple light signal (pick-to-light) or whether multimedia representations in the form of images, videos, and animations are possible. A corresponding distinction can be made for the type of information input. Information input can be manual by using actuators (e.g. buttons, switches), verbal by using voice input, or gesture-based by gesture recognition [10], or automation-based through the use of sensors that monitor the condition of the workpiece and thereby the status of the work process.

3   Productivity Management

Productivity management deals with the initiation and coordination of activities for improving productivity based on a company's business objectives, as well as with monitoring the effectiveness of such activities, in order to systematically improve the productivity of a company [12]. Productivity can be defined generally as the ratio of output to input. A distinction can be made between overall productivity and partial productivities [13]. While overall productivity aggregates all input factors in the denominator of the performance index, partial productivity relates the output only to a specific input factor [14]. For manual assembly, work productivity is a particularly relevant partial productivity because labor accounts for a high share of the total cost of manual assembly processes. Work productivity is always determined for a defined section of a company, e.g. for a cost center. In addition, this performance index must also relate to a defined period of time. To determine work productivity, the output quantity is divided by the number of working hours (attendance hours).

Models of productivity management are usually based on cybernetic control loops [15]. They distinguish the phases of productivity measurement, evaluation, planning, and improvement [14]. Concrete approaches to improving work productivity can be derived from the REFA time classification system [16]. For example, stochastically distributed times (allowances) can be reduced by eliminating unscheduled waiting periods and unscheduled activity times. Approaches include improved disposition of personnel, materials, and orders to avoid organizational problems. In addition, maintenance management may help to reduce the number and duration of technical problems. The scheduled basic time can be reduced by improving employee performance, which is determined by the employees' degree of familiarity with the task and their motivation. In addition, process standardization and optimization may also contribute to reducing the basic time.

4 Empirical Research

Work and time studies to determine the application potential of informational assembly assistance systems were conducted at two companies. One company is a manufacturer of customer-specific truck superstructures. The other company develops and produces machines in different variants for the wood-processing industry. Both companies have in common that they have large production areas in which their complex products are assembled by hand.

At the manufacturer of truck superstructures, the assembly is a sequential process that takes place at a series of work stations. At each station, usually one or two employees perform the assembly for a vehicle in a process that frequently takes more than an hour. The results of the work studies for one work system, the frame assembly station, are presented below. The actual truck superstructure is mounted on this frame at the subsequent work stations. A frame consists of two longitudinal members (rails) and usually more than ten cross-members. Between one and two employees work at the work station. The first step is job preparation. First, the employee asks their coworkers at the following work station which manufacturing job should be processed next. Then they obtain the relevant job documents. The order is logged into the ERP system by scanning a barcode that is printed on the job documents. Then the employee obtains an orientation for the job by viewing the BOM information in the job documents. In addition, the employee locates the relevant CAD drawing in the documents directory on the server. The employee opens the relevant document with the drawing to view the distance between individual cross-members, as well as additional job information (e.g. number and location of mounting brackets, number and location of bore holes). Specific information is noted on a sheet of paper. In a second step, the longitudinal members are cut to the required length through plasma cutter and reworked. In a third step, the employee prepares and fixes the rails on the assembly table. In a fourth step, the attachment points for the cross-members and the position of reinforcement plates and angle brackets are marked with the aid of a measuring tape. The fifth step includes the assembly, in which the cross-members and other elements are provided, aligned according to the markings, and bolted to the rails. In a sixth and final step, the finished frame is transferred to the next station, and the job status is changed to finished in the ERP system.

Potential for improvement with regard to the provision of information was identified especially for steps 1, 4, and 6 (see Fig. 3). For step 1, the problem is that it is not immediately clear which job should be performed next. Another problem is that the CAD drawing that needs to be opened contains data for the entire truck superstructure, so that the employee has to browse the complex overall drawing of the truck superstructure for the information that is relevant to their work station. In addition, the provided BOM also includes parts that are mounted at the next station, so that significantly more information is provided at this station than required. This increases the search effort. In addition, the information is usually documented on a sheet of paper to avoid having to view the drawing and the BOM repeatedly during the assembly process. The problem in step 4 is that the information from the model (CAD drawing) needs to be transferred to the actual workpiece by measuring distances and marking the object with a pen. Step 6 also offers potential for improvement, as the employee has to log in at the terminal to register the job as complete in the ERP system, which again takes several operating steps.

Introducing an informational assistance system could significantly improve the productivity at assembly steps 1, 4, and 6. When a job is completed, the system should automatically display the information for the next job. Only the information that is relevant for the assembly at the work station should be provided. The implementation could be such that the software of the assistance system accesses the CAD data directly and reads out the data relevant for the assembly. As a result, the employee is shown only those material numbers on a screen that are required for this station, thus enabling the employee to immediately begin commissioning and preparing the components. The fourth step, measurement and marking, could be eliminated with a projection-based assistance system that uses laser or light. Such a system can display the distances and specifications read out from the CAD data directly on the rails. An RFID tag on the frame could be used to simplify logging the completion of the job at the work station under consideration, as well as other work stations. According to the time studies performed, the mean actual total order time for a frame assembly is 54 min (n = 10). An informational assistance system would be able to reduce the total order time by about 7 min (13%).

The assembly area at the manufacturer of machines is divided into two sections. In the first area, individual functional units, referred to as modules, are assembled. The final assembly of the machines takes place in the second area. The final assembly involves the installation of the modules into the machine. The manufacturer offers a variety of machines. In addition, customers have numerous options for configuring machines to suit their requirements. The work system selected for observation at the machine manufacturer is in the module assembly section and produces pneumatic assemblies. The assemblies usually consist of a base plate, pneumatic valves, connecting elements, and hoses. In total, the work system produces 79 different pneumatic assemblies (modules) from 291 different components. The work system has several shelves, in which the components are provided in small load carriers. The work is

Actual assembly process	Target assembly process with projection-based assistance system
1. Job preparation	**1. Job preparation**
• Review paper-based job documentation and BOM • Open and log in job in the ERP system • Open and review drawing • Note information relevant for assembly	• Review digital BOM and information relevant for assembly on the screen
2. Preparatory tasks	**2. Preparatory tasks**
• Cut rails	• Cut rails
3. Provision of rails	**3. Provision of rails**
• Prepare rails on the assembly table	• Prepare rails on the assembly table
4. Transfer assembly locations	
• Mark attachment points on rails with a pen and measuring tape	
5. Execution of assembly	**4. Execution of assembly**
• Provide materials • Bolt rails to cross-members and other elements with the aid of the drawing and notes	• Provide materials • Bolt rails to cross-members and other elements according to the projected information
6. Job completion	**5. Job completion**
• Transport frame to next work station • Log out job in the ERP system	• Transport frame to next work station

Fig. 3. The introduction of a projection-based assistance system may significantly improve work productivity in vehicle assembly applications.

performed by a skilled technician. In a first step, the technician reviews the paper-based job documents and the attached BOM for each assembly. The selected job is then registered in the ERP system at a terminal. The second step is to obtain information about the module to be assembled. This is usually done by means of sample modules, in addition to hand-written notes and sketches in some cases. In a third step, the pneumatic module is assembled. In a fourth and last step, the finished module is transferred to the next work station and registered as completed in the ERP system.

The work system has significant potential for improvement with regard to the provision of information (see Fig. 4). In step 1, the problem is that incoming orders are not sorted by priority. As a result, the employee has to ask co-workers at the next process step which pneumatic assemblies will be required next. The subsequent job registration in the ERP system is done at a terminal located outside the work system, which requires the employee to walk there. The greatest potential for improvement is in steps two and three, because here the employee is supplied with inadequate information for the performance of the assembly tasks. In some cases, the provided sample modules do not match current product specifications, which increases the potential for errors. In addition, searching for the right sample module takes time. The records and sketches for each assembly are kept unsorted in a folder and likewise do not match the latest revision status. In step 4, the problem is that the employee needs to log in again at the terminal and perform several operations in the ERP system to set the job status to complete.

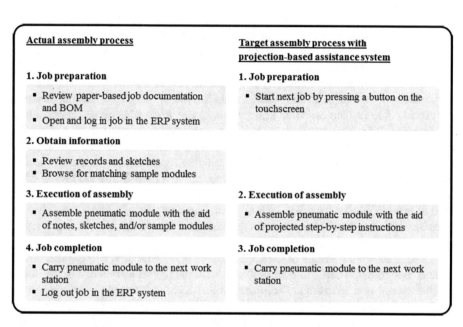

Fig. 4. The introduction of a projection-based assistance system may significantly improve work productivity in the assembly of modules (case study on machine production).

Introducing an informational assistance system could significantly improve the productivity at assembly steps 1, 2, and 4. Logging the job in and out – steps 1 and 4 – should be done directly at the assembly workplace using the touchscreen of the assistance system. Jobs should be sorted by priority, and the next job should open automatically when the previous one is completed. The second step can be eliminated with an assistance system that uses light projection to display relevant information on the workbench and on the workpieces themselves during the assembly process. The employee is guided through the assembly process step by step. A projection-based pick-by-light system shows which container holds the component that is required next. This procedure ensures that the employee is provided with all information relevant for the assembly in the work system. Time studies have yielded a mean actual total order time of 22 min for the assembly of a pneumatic module. The introduction of an informational assembly assistance system could reduce the total order time by up to 10 min (45%).

5   Productivity Increase Through Assembly Assistance Systems

The two case studies show that informational assistance systems are suitable for improving productivity in assembly operations. In order to identify basic approaches for improving information representation in manual assembly and be able to transfer the results of the two case studies to other manual assembly applications, a categorization of the observed problems of information representation is developed below. The classification is then used to derive design recommendations for informational assembly assistance systems.

The method of inductive category development [17] is applied for the classification of the observed problems. In this context, classification or category development means the organization of data based on criteria that appear reasonable either empirically or theoretically [17]. The method of inductive category development is an iterative approach that inspects data material according to a selection criterion. The selection criterion in this case is "problems with information representation in manual assembly." When the selection criterion is fulfilled for the first time during the review of the material, a first category is formulated as a term or brief phrase in consideration of the desired level of abstraction. For each addition identification of data that matches the selection criterion, the data is reviewed for whether it fits an already existing category (subsumption) or requires a new category.

Table 1 shows the results of the application of this method. We have been able to identify five different categories of problems with the representation of information in manual assembly applications. We observed that some information is missing in the work system (category 1), and that other information is presented but not required in the work system (category 2). Another problem is with the timing of the information delivery (category 3) – e.g., it is not beneficial for the worker if the large amount of information required for the performance of the overall task is not allocated to individual process steps. Since workers cannot be expected to retain the complete information about the job in their memory for the entire duration of the process, information should not be displayed before it is needed (step-by-step instructions). Additional problems are outdated information (category 4) and the compatibility of information (category 5). To ensure compatibility, information should be shown in the location where it is needed. In addition, the information should be prepared in a way that makes it easy for humans to absorb and process.

The principles for the design of digital assistance system presented in Table 1 contribute to a reduction of contingency allowances by making searches for required information unnecessary. In addition, optimizing workflows with the aid of assistance systems may reduce the basic time, e.g. by projecting information directly onto the workpiece and thereby into the employee's field of vision using light or laser beams. In addition, digital assistance systems may support the productivity management control loop, e.g. by providing data for productivity measurement and giving employees feedback about the current status of their work.

Table 1. System of basic problems of information representation in manual assembly and design recommendations for digital assembly assistance systems

Problems of information representation	Examples of observed problems of information representation	Examples of observed solution strategies of employees	Design recommendations for digital assistance systems
1. Lack of information within the work system	No information about next job to be processed within the work system	Ask co-workers at the subsequent process (internal customer) or supervisor	(Automatic) display of all information required for the performance of the task
2. Irrelevance of a subset of the information provided in the work system	Provision of a complete CAD model, even though only a subset of the provided information is required in the work system ("information overload")	Browse for and select relevant information	Display of only the information that is relevant within the work system
3. No process orientation: Presentation of information does not match process sequences	Review CAD model with numerous design details (contains significantly more information than a person is able to absorb in a short time)	Note all the information required in the subsequent assembly steps on a sheet of paper	Step-by-step assembly instructions: Display of information only as and when required
4. Outdated information	Assembly of modules based on outdated information, e.g. outdated samples or sketches	Obtain information about current revision from supervisor	Integration of the assistance system into IT systems with up-to-date product information
5. No compatibility of information representation with human modes of information processing	Representation of information in locations where it is not needed, and in ways that do not consider the characteristics of human information processing	Multiple reviews of BOM to make sure that the right components have been selected; measuring and marking assembly locations	Pick-to-light system to support commissioning processes; projection of information directly onto the workpiece ("in-situ projection")

6   Conclusion

Informational assembly assistance systems may contribute decisively to increasing work productivity in manual assembly applications. In two case studies, we have been able to identify typical problems of information representation that arise in manual assembly operations. At the same time, we have shown how these problems may be avoided, and productivity increased, through the use of informational assistance systems. Because the classification shown in Table 1 is only based on two case studies, additional research is required to verify and further define the design recommendations given here.

Acknowledgments. This publication is part of the "Montextas4.0" project for excellence in assembly funded by the German Ministry of Education and Research (BMBF) and the European Social Fund (ESF) (grant no. 02L15A260).

References

1. Hinrichsen, S., Riediger, D., Unrau, A.: Anforderungsgerechte Gestaltung von Montageassistenzsystemen. http://refa-blog.de/gestaltung-von-montageassistenzsystemen. 17 Apr 2017
2. Hinrichsen, S., Riediger, D., Unrau, A.: Montageassistenzsysteme – Begriff Entwicklungstrends und Umsetzungsbeispiele. Betriebspraxis & Arbeitsforschung. **232**, 24–27 (2018)
3. Lotter, B.: Einführung. In: Lotter, B., Wiendahl, H.-P. (eds.) Montage in der industriellen Produktion - Ein Handbuch für die Praxis, 2nd edn, pp. 1–8. Springer, Heidelberg (2012)
4. Reinhart, G., Shen, Y., Spillner, R.: Hybride Systeme – Arbeitsplätze der Zukunft. Nachhaltige und flexible Produktivitätssteigerung in hybriden Arbeitssystemen. wt Werkstattstechnik online, vol. 103, H. 6, pp. 543–547. Springer-VDI, Düsseldorf (2013)
5. Müller, R., Vette, M., Mailahn, O., Ginschel, A., Ball, J.: Innovative Produktionsassistenz für die Montage - Intelligente Werkerunterstützung bei der Montage von Großbauteilen in der Luftfahrt. wt Werkstattstechnik online, vol. 104, H. 9, pp. 552–560. Springer-VDI, Düsseldorf (2014)
6. Hinrichsen, S., Riediger, D., Unrau, A.: Assistance systems in manual assembly. In: Villmer, F.-J., Padoano, E. (eds.) Proceedings 6th International Conference on Production Engineering and Management, Lemgo, pp. 3–14 (2016)
7. Hinrichsen, S., Riediger, D., Unrau, A.: Development of a projection-based assistance system for maintaining injection molding tools. In: De Meyer, A., Chai, K.H., Jiao, R., Chen, N., Xie, M. (eds.) Proceedings of the 2017 IEEE International Conference on Industrial Engineering and Engineering Management, Singapore (2017)
8. Apt, W., Bovenschulte, M., Priesack, K., Weiß, C., Hartmann, E.A.: Einsatz von digitalen Assistenzsystemen im Betrieb. Forschungsbericht 502 des Bundesministeriums für Arbeit und Soziales (2018)
9. Wölfle, M.: Kontextsensitive Arbeitsassistenzsysteme zur Informationsbereitstellung in der Intralogistik. Diss. TUM, München (2014)
10. Schlick, C., Bruder, R., Luczak, H.: Arbeitswissenschaft, 3rd edn. Springer, Heidelberg (2010)
11. Geiser, G.: Informationstechnische Arbeitsgestaltung. In: Luczak, H., Volpert, W. (eds.) Handbuch Arbeitswissenschaft, pp. 589–594. Schäffer-Poeschel, Stuttgart (1997)
12. Saito, Y., Yokota, M.: Total productivity management. In: Zandin, K.B. (ed.) Maynard's Industrial Engineering Handbook, 5th edn, pp. 2.29–2.52. McGraw-Hill, New York (2001)
13. Smith, K.E.: The concept and importance of productivity. In: Zandin, K.B. (ed.) Maynard's Industrial Engineering Handbook, 5th edn, pp. 2.3–2.10. McGraw-Hill, New York (2001)
14. Sumanth, D.J.: Total Productivity Management: A systematic and quantitative Approach to compete in Quality, Price, and Time. CRC Press, Boca Raton (1998)
15. Dorner, M., Stowasser, S.: Das Produktivitätsmanagement des Industrial Engineering. Zeitschrift für Arbeitswissenschaft **66**, 212–225 (2012)
16. REFA - Verband für Arbeitsgestaltung, Betriebsorganisation und Unternehmensent-wicklung e.V. (ed.) Methodenlehre der Betriebsorganisation – Teil Datenermittlung. Hanser, München (1997)
17. Mayring, P.: Qualitative Inhaltsanalyse. In: Flick, U., von Kardorff, E., Steinke, I. (eds.) Qualitative Forschung - Ein Handbuch, pp. 468–475. Rowohlt, Reinbek bei Hamburg (2000)

Human Work Design: Modern Approaches for Designing Ergonomic and Productive Work in Times of Digital Transformation – An International Perspective

Martin Benter[✉] and Peter Kuhlang

MTM-Institut, Deutsche MTM-Vereinigung e. V. (DMTMV), Eichenallee 11,
15738 Zeuthen, Germany
{Martin.Benter,Peter.Kuhlang}@dmtm.com

Abstract. Enterprises face the challenge to design their processes in a rational way. This includes the reduction of physical loads in manual work, to name just one aspect. The German MTM Association (Deutsche MTM-Vereinigung e. V., DMTMV) develops methods and defines standards that support work design. Not least because of the demographic change in industrialized countries methods that consider ergonomic aspects in addition to purely time-related viewpoints gain more and more importance. Therefore, human work design should be based on both ergonomic and time-related aspects.

EAWS (Ergonomic Assessment Worksheet) is a method for the ergonomic assessment of work processes. This paper first describes the standardized dissemination of this method. It then explains how the method can be automated by using motion capture technologies. Finally, the new method MTM-HWD® is briefly introduced. With this method, an ergonomic and time-related assessment of work processes is possible simultaneously.

Keywords: Human work design · International standards
Ergonomic Assessment Worksheet · Motion capture

1 Internationally Uniform Training in EAWS

Well-designed work avoids abnormal biomechanical stress, sustains the health and performance of the employees, and increases their motivation. Holistic workplace design under consideration of ergonomic aspects is gaining more and more importance therefore [1].

To fulfil this task, enterprises normally apply methods for risk assessment, such as the Ergonomic Assessment Worksheet (EAWS). These methods aim at the identification of ergonomic risks for the employees. In addition, the assessment of risks caused by biomechanical stress is mandatory for employers in many European countries.

To ensure the worldwide uniform application of a method all future users of this method have to be trained identically as to scope and content. However, attention has to be paid to the fact that the general conditions may differ due to varying legal regulations with respect to ergonomics.

© Springer International Publishing AG, part of Springer Nature 2019
I. L. Nunes (Ed.): AHFE 2018, AISC 781, pp. 343–352, 2019.
https://doi.org/10.1007/978-3-319-94334-3_34

This requires high quality training throughout the world, which is achieved and guaranteed by globally uniform admission requirements, training materials, and syllabi. The German MTM Association (Deutsche MTM-Vereinigung e. V., DMTMV) guarantees a worldwide acknowledged training standard by, for example [2],

- clearly defined and worldwide comparable training measures (scope, materials, didactic means, examinations) with uniform admission requirements and degrees,
- guaranteed availability and multilingualism,
- clearly defined quality requirements on instructors (contents and accomplishment of the qualification for instructor), as well as, that these requirements are met, and
- the international publicity of the MTM methodology and the related training measures and degrees.

Accordingly, for example, the training for EAWS-Practitioner comprises two identical on-site attendance courses with 5 days each. In the training the participants acquire basic knowledge of EAWS and train the practical application of the method. To support the necessary calculations, the MTM software *TiCon4 EAWS* is used to analyse practical examples. Having successfully completed the training, the EAWS-Practitioner has to refresh and update her/his knowledge at three-year intervals [2].

2 EAWS Applied

EAWS was developed for use in industrial companies. Therefore, it focusses on typical physical loads imposed on the human body in an industrial environment. For this reason, it is used for the preventive and corrective ergonomic design of workplaces [3]. Assessment with EAWS is based on international (e.g. ISO) and national (e.g. DIN) standards, acknowledged assessment techniques, and relevant literature [4].

Consequently, the application of EAWS allows for the identification of ergonomic deficits at an early stage of the product engineering process (design and planning) and, thus, for the reduction of health risks from the start. It is the only method to record different types of biomechanical stress and aggregate them into a total load. The risks related to posture, action forces and handling of loads refer to biomechanical stress imposed on the whole body. In addition, EAWS can be used to assess the load imposed on the upper limbs by highly repetitive activities [4].

The EAWS user has to record the influencing factors defined for each of these types of load (cf. Table 1) [5]. The relevant influencing factors here are intensity (effort and posture) and the duration of biomechanical stress.

Apart from the types of loads listed above, the user collects general, organizational, and technical data (e.g. date of analysis, assessed workplace, cycle time) for the assessment of a work task [5].

Then she/he uses a point value to describe the ergonomic risk of the workplace for the employee. A higher physical load is visualized by a higher point value. By this, the user can classify the assessed work task into three categories, comparable to a traffic light. If the task is classified into the green area, it is ergonomically safe. If it is classified into the red area, measures to reduce biomechanical stress have to be taken [5].

Table 1. EAWS – risk areas and influencing factors [5]

Risk area	Influencing factors	Examples for the manifestation of the influencing factor	
Postures requiring little effort	posture	sit & walk in alternation	kneeling with the body bent forward
	duration of posture	20 % of cycle time	12 seconds per minute
Action forces	maximum force in posture	standing upright, lifting 245 N	kneeling bent forward, pulling 220 N
	load level	33.3 % of the maximum force	maximum force
	duration of load	10 % of the cycle time	9 seconds per minute
manual handling of loads	posture	upright upper body	stooping, or bending forward
	load level	10 kg	30 kg
	duration of load	25 reposition operations per shift	2,500 meters per shift
	means of transport	none	trolley
	operational conditions	little rolling resistance	checker plate or uneven floor
	gender	male	female
Additional negative influencing factors on the whole body		work area difficult to access	severe recoil forces
upper limbs low loads with many repetitions	postures of arm joints	inclined wrist	elbow rotated inwards
	type of grasp	cylindrical grasp	grasp with strong finger force
	load level	5 – 20 N	135 – 225 N
	duration of load in work task	10 actions per minute	20 seconds per minute
	duration of work task per shift	1.5 seconds per shift with breaks	5 hours per shift without breaks
	conditions for grasp	good	poor
	additional factors	unsuitable gloves	precision tasks

3 Using the AXS Suit to Automate EAWS

In industrial practice, methods for the assessment of ergonomic risks, such as the Ergonomic Assessment Sheet (EAWS), have gained acceptance. However, using EAWS requires expertise and manual efforts. Therefore, the practical application of this method is limited. On way to address this and to facilitate the application is the automation of the method. This is possible, for example, by using technologies for recording movements of the human body (motion capture).

3.1 Using the AXS Suit for Motion Capture

These technologies are capable to record human movements and process them into data that can be used digitally. Known representatives of these technologies are 3D cameras (e.g. Microsoft Kinect) or Motion Capturing suits (e.g. XSens). They record data about the positioning of human body parts and thus allow conclusions about postures and movements. Additionally, modern systems – like the AXS Suit – are equipped with sensors to record information about occurring forces. By automatically collecting this information, the AXS suit represents a great tool to partially automate ergonomic risk assessments like the EAWS [6].

The AXS suit was developed by AXS Motionsystem Korlátolt Felelősségű Társaság in collaboration with the German MTM Association. It allows for a mainly automated use of the EAWS analysis. The hardware part of the system consists of a suit with various sensors for the recording of movements and forces (cf. Figure 1a).

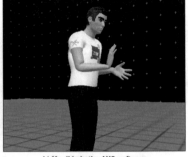

a) Assembly wearing the AXS suit   b) Manikin in the AXS software

Fig. 1. Using the AXS suit to record movements of the human body

3.2   Using the AXS Suit Together with EAWS

When the AXS suit is used to automate EAWS a worker puts on the suit and performs the work task that is to be assessed (cf. Fig. 1a). The software connected to the suit converts the recorded data into postures, force levels and force directions. The posture determined for the observed worker is shown in Fig. 1b.

Using the AXS suit significantly reduces the time required to gather the majority of data necessary for an EAWS analysis. Additional required information has to be entered manually by the user. At this point, the AXS suit records posture and forces involved, as well as, the related durations [6].

However, part of the time-related information (e.g. duration of load per shift) cannot be determined completely with the suit as it is not intended for permanent use. The suit can, for example, determine the time required to perform a task once, but not how often this task occurs during one shift. Also it is not able to determine additional information, such as tools used (e.g. gloves) or conditions (e.g. slippery floor).

3.3   Evaluation of the Use of the AXS Suit

Next, we will present a work task that was assessed with EAWS. First, this assessment was done in the classical way, i.e. manually by an EAWS expert. Then it was done with support by the AXS suit. Having described the work task, the results of both analyses will be compared.

At the Institute of Production Systems of the Technical University of Dortmund the assembly of a gearbox was observed. Figure 2 presents the workplace and the observed work task.

For the manual EAWS analysis, all general information was recorded first. Then all influencing factors in the various risk areas (cf. Table 1) were determined; from these the point value of the ergonomic risk was derived.

For the analysis with the AXS suit one work cycle was recorded (cf. Figure 1). Then the user collected the missing information. Therefore, this information is identical in both analyses and will not cause any deviations.

No.	Operation step
1	Place housing into mounting device
2	Place gasket on housing
3	Assemble housing lid to housing
4	Turn on 4 screws by hand
5	Tighten 4 screws with cordless power screwdriver
6	Turn housing
7	Place shaft into mounting device
8	Place 2 gear wheels onto shaft
9	Assemble 2 sealing rings with pliers
10	Place 2 ball bearings onto shaft
11	Place AU ǐshaftǐ into housing
12	Aside housing

a) Workplace   b) Process description

Fig. 2. Work task for the evaluation of the EAWS analysis with the AXS Suit

Table 2 presents the results of the analyses and compares them [6]. The critical aspects of the work task observed are, in particular, the manual handling of weights (e.g. the transport of gearboxes) and the repetitive activities of the upper limbs (e.g. the use of the screwdriver). The comparison reveals that postures with little physical effort and action forces were identically analysed in both analyses. However, for the manual handling of weights and activities of the upper limbs the automated analysis yielded a higher point value than the classical one. The additional influencing factors for the whole body were determined manually for both analyses and are, therefore, identical. The total score of the automated analysis is 24.0 points and, thus, exceeds the result of the classical analysis by 5 points (26%).

Table 2. Comparison of the EAWS analyses (manual and automated)

Risk area	Point value of the EAWS analysis			
	manual analysis	automated analysis	deviation	
Postures requiring little effort	2.0	2.0	0.0	0 %
Action forces	0.0	0.0	0.0	0 %
Manual handling of loads	17.0	22.0	5.0	29 %
Additional influencing factors – whole body	0.0	0.0	0.0	0 %
Total for whole body	19.0	24.0	5.0	26 %
Upper limbs	17.0	17.3	0.3	2 %

The comparison shows that the analysis with the AXS suit is not yet identical with the classical analysis. The deviations are caused by different ways of recording [6]:

- In the manual analysis the time was determined by means of an MTM-1 analysis. Therefore, the results provide reliable data for the average durations of the individual process steps. In the analysis with the AXS suit one cycle was recorded. In this way, individual steps may influence the risk assessment disproportionally if these steps take longer than intended.

- The suit does not impede the movements of the whole body; yet the gloves with the sensors obstruct the hand movements. Therefore, it is possible that process steps are performed irregularly.
- Faulty sensors or poor sensor calibration may lead to faulty information. For example, the determined forces may be inaccurate.

In spite of the detected deviations and their possible causes, the comparison has shown that the suit at its present status of development is a suitable support for EAWS analyses. Moreover, the revealed differences will be integrated into the further development of the suit.

4   The Assessment of Ergonomics and Time with MTM-HWD®

The ergonomic design of work processes is of increasing importance in the industrial work environment. Nonetheless, the time-related assessment of processes remains relevant for work design, as it gives significant input to production planning on the one hand and is partly the basis for remuneration on the other hand. The new process building block system Human Work Design (MTM-HWD®) is an integrated approach to the recording of factors influencing both time and the ergonomics of human movements [7–10]. It thus allows for the design or productive and ergonomic work processes.

4.1   How to Use MTM-HWD®

With MTM-HWD® the influencing factors affecting parts of the human body are noted down on a record sheet. The record sheet makes the method easy to understand and ensures that the user records all influencing factors completely and systematically [9].

The assembly of a gearbox shall exemplary explain the main steps in applying MTM-HWD®. The example concentrates on one step in the assembly: placing the housing at the assembly workplace. Before that the worker has to pick up the housing and only after placing it will the actual assembly take place. However, picking up the housing and the actual assembly will not be considered here.

4.2   Actions

The user of MTM-HWD® first records which actions the worker performs with which parts of her/his body and which object she/he uses in doing so [10]. In the given example, she/he uses both hands to transport the housing to the assembly workplace and place it there (DEPOSIT). Further possible actions are, for example, picking up the housing (OBTAIN) or checking components (CHECK). Objects could be, for example, tools or actuators. Moreover, the worker may use only one hand or foot.

Depending on the selected action the user then records various influencing factors.

First the user observes the workers' lower limbs. The relevant influencing factors here are the path taken, the conditions of execution, and leg posture (cf. Table 3). Another influencing factor is stability, which will not be described here [9, 10].

Table 3. MTM-HWD® influencing factors: lower limbs

Influencing Factor	Path		Operational conditions		Basic posture		Leg posture (left)		Leg posture (right)	
Housing to mounting device	1 Step	1 side step		unimpeded		stand		stretched		stretched

4.3 Influencing Factors: Lower Limbs

The path describes the distance covered by the worker. The description differentiates between walk, side step, and crawl. For each of these variants the number of meters covered or steps performed has to be stated. In the example the worker performs one side step to place the housing.

The influencing factor Conditions of Execution describes the floor conditions in the work area, i.e. the degree of control required during walking. In the example the conditions are good (even floor, no obstacles).

The basic position indicates whether the worker sits, squats, or kneels at the end of the movement. In addition, leg posture indicates for each leg whether it is stretched or bent. After placing the housing, the worker stands with her/his legs stretched.

4.4 Influencing Factors: Trunk and Head/Neck

Subsequent to the lower limbs the trunk and the head-neck section are described. The influencing factors Trunk Flexion, Trunk rotation and Trunk Inclination as well as head posture are of particular relevance in this connection (cf. Table 4) [9, 10]. However, head posture is not explicitly dealt with here.

Table 4. MTM-HWD® influencing factors: trunk

Influencing Factor	Trunk flexion		Trunk rotation		Trunk inclination	
Housing to mounting device		upright		no trunk rotation		no trunk inclination

The influencing factors Trunk Flexion, Trunk Rotation, and Trunk Inclination represent the trunk posture at the end of the movement. They are assessed independent of the lower limbs. Thus, the trunk may not be twisted at all, or strongly twisted – independent of whether the worker sits, stands, or squats. In the given example the worker's body is in the neutral position at the end of the placing movement. In other words, the trunk is neither bent, nor rotated, nor inclined. The influencing factor Head Posture, which is not included here, analogously describes head flexion, head rotation, and head inclination.

4.5   Influencing Factors: Arm and Force

In the next step, the user records arm posture and the necessary forces (cf. Table 5). This is done for each arm separately [9, 10]. The given example concentrates on the right arm.

Table 5. MTM-HWD® influencing factors: arm and forces

Influencing Factor	Upper arm posture	Hand position	Arm stretching	Hand posture
Housing to mounting device	Angle 20°<x<60°	below shoulder	40%<x<80%	Side rotation

The influencing factor Upper Arm Posture describes the deflection of the arm relative to the neutral position (hanging to the side). In the example, placing the housing into a mount, the right arm is slightly deflected.

Hand position describes the height of the worker's hand relative to her/his shoulder. In the example the hand is below shoulder height.

Arm extension refers to the distance between hands and shoulder joint. Possible differentiations here are: close, half stretched, and stretched. The placing action requires the worker to lift her/his arm into a half-stretched position.

Hand posture describes, analogous to trunk posture (cf. Table 4), whether the worker rotates, bends, or inclines the wrist. While placing the housing she/he inclines her/his right hand to the right, but does not bend or rotate it.

The time-related and ergonomic assessment of activities does not depend on postures only, but also requires knowledge of the efforts to be exerted by the worker. These are described with the influencing factor Weight/Force. In the given example the worker has to carry the housing (weight) while transporting it (3 kg).

4.6   Influencing Factors: Hands

Finally, the user has to record the influencing factors related to hand movements. These include Covered Distance, Type of Supply, Accuracy of Place, Mounting Position, Positioning Conditions, Grasp Motion, and Type of Grasp (cf. Table 6). The influencing factor Vibration is not considered here [9, 10].

The influencing factor Distance Class describes the path covered. There are various degrees in distance. In the given example, the worker moves her/his hand over 40 cm to the mount while placing the housing.

Type of Supply refers to the arrangement or position of the objects to be grasped. This influencing factor occurs only with the action OBTAIN and is, thus, not relevant for the placing process.

Table 6. MTM-HWD®-influencing factor: hands

Influencing Factor	Distance Class	Supply / accuracy of Place + mounting position	Positioning conditions	Grasp movement + type of grasp
Housing to mounting device	40 up to 40 cm	tight Place, with adjusting	Grasping distance obstructed view	Cylindrical grasp

Accuracy of Place, Mounting Position, and Positioning Conditions describe the accuracy required in placing an object, i.e. whether the worker has to pay attention to the symmetry, shape, characteristics, or weight of the object. To place the housing into the mount the worker has to exert slight pressure (tight) and at the same time has to align the housing. In addition, the sight of the mount is obstructed and placing is complicated by the position of the housing handle.

The influencing factors Grasp Motion and Type of Grasp finally describe the finger posture necessary to gain control of the object, or adjust the control. While placing the housing the user maintains control over the object by a cylindrical grasp.

5 Result

Having recorded all influencing factors, the user can assess both the time required to perform the actions and their ergonomic status. The time-related assessment is based on MTM-1, the approved MTM Basic System. Thus, reliable times are available for the individual steps. The ergonomic assessment is done with the EAWS technique, which can be done automatically with the gathered information. Thus, the essential advantage of MTM-HWD® is the possibility to determine reliable times for and the ergonomic risk of a process in one go. This also reduces the time and costs required for the assessment of the work task.

References

1. Schlick, C., Bruder, R., Luczak, H., Mayer, M., Abendroth, B.: Arbeitswissenschaft, 3rd edn. Springer, Berlin (2010).
2. Ostermeier, M., Kuhlang, P.: Internationale Methoden- und Ausbildungsstandards zur Bewertung ergonomischer Risiken. In: TBI 2017 (2017)
3. Lavatelli, I., Schaub, K., Caragnano, G.: Correlations in between EAWS and OCRA Index concerning the repetitive loads of the upper limbs in automobile manufacturing industries. Work **41**, 4436–4444 (2012)
4. Schaub, K., Caragnano, G., Britzke, B., Bruder, R.: The European assembly worksheet. Theoret. Issues Ergon. Sci. **14**(6), 616–639 (2013)
5. Deutsche MTM-Vereinigung e. V.: Lehrgangsunterlage – EAWS-Praktiker. Deutsche MTM-Vereinigung e. V., Hamburg [2914]

6. Benter, M., Rast, S., Ostermeier, M., Kuhlang, P.: Automatisierung von Ergonomiebewertungen durch Bewegungserfassung am Beispiel des Ergonomic Assessment Worksheet (EAWS) Submitted at: Gesellschaft für Arbeitswissenschaft e.V. (ed.): Gestaltung der Arbeitswelt der Zukunft - 64. Kongress der Gesellschaft für Arbeitswissenschaft, Dortmund (2018)

7. Finsterbusch, T., Wagner, T., Mayer, M., Kille, K., Bruder, R, Schlick, C., Jasker, K., Hantke, U., Härtel, J.: Human Work Design - Ganzheitliche Arbeitsgestaltung mit MTM. In: Gesellschaft für Arbeitswissenschaft e.V. (ed.): Gestaltung der Arbeitswelt der Zukunft - 60. Kongress der Gesellschaft für Arbeitswissenschaft, pp. 324–326. GfA-Press, Dortmund (2014)

8. Finsterbusch, T., Kuhlang, P.: A new methodology for modelling human work – evolution of the process language MTM towards the description and evaluation of productive and ergonomic work processes. In: Proceedings of 19th Triennial Congress of the IEA, 9–14 August 2015, Melbourne (2015)

9. Finsterbusch, T., Petz, A., Haertel, J., Faber, M., Kuhlang, P., Schlick, C.: A comparative empirical evaluation of the accuracy of the novel process language MTM-human work design. In: Schlick, C., Trzcieliński, S. (eds.) Advances in Ergonomics of Manufacturing: Managing the Enterprise of the Future. Advances in Intelligent Systems and Computing 2016, vol. 490, pp. 147–155. Springer, Switzerland (2016).

10. Kuhlang, P., Finsterbusch, T., Rast, S., Härtel, J., Neumann, M., Ostermeier, M., Schumann, H., Mühlbradt, T., Jasker, K., Laier, M.: Internationale Standards zur Gestaltung produktiver und ergonomiegerechter Arbeit. In: Dombrowski, U., Kuhlang, P. (eds.) Mensch – Organisation – Technik im Lean Enterprise 4.0, pp. 91–154. Shaker, Aachen (2017)

Evaluation and Systematic Analysis of Ergonomic and Work Safety Methods and Tools for the Implementation in Lean Production Systems

Uwe Dombrowski, Anne Reimer[✉], and Tobias Stefanak

Institute for Advanced Industrial Management,
Technische Universität Braunschweig,
Langer Kamp 19, 38106 Brunswick, Germany
{u.dombrowski,anne.reimer,
tobias.stefanak}@tu-braunschweig.de

Abstract. According to the VDI guideline 2870, Lean production systems represent an effective structure for cascading strategic objectives into methods and tools. The primary focus in a Lean production system is the elimination of waste. Besides the known lean methods, human factors and ergonomics (HFE) methods can contribute to this focus. Therefore, HFE methods have to be implemented into the structure of LPS which is achieved by designing a new principle. In this paper, relevant methods in the field of HFE for the new principle are identified and systematically structured. Therefore, a literature research was conducted to identify the relevant methods. As a result, 242 HFE methods and tools were identified. However, the further analysis of the methods and tools showed that only 90 of those were applicable in a LPS. Because of the high number of the methods and tools, a classification was done to give the methods a structure.

Keywords: Lean production systems
Human factors and ergonomics methods · Ergonomics and work safety

1 Introduction

Nowadays, manufacturing companies face an uncertain market environment leading to volatile markets in which they have to stay competitive. The need for high return on investment and more and more flexibility in production in order to obtain competitiveness is obvious. As to achieve these aims, manufacturing companies use lean production systems (LPS) as a possibility for the cascading of strategic objectives into specific methods and tools. A lean production system is defined as enterprise-specific, methodical system of rules for the continuous orientation of all enterprise processes to the customer [1]. There exist many opportunities on the fields of action in an enterprise. One central aspect in an enterprise is the employee. The field of human factors and ergonomics (HFE) focuses on the well-being of employee and the enterprises overall efficiency and can contribute to the LPS aims by the integration of a new principle

© Springer International Publishing AG, part of Springer Nature 2019
I. L. Nunes (Ed.): AHFE 2018, AISC 781, pp. 353–362, 2019.
https://doi.org/10.1007/978-3-319-94334-3_35

called "ergonomics and work safety" [2, 3]. With this, many use cases show a positive impact on the return on investment [4]. However, many HFE methods and tools are not known or are summarized within another principle in a lean production system [2]. Moreover, by implementing isolated methods and tools from the overall methodical system, there is a high chance of failure of the method or tool resulting in a loss of time and increasing costs [5, 6]. As a consequence, it is necessary to identify relevant HFE methods and tools for the new principle and to design a catalogue for manufacturing companies. The aim is to integrate these identified methods and tools in the overall structure and, as a result, optimize the work conditions for employees and the overall efficiency of a company. In addition, all activities should be harmonized, so there are no negative interactions between the different methods.

2 Lean Production Systems in Combination with Human Factors and Ergonomics

Nowadays, it is crucial for manufacturing companies to organize all processes coordinated and efficiently. LPS offer one possibility for the process structure and take into account aspects of technology, organization and humans [7]. The definition of a LPS is stated in the VDI 2870 as "an enterprise-specific, methodical system of rules for the continuous orientation of all enterprise processes to the customer in order to achieve the objectives set by the enterprise management." [1, 8]. Therefore, as to reach higher efficiencies and, consequently, to be more competitive, elimination of wasteful activities in all company processes is the major target in LPS [1, 9]. In this context, waste can be seen as all kinds of non-value-adding activities that do not contribute to the product in terms of increasing the customer value. According to VDI 2870, there can be seven different types of waste distinguished:

- Over production
- Waiting time
- Transportation
- Over processing
- Inventory
- Motion
- Defects and touch up [1].

These different types of waste lead to a loss in value-adding and have to be eliminated or reduced. This can be accomplished by the use of different methods and tools, e.g. Kanban or milkrun. However, the implementation of several individual methods and tools do not necessarily lead to an overall optimum. Therefore, a structure for LPS had been determined which organizes the use of methods and tools to avoid or reduce waste.

As shown in Fig. 1, an LPS consists of different elements. On the first level, an enterprise has to define objectives. In most cases, target dimensions stand for the strategic objectives of an enterprise and are quality, costs and time. Since strategic objectives affect the entire organizational structure they need to be referred to all enterprise processes. Enterprise processes are the second element in a LPS. Within the third element of a LPS, the strategic objectives are executed. On this level, an LPS consists of different Principles that define a coherent overall framework. Each principle leads to defined methods and tools which can be used in order to achieve the objectives. Methods and tools are the fourth element in a LPS [1].

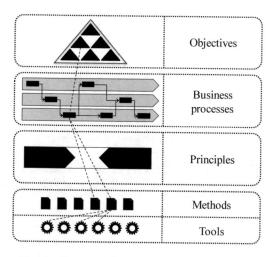

Fig. 1. Structure of a lean production system [1]

In the following, the structure of a LPS is explained with an example. The top target of a manufacturing company is to improve the quality. Therefore, Sub-objectives are sustainable process mastery in manufacturing and assembly-friendly product design. For this purpose, relevant manufacturing processes need to be defined, e.g., for turning, milling or grinding. As to achieve the strategic target, a suitable principle is "zero-defects-production". The principle combines methods and tools that are used to re-duce the number of defects that are passed to the next production step and to ensure a high product and process quality. Especially Six Sigma, automation, Poka Yoke and 5x Why are methods of this principle [1].

As it is mentioned above, LPS are used to comprehensively and continuously design enterprise processes. The processes are optimized to lower costs, save time or improve quality. Nowadays, these objectives should be revised. Especially the demographic change and accompanied change in the age structure of the German population as well as technological advances and digitalization are trends which enterprises have to consider [10]. The employee becomes a central aspect in an enterprise and are crucial for the success. Therefore, employees have to be bound to the company as long as possible.

However, methods of human factors and ergonomics are not consequently implemented in all elements of an LPS. Mostly, they are just used as methods and tools in different principles. As an example, the 5S-Method includes the cleaning of the work station. This is an important prerequisite for Human Factors and Ergonomics because it focuses the safety of the work station. However, it is not perceived as part of Human

Factors and Ergonomics but to LPS. Therefore, it is assumed that there is only little awareness for Human Factors and Ergonomics in manufacturing companies [2].

In order to reach a change in mind considering HFE, the discipline has to be integrated into the structure of an LPS. As it was shown with the structure of LPS, the integration of human factors and ergonomics into an LPS is possible on four levels. A previous evaluation of the possibilities showed that the integration of ergonomics and work safety as a new design principle is favored [3]. With this specification, the new design principle includes all important technical, organizational and human aspects and, therefore, creates an overall framework for the integration of methods and tools. The extension of LPS principles is shown in Fig. 2.

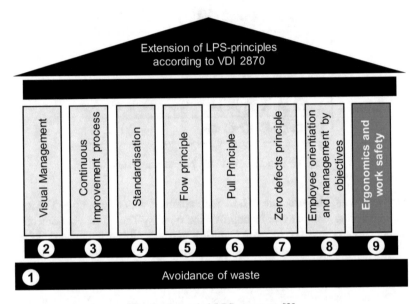

Fig. 2. Extended LPS structure [3]

The purpose of the new principle is to extract methods considering HFE and sum up those into one principle. The first principle focusing the avoidance of waste represents the basis for all principles. Therefore, the new principle is visualized above the first principle, according to the other seven principles. It can be stated that the principle "Employee orientation and management by objectives" and the new principle "Ergonomics and work safety" have a similar focus. However, the new ninth principle distinguishes itself to the eighth principle "employee orientation and management by objectives" with the focus on employee's health. In comparison, the eighth principle focuses on methods for the culture of avoidance of defects and waste among employees [1].

The integration of a new design principle depends also new methods which focus on the improvement of employee's health. The next section, a variety of methods and tools is identified and analyzed for the suitability for the new design principle.

3   Literature Research of Methods and Tools for Human Factors and Ergonomics

As mentioned in the previous section, methods and tools for the design principle ergonomics and work safety are often assigned to other disciplines. Because of the lack in distinguishing the different methods and tools to its scope, there is no classification regarding HFE. Therefore, a literature research was conducted, to identify methods and tools which contribute to the new design principle. A method is defined according to VDI 2870, describing a "specific standardized procedure that is assigned to a principle and is used to achieve enterprise targets" [1, 11]. A tool is a physical object which is necessary for the implementation of a method. Therefore, methods and tools are closely related [2]. For this research, literature focusing on engineering aspects were evaluated due to the implementation in LPS. The approach for the systematic analysis is shown in Fig. 3.

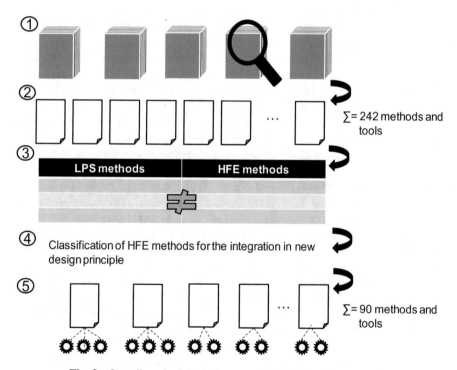

Fig. 3. Overall methodological approach for the literature research

In the first step, relevant literature was identified. For this, several search engines were used to research national as well as international resources. The focus for relevant literature lay on monographs. Besides, monographs include a higher number of HFE methods and introduce different characteristics of the methods. In addition to those monographs, several papers were considered likewise, in order to strengthen the relevance of the identified methods. However, most papers only focused and explained one specific HFE method or tool in depth. Therefore, the papers could not all be listed in the comparison because of the single focus. However, the relevance of most identified methods was strengthened with those papers. In order to establish a comparison between methods and tools in the field of HFE, research terms had to be adapted. In an international context, HFE has a broader scope of topics e.g. including work and society or body functions [12]. In comparison, the national literature focuses on work tasks and places, personal actions or Forms of cooperation in working groups. For the identification of relevant HFE methods and tools, there is no exclusion of methods in order to have a holistic overview of available methods and tools. However, for the systematization of the identified methods and tools, the difference of scope has to be taken into account.

In total, five relevant monographs were identified. It can be stated that international monographs focus on HFE methods and tools. Most methods were listed and explained in Stanton et al., stating around 90 recognized methods and tools for HFE [13]. The range of methods included physical, psychophysical, behavioral and cognitive, team, environmental and macroergonomic methods. In addition to this monograph, Stanton et al. reviews 12 additional ergonomic methods in another monograph. In particular, the focus is put on the design for human use [14]. Another monograph was introduced by Letho et al., giving an introduction to HFE for engineers [15]. Here, several methods were introduced in the context of basic knowledge. Compared to Stanton et al. and Stanton et al., most methods were already known from these sources. Compared to these identified, highly relevant monographs, national literature was also evaluated. For this, two relevant methods catalogue were identified. One methods catalogue resulted from a research project which focused on the improvement of production relevant indices and increase of employee competences. In this catalogue, 58 methods are introduced and evaluated for different criteria. [16]. The second methods catalogue is given in the VDI guideline 2870 "Lean Production Systems". This catalogue contains 35 methods which focus on the elimination of waste in production processes [1].

In the second step of the literature research, all the methods were summed up in one list. Methods which were mentioned in more than one monograph or paper were only counted once in the methods list. Furthermore, methods referring to an identical approach but differently named were also counted as one method. All in all, taking into account the two assumptions, there are 242 methods and tools which consider HFE. Examples of the methods are applied cognitive work analysis, Indices for heat or cold stress or ergonomic assessment methods like NIOSH or OCRA.

The third step represented the sorting of the identified methods. This step is crucial to further systematization due to the suitability for production systems. Hence, the first part of the literature analysis was the comparison of the HFE methods list with the VDI guideline 2870. The starting point of the research was chosen, because of the implementation of the methods in LPS. In the VDI guideline 2870, as mentioned above, there are 35 different methods presented, all of which aim to eliminate waste in the production process. Those methods can be extended by their fundamentals, e.g. standardization or visualization [1]. Examples of the methods mentioned in the VDI guideline are 5S, Poka Yoke, Six Sigma, Kanban, Total Productive Maintenance, PDCA, Hancho and more [1]. The comparison of the identified HFE methods and LPS methods and fundamentals showed an overlap of 48 methods. For this reason, those methods were not included in the new design principle ergonomics and work safety because of the integration in other parts of the VDI guideline. As a result of this third step, there are 194 methods which are applicable in production processes.

Now that a methods list is revised, it is necessary in the fourth step to evaluate the methods for the integration into the VDI guideline. Hence, criteria for this evaluation have to be defined. The definition of HFE shall be used to derive criteria for the integration in the VDI guideline. The International Ergonomics Association (IEA) defines HFE as followed "Ergonomics (or human factors) is ... concerned with the understanding of interactions among humans and other elements of a system, ... in order to optimize human well-being and overall system performance" [2, 17]. This definition represents the international understanding of HFE. For the purpose of the methods integration in the national VDI guideline, a national definition was also analyzed. Schlick et al. state that the aim of HFE is to analyze present working conditions, to systematically process the gained knowledge and derive design rules [12]. In combination with the above mentioned definition of LPS, the following criteria for exclusion of a method were identified and will be explained:

1. Single economic focus
2. Creative techniques
3. Product ergonomics methods/customer focus
4. Methods with medical focus

The list of methods showed several methods with an (1) single economic focus such as cost-benefit analysis or discounted cash-flow method. Compared to the definitions mentioned above, those kind of methods do not support the aim of HFE with the human well-being and overall system performance in focus. Therefore, those methods were excluded from the list. In addition to those methods, (2) creative methods focusing on the generation of new knowledge and ideas were excluded due to the reason that those are already implemented in continuous improvement processes. Examples of this kind of methods are the 6-3-5 Brainwriting or mind mapping. (3) Another reason for the exclusion of methods was the customer focus. HFE methods which have the aim to improve the final product were not applicable for the integration in the VDI guideline because methods as such do not add to employee's improvement of well-being. Lastly, according to the broader scope of HFE in international literature, many (4) medical methods for the improvement of ergonomics were mentioned in the methods lists. However, the prerequisite for the suitability of the identified methods is the application into production processes. Therefore, HFE methods regarding and evaluating bodily functions are excluded from the methods list. Examples are Electromyogram or an Electroencephalogram. Those methods involve medical devices to determine data, like heart rate or brain activity, which cannot be applied in a production process.

Consequently, applying the criteria to the methods list, a number of 104 methods have to be excluded from the application in the VDI guideline. Hence, a total number of 90 methods and tools can be used for the next step of the research process. The last step is the classification of the methods.

4   Systematization of HFE Methods and Tools

Due to the number of identified methods and tools, a classification is crucial for the systematic implementation in a production process. First of all, methods which can be merged within an overall method characterization are summarized into a classification. This determination is exemplarily explained with ergonomic assessment methods.

The list of methods showed numerous approaches for the ergonomic assessment of work. According to the definition of a method and a tool mentioned in a previous section, the approaches for ergonomic assessment belong as methods. The reason is that the methods give an approach on the methodological assessment of work. However, all assessment methods have a different focus on evaluating work. Because of this, the approaches are classified as ergonomic assessment methods. Examples are Rapid upper limb assessment (RULA), Occupational repetitive action (OCRA), Ovako Working Posture Analysing System (OWAS) and more. In total, 12 different ergonomic assessment methods are merged into this classification, leading to a reduction in method complexity.

For the classification of methods, a general approach is needed. Schlick et al. introduce a general model of a work system which is used for the classification of methods [12]. This model is best applicable because it takes into account national requirements and has been revised by experts.

The different classifications are shown in Fig. 4.

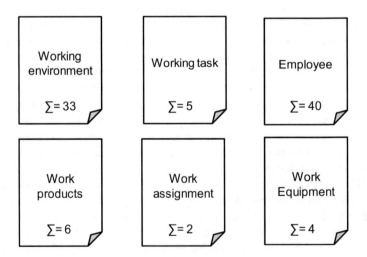

Fig. 4. Classification of HFE methods into work system model

As a result, six different classifications were identified. For the classification of the methods and tools, the aim of these is used to determine the classification. For example, the tool heat stress index is classified into the class employee. The aim of the tool is the determination of exposure to heat and derive actions e.g. heat breaks or enough hydration. Therefore, it is used to improve the well-being of the employee.

In total, the classification showed that most methods and tools focus the employee and working environment.

5   Conclusion

LPS are often implemented into manufacturing companies to structure processes and, consequently, reach the strategic objectives set by the enterprise. Therefore, a structure for the cascading of the objectives down to methods and tools is presented in the VDI guideline 2870. There are four different elements: objectives, business processes, principles and methods and tools. In order to achieve the objectives, elimination of waste in the production process is the fundamental approach. Waste are all actions which do not add to the value of a product. One wasteful activity is non-ergonomic work because it can lead to a long absence of work of employees. Hence, the introduction of the new wasteful activity of non-ergonomic work was presented, leading to the introduction of a new design principle: ergonomics and work safety. Consequently, new methods and tools had to be identified. For this, a literature research was conducted by using national as well as international literature. As a result, 242 methods and tools were identified. However, not all methods and tools were applicable in LPS, so a classification was applied to sort the methods and tools. In total, 90 methods could be identified which can be used within an LPS. It was shown that especially methods focusing the employee and work environment were primarily used to overcome non-ergonomic work. Further research has to be done on methods extending the classifications or regarding other classifications. For this, papers have to be used because monographs only give an overview of existing basic methods.

References

1. VDI: Guideline 2870-Part 1: Lean Production Systems-Basic Principles, Introduction, and Review, vol. 2870-1. Beuth-Verlag, Berlin (2014)
2. Dul, J., Bruder, R., Buckle, P., Carayon, P., Falzon, P., Marras, W., Wilson, J.R., van der Doelen, B.: A strategy for human factors/ergonomics: developing the discipline and profession. Ergonomics **55**, 377–395 (2012)
3. Dombrowski, U., Reimer, A., Wullbrandt, J.: An approach for the integration of non-ergonomic work design as a new type of waste on lean production systems. In: Nunes, I.L. (ed.) Advances in Human Factors and System Interactions. Proceedings of the AHFE 2017 International Conference on Human Factors and Systems Interaction, The Westin Bonaventure Hotel, Los Angeles, California, USA, 17–21 July 2017 (2017)
4. Dul, J., Neumann, W.P.: Ergonomics contributions to company strategies. Appl. Ergon. **40**, 745–752 (2009)
5. Koukoulaki, T.: The impact of lean production on musculoskeletal and psychosocial risks: an examination of sociotechnical trends over 20 years. J. Appl. Ergon. **45**, 198–212 (2013)

6. Dombrowski, U., Reimer, A., Ernst, S.: Occupational stresses on employees in lean production systems. In: Goossens, R.H.M. (ed.) Advances in Social and Occupational Ergonomics, Advances in Intelligent Systems and Computing 487, Proceedings of the AHFE Conference International Conference on Human Factors and Systems Interaction, 27–31 July 2016, Orlando, USA, pp. 215–226 (2016)
7. Spath, D.: Ganzheitlich produzieren. LOG_X Verlag, Stuttgart (2003)
8. Dombrowski, U., Mielke, T. (eds.): Ganzheitliche Produktionssysteme. Aktueller Stand und zukünftige Entwicklungen. Springer, Berlin (2015).
9. Ohno, T.: The Toyota Production System - Beyond Large-Scale Production. Productivity Press, Portland (1988)
10. Winkelhake, U.: The Digital Transformation of the Automotive Industry. Catalysts, Roadmap, Practice. SpringerVieweg, Wiesbaden (2017).
11. Liker, J.K.: Der Toyota Weg: Das Praxishandbuch. FinanzBuch Verlag, München (2007)
12. Schlick, C., Bruder, R., Luczak, H.: Arbeitswissenschaft. Springer, Berlin (2010).
13. Stanton, N., Hedge, A., Brookhuis, K., Salas, E., Hendrick, H.: Handbook of Human Factors and Ergonomics Methods. CRC Press, Boca Raton (2005)
14. Stanton, N., Young, M.S., Harvey, C.: Guide to Methodology in Ergonomics. Designing for Human Use. CRC Press, Boca Raton (2014)
15. Letho, M., Landry, S.J.: Introduction to Human Factors and Ergonomics for Engineers, 2nd edn. CRC Press, Boca Raton (2014)
16. Institut für Fabrikanlagen und Logistik (IFA): Methodenkatalog des Forschungsprojektes prokoMA. IFA, Hannover (2017)
17. IEA Council: The Discipline of Ergonomics. International Ergonomics Society (2000)

Indicators and Goals for Sustainable Production Planning and Controlling from an Ergonomic Perspective

Maximilian Zarte[1]([⊠]), Agnes Pechmann[2], and Isabel L. Nunes[1,3]

[1] Faculty of Science and Technology, Universidade Nova de Lisboa,
Campus Caparica, 2829-516 Caparica, Portugal
m.zarte@campus.fct.unl.pt, imn@fct.unl.pt
[2] Department of Mechanical Engineering,
University of Applied Sciences Emden/Leer, Constantiaplatz 4,
26723 Emden, Germany
agnes.pechmann@hs-emden-leer.de
[3] UNIDEMI, Campus de Caparica, 2829-516 Caparica, Portugal

Abstract. The need for sustainable development has been widely recognized by governments and enterprises and has become a hot topic of various disciplines including ergonomics and industrial engineering. Sustainable development can only be achieved if impacts and benefits on current and future generations of humans are focused according to the three dimensions of sustainability. The paper provides an overview on indicators and goals for sustainable production planning and controlling (sPPC) processes in manufacturing companies from an Ergonomic perspective. This was done based on a root cause analysis. For the analysis, the main stakeholder groups for sPPC are identified, and sustainable indicators and goals are selected through a text review according to the interests of the stakeholder groups. The results can be used as basis to support and improve the outcome of the sPPC process discussions with company managers, experts and employees' representatives.

Keywords: Sustainable Development · Human Centered Development

1 Introduction

The need for sustainable development has been widely recognized by governments and enterprises and has become a hot topic of various disciplines including ergonomics and industrial engineering [1]. For the terms sustainability or sustainable development, it exists no standard definition or universal truth. The nature of sustainability can be varied from the considered system (e.g. household, industry, country, world), effected stakeholder group (e.g. employees, customers, suppliers, etc.), and/or perspective and scientific background of the researcher. Therefore, researchers need always to define what is meant by sustainability in relation to their presented research approaches, which leads to multiple theories about the understanding of sustainable products and processes which are all truth for a specific case.

© Springer International Publishing AG, part of Springer Nature 2019
I. L. Nunes (Ed.): AHFE 2018, AISC 781, pp. 363–373, 2019.
https://doi.org/10.1007/978-3-319-94334-3_36

The Brundtland Commission published a widely accepted definition for sustainable development: "development that meets the needs of the present without compromising the ability of future generations to meet their own needs" [2]. From a more human perspective, an United Nation (UN) report defines sustainable development as: "Human beings are at the center of concerns for sustainable development. They are entitled to a healthy and productive life in harmony with nature" [3]. Moreover, sustainable development is bonded by the triple bottom line which balances three dimension of sustainability: economic goals, environmental cleanness, and social responsibility [4]. According to these definitions, sustainable development can only be achieved if impacts and benefits on current and future generations of humans are focused according to the three dimensions of sustainability.

Ergonomics is "the scientific discipline concerned with the understanding of the interactions among humans and other elements of a system and the profession that applies theory, principles, data and methods to design in order to optimize human wellbeing and overall system performance" [5]. Obviously, this definition for Ergonomics already matches two dimensions of sustainability: social responsibility (human well-being), and economic success (system performance). Moreover, Thatcher (2013) [6] investigates the relationship between ergonomics and the environmental dimension of sustainability. This author discusses the term "green ergonomics" and presents ideas on the design of low resource systems and products, design of green jobs, and design of processes and products for behavior change to meet sustainable goals. The change of behavior in human-system interaction plays an important role in sustainable development and industrial engineering. To measure the change, sustainable indicators are required, which measuring the distance to a goal to get the appropriate information on the current status or trend [7]. An indicator set is a group of indicators that comprise a holistic view of sustainability. Combining indicators from the environmental, economic, and social dimensions and evaluating those indicators together is a practice to measure the sustainability on a much larger scale than individual indicators [8].

In recent years, numerous reports and studies have been published which presents sets of indicators and procedures for the evaluation of sustainability by countries and businesses while being applicable to the evaluation of products and processes as well. Examples for sustainable indexes are the Ecological Footprint [9] and the Living Planet Index [10], which provide several indicators for the evaluation of sustainability comparing products, processes, and investment projects. Joung et al. reviewed a set of publicly available indicator sets for corporate social responsibility (CSR), and provided a categorization framework and a database of indicators that are quantifiable, qualifiable, and clearly related to sustainable manufacturing [8]. Popvic et al. reviewed existing scientific articles to identify indicators for quantitative assessment of social sustainability in supply chains. Through statistical analysis, the relevance of the identified indicators was investigated [11].

Previous developed indexes and set of indicators for sustainable development are focused on single dimension of sustainability (social, environmental) and/or consider specific enterprise processes (production, supply chain). Moreover, the procedures for the identification of relevant sustainable indicators neglect the role and interests of humans (so called stakeholder groups). Recent studies (see [12, 13]) argue strongly that ergonomics should be an integral part of finding solutions to the current global challenges for sustainable development. Initiatives in these domains will be more successful if they follow an integrated approach.

The goal of this paper is to provide an overview of indicators, and goals for sustainable production planning and controlling (sPPC) in enterprises from an Ergonomic perspective. To identify sustainable indicators, a root cause analysis has been made. For the root cause analysis, the main stakeholders for sustainability in manufacturing enterprises are identified. For these main stakeholders, the causes in form of indicators and goals for sPPC are presented in a root cause diagram. The indicators and goals are selected according to defined criteria. With the aid of the criteria, a text review of an existing database for sustainable indicators used in manufacturing companies, and political goals for sustainable development was made.

After this introduction, Sect. 2 presents the methodology, which was used for the identification of sustainable goals and indicators according to the stakeholder groups. Section 3 presents the results and a discussion of the root cause analysis. Section 4 offers the conclusions and future works, followed by references.

2 Methodology and Methods

A root cause analysis was performed with the purpose of identifying sustainable indicators and goals contributing for sPPC from an Ergonomic perspective. In fact, the goal of the root cause analysis methodology is to reveal key relationships for a specific topic among various main variables, and possible causes, providing additional insight into process behaviors [14]. The procedure for a root cause analysis can be structured in three steps:

1. Determination of the specific topic,
2. Definition of main variables related to the specific topic,
3. Identification of causes which influence the main variables.

The specific topic is sPPC integrating also an Ergonomics' perspective, as already discussed in the introduction.

In the second step, the main variables related to sPPC must be defined. According to the above-mentioned definition for sustainability, the main variables for sustainable development can be defined as present and future generations, which are affected from enterprises. These generations are already identified in related studies (see e.g. [15, 16]) and can be categorized in six main stakeholder groups:

- Company owners
- Employees
- Costumers
- Suppliers
- Local Community
- Global Community.

The six main stakeholder groups represent the main variables for root cause analysis related to sPPC. *"Company owners"* are individuals with a financial investment in the business and responsible for the strategic orientation of the company.

"Employees" are individuals who provides their skills to a firm, usually in exchange for a monetary wage. *"Costumers"* can be viewed as any end-users of products, services, or processes. *"Suppliers"* provide goods and/or services to the company. *"Local community"* are spatially-related groups of individuals utilizing a shared resource base within which a company enterprise exists. *"Global community"* are communities outside the boundaries of local communities, e.g. state, national, and international entities.

In the third step, the causes must be identified related the main variables. The causes are represented as indicators and goals which can be controlled and achieved through sPPC. For the identification of the causes, the German Government's Sustainable Development Strategy [17] and the indicator framework for sustainable manufacturing provided by NIST's Sustainable Manufacturing Indicator Repository (SMIR) [18] were analyzed through a text review.

The new German Sustainable Development Strategy was developed in 2016 [17] based on previous developed and extended Government's Sustainable Development Strategies from 2002 to 2012 and according to the United Nation (UN) sustainable development goals (SDGs) from 2015 [19]. A total of 17 key goals are defined with varying number of sub-goals for the sustainable development of Germany.

SMIR provides a set of indicators that are quantifiable and clearly related to manufacturing. The database was developed through an analysis of eleven publicly available indicator sets. The indicators are structured in five categories: environmental stewardship, economic growth, social well-being, performance management, and technological advancement management. The indicator categorization work is also intended to establish an integrated sustainability indicator repository as a means to providing a common access for manufacturers, as well as academicians, to learn about current indicators and measures of sustainability [8, 18].

Through a text review of the German Sustainable Development Strategy, applicable goals are identified, which are affected though companies. With aid of the sustainable indicator database SMIR, sustainable indicators are identified to control the progress of the goals in manufacturing enterprises. To identify and select sustainable goals and indicator, a set of criteria was defined for sPPC:

- The goal or indicator must have a predictable quantitative impact on short and mid-term production planning and/or controlling;
- The goal or indicator must be relevant regarding sustainable manufacturing and to the sustainable goal;
- The goal or indicator must be understandable and easy to interpret by the company;
- The goal or indicator must be allocable to one (or more) stakeholder groups supporting their interests.

The results of the analysis are presented as root cause diagram. Figure 1 presents the general structure of the diagram presenting the results.

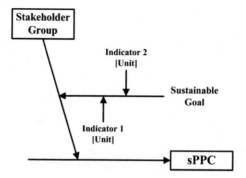

Fig. 1. General structure of the results presented as root cause diagram.

3 Results and Discussion of the Root-Cause Analysis

3.1 Results of the Root-Cause Analysis

Figure 2 presents the results of this paper as root cause diagram.

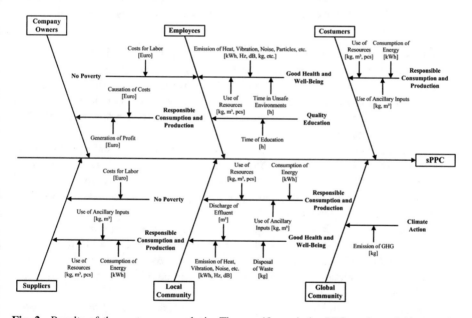

Fig. 2. Results of the root cause analysis: The specific topic is sPPC, main variables are the main stakeholder groups for sustainable manufacturing, and the causes are represented as sustainable goals and indicator.

Sustainable goals and indicators (Causes) have been allocated to each main variable (stakeholder group). The following section presents the interests of the stakeholder groups from an enterprise perspective considering sustainable aspects. For each stakeholder group, sustainable goals according to these interests are identified, and related sustainable indicators to measure the status and trend of the goals are presented.

Company Owners are interested in the economic value creation of the company. Therefore, the identified sustainable goal is "Responsible Consumption and Production" to ensure the long-term survival of the company on the market against competitors. This goal can be controlled through following sustainable indicators:

- Causation of Costs [Euro]: Costs caused by an enterprise focusing on the categories e.g. costs for material, labor, energy, waste disposal, effluent disposal, delivery. The indicator can be normalized on processes, costumer orders and/or products, and compared with historical costs (internal benchmarking), and national and international competitors (external benchmarking).
- Generation of Profit [Euro]: Profit generated by an enterprise focusing on the categories e.g. financial profits, and non-financial profits. The indicator can be normalized on processes, costumer orders and products, and compared with historical profit generations (internal benchmarking) and previous set financial goals.

Employees are interested in safe and healthy workplaces, opportunities for trainings and qualification activities, and fair salaries. Therefore, the identified sustainable goals are "Good Health and Well-Being", "Quality Education", and "No Poverty" to ensure long-term wellbeing of the employees. These goals can be measured through following sustainable indicators:

- Emission of Heat, Vibration, Noise, Particles, etc. [kWh, Hz, dB, kg, etc.]: Emission of heat, vibration, noise, etc. emitted by an enterprise focusing on the categories e.g. hazard emissions, non-hazard emissions. The indicator can be normalized on processes, costumer orders and products, and compared with set of limits for health and safety at the workplace (see e.g. European directives on safety and health at work [20]).
- Use of Resources [kg, m^3, pcs]: Amount of used materials by an enterprise focusing on the categories e.g. hazard materials, radioactive materials, heavy metals, toxic chemicals. The indicator can be normalized on processes, costumer orders and products, and compared with historical consumptions of these resources (internal benchmarking) and previous set finical goals.
- Time in Unsafe Environments [h]: Time of employees in unsafe environments focusing on the categories e.g. working with hazard materials, exposure to toxins, working in high. The indicator can be normalized on processes, costumer orders and products, and compared with historical times in unsafe environments (internal benchmarking).
- Time of Education [h]: Time of employees for education focused on the categories e.g. education for security, human rights, corruption, career development, emergency drills. The Indicator can be compared be compared with historical times for education.
- Costs for Labor [Euro]: Costs for labor paid by an enterprise focusing on the categories e.g. job type, gender, age. The indicator can be normalized on processes, costumer orders and products, and compared with local and national competitors (external benchmarking).

Costumers are interested in consuming sustainable products to satisfy their needs. These products should be produced as sustainable as possible, through e.g. using renewable and recycled materials, green energy, and environmental compatible ancillary inputs. Therefore, the identified sustainable goal is "Responsible Consumption and Production" to ensure the long-term satisfaction of the costumers. This goal can be measured through following sustainable indicators:

- Use of Resources [kg, m^3, pcs]: Amount of used resources by an enterprise focusing on the categories e.g. source of the resources, renewable, recycled, fossil. The indicator can be normalized on processes, costumer orders and products, and compared with historical consumptions (internal benchmarking), and national and international competitors (external benchmarking).
- Use of Ancillary Inputs [kg, m^3]: Amount of used ancillary inputs by an enterprise focusing on the categories e.g. cleaners, lubricants, oils, coolants. The indicator can be normalized on processes, costumer orders and products, and compared with historical consumptions (internal benchmarking), and national and international competitors (external benchmarking).
- Consumption of Energy [kWh]: Amount of consumed energy by an enterprise focusing on the categories e.g. source of the energy, renewable, reuse, fossil. The indicator can be normalized on processes, costumer orders and products, and compared with historical consumptions (internal benchmarking), and national and international competitors (external benchmarking).

Suppliers are interested in offering sustainable resources to the company These resources should be produced using renewable and recycled materials, green energy, environmental compatible ancillary inputs. Moreover, the suppliers should consider social requirements, e.g. paying fair salaries to contractor and employees. Therefore, the identified sustainable goals are "Responsible Consumption and Production", and "No Poverty" to ensure sustainability considering also the environmental and social impacts of the extraction of resources. These goals can be measured through following sustainable indicators:

- Use of Resources [kg, m^3, pcs]: Amount of resources used by the supplier to produce and/or extract resources focusing on the categories: sources of the materials, renewable, recycled, fossil. The indicator can be normalized on processes, costumer orders and products of the enterprises, and compared with historical consumptions by the supplier and other suppliers.
- Use of Ancillary Inputs [kg, m^3]: Amount of ancillary inputs used by the supplier to produce and/or extract resources focusing on the categories: sources of the materials, renewable, recycled, fossil. The indicator can be normalized on processes, costumer orders and products of the enterprise, and compared with historical consumptions by the supplier and other suppliers.
- Consumption of Energy [kWh]: Amount of energy consumed by the supplier to produce and/or extract resources focusing on the categories: sources of the materials, renewable, recycled, fossil. The indicator can be normalized on processes, costumer orders and products of the enterprise, and compared with historical consumptions of the supplier and other suppliers.

- Costs for Labor [Euro]: Costs for labor paid by supplier to produce and/or extract resources focusing on the categories e.g. job type, gender, age. The indicator can be normalized on processes, costumer orders and products of the enterprises, and compared with historical pad salaries of the supplier and other suppliers.

The *local community* share the environment with the enterprise and is interested in healthy and safe living environment. Therefore, the identified sustainable goals are "Good Health and Well-Being", and "Responsible Consumption and Production" to ensure long-term wellbeing of the local community. These goals can be measured through following sustainable indicators:

- Use of Resources [kg, m^3, pcs]: Amount of used resources by an enterprise focusing on local extracted and/or purchased resources. The indicator can be normalized on processes, costumer orders and products, and compared with historical consumptions (internal benchmarking), and with limits for local available renewable resources.
- Use of Ancillary Inputs [kg, m^3]: Amount of used ancillary inputs by an enterprise focusing on local extracted and/or purchased resources. The indicator can be normalized on processes, costumer orders and products and, compared with historical consumptions (internal benchmarking), and with limits for local available renewable ancillary inputs.
- Consumption of Energy [kWh]: Amount of consumed energy by an enterprise focusing on local produced energy. The indicator can be normalized on processes, costumer orders and products, and compared with historical consumptions of local energy (internal benchmarking), and with limits for local available renewable energy.
- Discharge of Effluent [m^3]: Volume of discharged effluent by an enterprise focusing on the categories e.g. hazard effluents, non-hazardous effluents. The indicator can be normalized on process, costumer order and product, and compared with historical consumptions of local energy (internal benchmarking), and with local limits for the discharge of effluent.
- Disposal of Waste [kg]: Amount of disposed waste by an enterprise focusing on the categories: hazardous waste, non-hazardous waste. The indicator can be normalized on processes, costumer orders and products, and compared with historical consumptions of local energy (internal benchmarking), and with local limits for the discharge of effluent.
- Emission of Heat, Vibration, Noise, etc. [kWh, Hz, dB]: Emission of heat, vibration, noise, etc. emitted by an enterprise, focusing on emissions which impacts the local environment. The indicator can be normalized on processes, costumer orders and products, and compared with historical emissions in the local environment (internal benchmarking), and with local limits for the emissions.

The *Global Community* is interested in a healthy and safe world. Sustainable impacts must be considered which have direct effects on communities all over the world. Therefore, the identified sustainable goal is "Climate Action" to ensure healthy living environment and long-term wellbeing of the international community. This goal can be measured through following sustainable indicator:

- Emission of GHG [kg]: Emission of GHG emitted by an enterprise focusing on the categories e.g. CO_2, CH_4, N_2O, CFC, NO_x, SO_x. The indicator can be normalized on processes, costumer orders and products, and compared with historical emissions of GHG (internal benchmarking) with international limits for GHG (see e.g. Paris agreement [21]).

3.2 Discussion of the Root-Cause Analysis

The analysis identified five national SDGs which can be affected by considering sustainable aspects in PPC processes. For the identified SDGs, sustainable indicators are selected to control the status of the identified goals in manufacturing companies. According to the interests of the stakeholder groups, the meaning of the indicator is different.

For example, the sustainable goal "Responsible Consumption and Production" is allocated to multiple stakeholder groups company (owners, customers, suppliers, and local community) using similar and/or the same sustainable indicators. Local communities are interested in protecting of local resources and the sustainable indicators are focused on the availability local resources for manufacturing companies. However, costumers are interest in products which were produced using renewable and recycled materials, green energy, and environmental compatible ancillary inputs. For both stakeholder groups the sustainable indicator "Use of Resources" is selected, but with different focused on type of resources which are controlled.

Moreover, conflicting indicators must be identified for sustainable development. The company owner is interested in generating profits to ensure a long-term survival of the company. In case that, the use of renewable resources is more expensive than the use of non-renewable resources, the production costs can be decreased using non-renewable materials. But the use of non-renewable resources meets not the interests of the costumers who prefer renewable resources. The use of renewable and non-renewable resources must be in balanced meeting customer requirements, finical goals, and protection the sources of non-renewable resources. This balance should also be controlled at suppliers to ensure sustainability considering the whole life cycle of products. This principle is also valid for recyclable resources.

The sustainable goal "Good Health and Well-Being" is allocated to the stakeholder groups employees and local community. The focus of the sustainable indicator is related to the interests of the stakeholder group, but also on the definition of the system boundary of the selected sustainable indicators. For example, the discharge of toxic effluent has direct impacts on the local community but no direct impacts on the employees. Therefore, it is also very important to define the considered system which is controlled by the sustainable indicator.

4   Conclusion and Future Work

The paper provides an overview on indictors and goals for sPPC process for manufacturing companies, referring also the benefits of integrating ergonomic factors in this process. This was done based on an analysis of the existing set of indicators for

sustainable manufacturing and national sustainable development goals. The key findings of the analysis are:

- The relation of the indicators to the goals must always be described according to the
  interests of the affected stakeholder groups.
- Conflicting indicators controlling the same goal must be in balanced meeting
  requirements and interest of all affected stakeholder groups.
- For the selected sustainable indicators, the system boundaries must be defined
  according to the requirements and interest of the affected stakeholder groups.

The results of the analysis can be used as basis to support and improve the outcome
of the PPC process discussions with company managers, experts and employees'
representatives. The integration of the Ergonomics' perspective ensures that the sustainable PPC process, is guided by principles of social responsibility (human
well-being), green system design (environmental cleanness), and economic success
(system performance).

In future work, statistical analysis should be performed to investigate the relevance
of the identified sustainable indicators and goals. The most relevant indicators must be
implemented in sPPC systems to test the feasibility and practicability in manufacturing
enterprises.

References

1. Radjiyev, A., Qiu, H., Xiong, S., Nam, K.: Ergonomics and sustainable development in the
   past two decades (1992-2011): research trends and how ergonomics can contribute to
   sustainable development. Appl. Ergon. **46**, 67–75 (2015)
2. WCED: Report of the World Commission on Environment and Development: Our Common
   Future (1987)
3. United Nations: Report of the United Nations Conference on Environment and Development. In: Rio Declaration on Environment and Development, no. A/CONF.151/26, vol.
   1 (2012)
4. Elkington, J.: Cannibals with Forks: The Triple Bottom Line of 21st Century Business.
   Capstone, Oxford (1997)
5. IEA: IEA Executive Defines Ergonomics. On-Site Newsletter for the IEA 2000/HFES 2000
   Congress (2000)
6. Thatcher, A.: Green ergonomics: definition and scope. Ergonomics **56**(3), 389–398 (2013)
7. Moldan, B., Janoušková, S., Hák, T.: How to understand and measure environmental
   sustainability: indicators and targets. Ecol. Indic. **17**, 4–13 (2012)
8. Joung, C.B., Carrell, J., Sarkar, P., Feng, S.C.: Categorization of indicators for sustainable
   manufacturing. Ecol. Indic. **24**, 148–157 (2013)
9. Global Footprint Network: Ecological Footprint. https://www.footprintnetwork.org/our-
   work/ecological-footprint/. Accessed 08 Feb 2018
10. ZSL: Living Planet Index. http://www.livingplanetindex.org/home/index. Accessed 08 Feb
    2018
11. Popovic, T., Barbosa-Póvoa, A., Kraslawski, A., Carvalho, A.: Quantitative indicators for
    social sustainability assessment of supply chains. J. Clean. Prod. **180**, 748–768 (2018)
12. Coskun, A., Zimmerman, J., Erbug, C.: Promoting sustainability through behavior change: a
    review. Des. Stud. **41**, 183–204 (2015)

13. Thatcher, A., Waterson, P., Todd, A., Moray, N.: State of science: ergonomics and global issues. Ergonomics 1–33 (2017)
14. Wilson, P.F., Dell, L.D., Anderson, G.F.: Root Cause Analysis: A Tool for Total Quality Management. ASQC Quality Press, Milwaukee (1993)
15. Benoît-Norris, C., et al.: Introducing the UNEP/SETAC methodological sheets for subcategories of social LCA. Int. J. Life Cycle Assess. 16(7), 682–690 (2011)
16. Sutherland, J.W., et al.: The role of manufacturing in affecting the social dimension of sustainability. CIRP Annals – Manuf. Technol. 65(2), 689–712 (2016)
17. Deutsche Bundesregierung: Deutsche Nachhaltigkeitsstrategie
18. NIST: Sustainable Manufacturing Indicator Repository. http://www.mel.nist.gov/div826/msid/SMIR/index.html. Accessed 07 Feb 2018
19. United Nations: Sustainable Development Goals: 17 Goals to Transform our World. http://www.un.org/sustainabledevelopment/sustainable-development-goals/. Accessed 08 Feb 2018
20. European agency for safety and health: European directives on safety and health at work. Accessed 23 Feb 2018
21. European Commission: Paris Agreement. https://ec.europa.eu/clima/policies/international/negotiations/paris_en. Accessed 23 Feb 2018

Environment-Integrated Human Machine Interface Framework for Multimodal System Interaction on the Shopfloor

Katrin Schilling[✉], Simon Storms, and Werner Herfs

Laboratory for Machine Tools and Production Engineering (WZL),
RWTH Aachen University, Steinbachstr. 19, 52074 Aachen, Germany
{k.schilling, s.storms, w.herfs}@wzl.rwth-aachen.de

Abstract. Today's production systems evolve into an ecosystem of smart automation technology and networking. Algorithms lead the processes that are derived from digital data over the system lifecycle. In this context the human worker is still an important element despite the significant role of automation. But the physical world gives only limited access to the digital part of the production system. A natural and intuitive user interface between the real and the digital world is a crucial factor. Moreover, the shopfloor itself should become a smart environment where the digital information is merged into the physical world. That requires a flexible framework for various interaction technologies, regarding the data interface, a multi-device infrastructure and a concept to adapt the presentation of information.

Keywords: Environment-Integrated user interface
Cyber Physical Production System · Human system interaction

1 Introduction

The present trends in industrial production are characterized by flexible automation and a high level of networking: Machines communicate among each other and with upper systems, self-learning algorithms manage the work order and artificial intelligence controls the process. The plant becomes a so-called Cyber Physical Production System (CPPS) that enables the manufacturer to fulfill customer-specific requirements, react to short-term adjustments and increase efficiency. [1] But this also leads to a significant increase of dynamic and complexity. To handle such a system, transparency is most important. Clear traceability helps to avoid failure and supports optimization. While the human factor is often seen as a cause of failure in automation, he also is the most creative and adaptable actor when it comes to sense and solve unexpected problems. In that, the human worker must not be excluded from the CPPS but the User Interface (UI) has to empower him to stay on top of the process and lead the system.

The crucial problem is, that the worker in the real world cannot see the data and algorithms that operate a plant. Today, at best, those are visualized in mostly graphical UIs within an operating system or a mobile app. But while monitoring them, the worker is focused on computer screens. Even with use of mobile devices, the world behind

© Springer International Publishing AG, part of Springer Nature 2019
I. L. Nunes (Ed.): AHFE 2018, AISC 781, pp. 374–383, 2019.
https://doi.org/10.1007/978-3-319-94334-3_37

these screens stands beside the real processes and separates them. The user has to associate the two systems based on his knowledge. He has to learn how to read the machine control and monitoring tools. With the increasing weight of digital content, the operator's attention might drift into the virtual world as we already can observe in social life, where messaging with mobile devices can rapidly wipe out a face-to-face conversation. Or else, the worker observes the physical process and misses incipient data trends that are hidden behind other windows or in the depth of the operating system. But the shopfloor-worker must equally concentrate on both: digital *and* physical procedures. Moreover, he has to understand its relation and must immediately identify the need for action. In doing so he needs access to various information about the CPPS, not just the current production process but the whole system lifecycle.

2   Interaction in the Cyber Physical Production System

2.1   The Role of the Human Worker and His Tasks

With the technical transformation to intelligent, self-controlled manufacturing systems the human role in the CPPS changed from a machine operator and programmer to the observer and supervisor of the system. The tasks shift mainly to process management, maintenance tasks and multiple machine operation. Some tasks that cannot be auto-mated economically (usually caused by a high level of agility and individualization) must be performed by a human worker. While the computer is able to handle large quantities of data and optimize complex process in a defined setting, the human worker reacts more flexible and creative when it comes to unexpected situations. To do so he has to be able to collaborate with other process participants and needs methodologic and professional skills [2].

2.2   The Human Machine Interface on the Shopfloor

The Human Machine Interface (HMI) is the technical aspect of the sozio-technical system. It provides the functionality of machines and devices to the operator and forms the tool to fulfill his tasks. Moreover, the HMI is the "face" of the technical system which means that the HMI design rules the way of the user's understanding and relation to the whole system. The increasing weight of networking in production leads to a shift of the HMI's function and role. Mobile devices from the consumer market, such as tablets and smartphones, complement or replace the native HMI of modern field devices and machines. More and more manufacturers provide mobile apps for their products. One device can now hold various applications to monitor, control and analyze multiple machines and related job data. So, the user can interact with the whole system by use of one single device. That gives him a better overview of the CPPS and makes him independent from local panels.

The idea of *ubiquitous computing,* where users are surrounded by numerous computers and interact with a smart environment, was described by Mark Weiser in 1991 [3]. At that time mobile computer were still very bulky and the implementation of a comprehensive network infrastructure was associated with a big effort. But today we have both: small and low-cost computers that can be integrated in mostly every device and a data infrastructure consisting of standards in communication protocols and data description. The vision of ubiquitous computing turns into the present technologies for a smart home, smart cities and the smart factory.

Smart Devices. The success of smart devices, such as smartphones and tablets, in daily life has improved the advances in mobile interaction on the shopfloor. The high-performance devices from the consumer market became affordable. They are easy to use and bring a broad setting of ready-to-use sensors and actors, in particular the camera is useful for various applications. With the establishment of a few operating systems (Android, Windows and iOS) even the development of applications for mobile devices became more efficient. Nowadays, most decision-makers see the benefits of mobile devices on the shopfloor. The former main challenges like device management and integration into the IT infrastructure decrease with the broad acceptance of smart devices in the context of manufacturing.

Augmented Reality. Augmented reality (AR) is a promising technology when it comes to merge the digital and real world [4]: You can see the virtual content embedded in the real world through a tablet, glasses or with use of projection. AR has a lot of advantages and seems to become a key technology in the future. But there are still some restrictions: AR on the tablet is easy to implement, but it needs free hands to hold the device. And due to the video-based visualization the user still focuses on a screen. It is a good option for short-term scenarios, such as remote assistance or instructions for nonprofessionals. With their increasing weight and better comfort, Head Mounted Displays (HMD) become a promising class of AR-technologies, especially in the context of a production plant. A characteristic feature of AR on HMD is, that the virtual content is a personal information. That privacy can be an advantage, e.g. in training situations, but in relation to collaboration it becomes a disadvantage. Even if the technology allows to share information and virtual objects (see shared experiences on Microsoft HoloLens), the device remains a personalized item. It forms a bubble around the user that divides him from the others by the accessibility to the virtual content (Fig. 1).

Fig. 1. Augmented reality application that shows embedded sensor values in the plant [4].

Nowadays, mobile devices and AR form a good approach to realize the idea of mobile and ubiquitous computing. But there is still a lack to the vision of a holistic and immersive system interaction where the user no longer distinguishes digital content from the real world and interacts with the whole CPPS.

3   A Vision of the Human Machine Interface on the Shopfloor

The purpose of the CPPS user interface is to provide the required information at the right time and place, to support the user in decision-making and give access to the technical functionality - that all in an easy and efficient way. Because of the growing ratio of software in contrast to mechanical function elements in manufacturing systems, most properties and dependencies become (physically) invisible. Today, experienced machine operators are able to perceive the plant condition by its sound, its behavior or even the smell. They know the typical impression of the shopfloor and rapidly get an instinctive feeling of normal or unusual machine behavior. Their assessment is not based on a single feature, but on the overall characteristic impression regarding all senses. The UI for the CPPS has to bring back the visibility of processes and system information to ensure transparency. The guiding principles to design such a HMI-framework follow the idea of ubiquitous computing and meet the requirements of a CPPS.

Immersive Interaction – Forget About the Tools. Immersion means the way of user perception where the user forgets about the interacting activity itself and interacts with the technical system in such a natural way, that the devices seem to disappear. Technologies such as gesture control, AR/VR or natural speech processing enable a natural interaction that reduces more and more the distance between the user and technology. Figure 2 shows an example for the growing level of immersion related to the inspection of a three-dimensional part: The technical drawing shows a limited number of views, the part is reduced to its contour lines. The visualization is based on rules and conventions. The reader has to know those rules und transfer the drawing into a mental imagination of the object. The same drawing in a CAD viewer is much closer to the real object. The user observes the virtual object via an imaginary camera, he can move the camera with a (programmable) key/mouse combination. More intuitive is the presentation on the tablet computer. Based on the internal sensors and motion tracking, the user can rotate and move the viewport by physically moving the device (see *magic lens metaphor* [5]). The most immersive experience of the virtual part is a three-dimensional visualization in space, which adapts the user's head position (e.g. holograms with Microsoft HoloLens). The user can observe the virtual object in the same way as he observes a physical object. He does not need any knowledge of the navigation or encoded information, he just has a natural understanding. This idea works for most information: the more the digital content feels like a physical object, the better we understand it. In that way the shopfloor HMI should provide a user experience that enables the worker to interact with the digital aspect of the manufacturing system in a natural and intuitive way – until he forgets about the partition between *reality* and *digital data*.

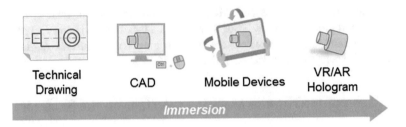

Fig. 2. Examples for the level of immersion: inspection of a three-dimensional part.

Spatial Interaction – Walk Through the Data. Spatial orientation dominates natural perception and orientation in the physical world. This powerful ability should be used to support the orientation and navigation in digital content. A spatial understanding of abstract data models improves a better understanding. Various memory techniques are based on the linkage between information and well-known spaces. But on the one hand it is common to organize data in grouped segments and sorted tables, accessible with ids and keywords. On the other hand, the shopfloor forms a well-known space that already has an abstract association to the digital production data. Today, those data are accessible via conventional software applications on computers that are located next to machines or production cells. The user has to find the wanted information in menus and dialogs. Instead, the shopfloor itself should become part of the interface. Digital content must be linked to the physical resources and the user context so that the worker can call the information directly. The space becomes a tool for structuring information and the user no longer needs to remember the id or keyword.

Scalability of Information – From Distributed to Reduced Interfaces. Within a holistic UI concept, the single devices become less important. They form an interaction network instead, that is seen as one interactive environment. The UI is distributed over various terminals and mobile devices, even the plant facility and the shopfloor environment themselves become a representation of the digital objects. Those different devices and their individual properties require an adaption of the interface design. In web design this is solved with a device and platform independent design concept, such as responsive or adaptive design. A model-based layout definition allows the GUI to scale from small smartphone screen to large computer displays. This concept can be transferred to the greater extent of a network of multi-modal HMI devices in an interactive environment. The information density expands with the display possibilities, from a large dashboard or a smaller terminal to a mobile device like tablets as well as a smartwatch and even a simple light element. Figure 3 shows how the UI can adapt the display capability.

Thereby the worker can interact with the overall system, no matter where he is situated. Moreover, it is possible to sense the plant health by instinct without the observation of a concrete screen.

Multi-Modal Interaction – The Environment Becomes Alive. In addition to the graphical user interface as the most common UI, the interaction can address much more modalities. The visual perception is the most important human sense for information

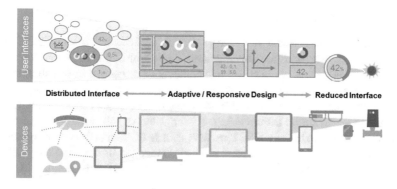

Fig. 3. Scaled information depending on the devices.

intake. Integrated lights in machines and devices can support subtle information that the worker can notice in passing. The optical impression of a plant is enriched with acoustic feedback like sounds up to speech output. At least haptic feedback complements the multi-modal interface, e.g. in a wearable or (if possible) integrated vibration motors. Based on various such interaction elements, even though every single one seems to be inconspicuous, the environment becomes a holistic natural interface that communicates with the worker. He is able to get a feeling of the CPPS status, even without starting any application.

Scalable Platform – The HMI Grows up with the System. As well as the font-end is integrated into the production plant, the back-end concept should fit in the system architecture. Therefore, a framework is needed, that defines the boundary conditions but also gives enough flexibility to add new functions continuously. To ensure compatibility the use of standards and common technologies in data modelling and communication is important. In doing so, the framework should follow the common design paradigms of the Internet of Things (IoT), such as modularization and loose linkage, platform independency and semantic description. In that ecosystem the HMI can be initialized with some basic applications, it expands over time and adapts the transformation of a CPPS over lifetime.

4   A Technical Approach to a HMI for the CPPS

The HMI for the CPPS is based on a framework that defines an open and extendable data model, a back-end infrastructure to share those models between various devices and a multi-modal front-end concept. The technical approach is based on common (web-)technologies to provide a universal interface to the CPPS, especially the IoT devices that control and connect machines, sensors and field devices. The concept is designed to ensure high flexibility and scalability, so that new applications can be added continuously.

4.1   The Data Item Definition

Every content such as a sensor value, an executable function or a media file is described as a *data item*. It integrates information from applications, machines and devices. The data item definition is shown in Fig. 4 and contains some general information about the content, a description of the data source and optional parameter for specific applications. The data source represents either some static content (e.g. text, picture or video) or is linked to a dynamic resource (e.g. sensor value or job progress). The dynamic content can be integrated via REST as well as a MQTT interface. Furthermore, this source-type definition can be extended to other connection information (e.g. OPC UA) if needed. The app-data element holds optional information that is required for specific applications. As an example, an AR application needs the coordinates for the position in space where the data item is allocated and the id of a related area description.

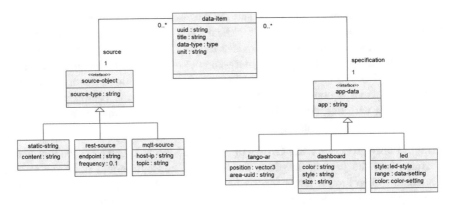

Fig. 4. The model definition of a data item.

4.2   Shared Data Items – Information *To Go*

When a data item was created it can be saved (as JSON serialization) within an application. But the primary purpose of these definition comes with sharing data items between multiple application and devices. To do so the framework provides a sharing feature that organizes available devices or rather the participating applications. In use of this feature a data item can be send from one application to another, for example from a dashboard to a mobile device. This allows the user to take away relevant data items or transmit them to colleges as well as machines.

Another way to transfer data items without the need of a sharing service is the encoding of a data item in a QR code or RFID tag. Those can easily be attached to machines, products, tools or supply lines. An app on a mobile device can read the tag with use of the camera or NFC sensor and present the information immediately.

4.3 Widgets – the Graphical Representation

A data item can be visualized by a graphical object that is called *widget*. A widget is a container which presents the static or dynamic content of the data item. Figure 5 shows some examples of different styles, such as a textual representation, a trend chart or a diagram. The widgets can form a regular GUI, e.g. on a dashboard or on a mobile device, or they are embedded in the environment as AR application (see Fig. 6). The interpretation of the data item is up to the individual application. The specification of optional app-data in the definition allows to store specific information that are ignored by other applications.

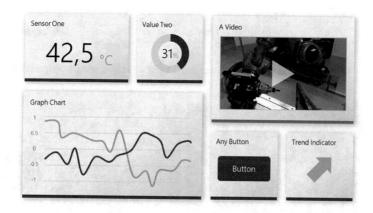

Fig. 5. Various styles for a widget.

4.4 The Environment-Integrated Representation

Even while the graphical representation might be familiar and most expressive, it always requires a screen and a HMI device. But the presentation of a data item is not limited to displays. Some data are also suitable for special interface elements: A simple but powerful feature are integrated lights. For example, a LED stripe can easily be mounted in the housing of a production cell as shown in Fig. 7. While a signal light for a single machine is not a new thing, as part of the CPPS it becomes a powerful interface. It can present the machine status as well as the job priority with use of color. A progress bar can show the remaining time, or a slight pulse indicates that the system is loading some data from the cloud. This kind of interface can be perceived in passing even when the worker is far away from the next HMI terminal. Of course, the scope of such an interface must be reduced to some basic information to avoid an overload.

The technology for such a light element is just the same concept as the widgets: The information (e.g. job progress or system state) is represented by a data item. A customizable configuration file on the controller defines the mapping from a value to a light pattern. The configuration follows a simple *if-this-then-that* logic. In the same way a data item can be represented by acoustic or haptic feedback.

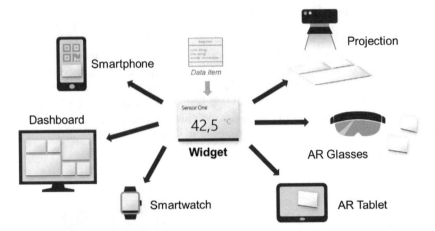

Fig. 6. Representation of the data item in form of a widget within various devices.

Fig. 7. Integrated light elements in the smart automation lab and a HMI terminal.

5 Conclusion and Further Work

The important role of digital data in nowadays manufacturing systems requires a strong focus on the design of a holistic and immersive UI. At the same time there are established information and communications technologies to integrate interactive elements in the environment that form a distributed UI network. It is based on a framework that defines a data model and the back-end concept. The overlaying front-end concept can be applied on application for mobile devices, augmented reality as well as reduced feedback elements.

The next phase of the framework design will concentrate on the linkage of data items among each other, the association with context data and a semantic description that refers to the physical environment. The concept will be oriented towards the idea of a semantic web and linked data, that will be applied on the HMI framework.

Acknowledgments. This research and development project is funded by the German Federal Ministry of Education and Research (BMBF) within the Program "Innovations for Tomorrow's Production, Services, and Work" (funding number 02P14B145) and managed by the Project Management Agency Karlsruhe (PTKA). The author is responsible for the contents of this publication.

References

1. Kagermann, H., Wahlster, W., Helbig, J.: Recommendations for implementing the strategic initiative INDUSTRIE 4.0. Final report of the Industrie 4.0 Working Group. Frankfurt/Main (2013)
2. Gorecky, D., Schmitt, M., Loskyll, M.: Mensch-Maschine-Interaktion im Industrie 4.0-Zeitalter. In: Bauernhansl, T., ten Hompel, M., Vogel-Heuser, B. (Hrsg.) Industrie 4.0 in Produktion, Automatisierung und Logistik, pp. 525–542. Springer Fachmedien, Wiesbaden (2014)
3. Weiser, M.: The computer for the 21th century. In: Scientific American, Special Issue on Communications, Computers, and Networks (1991)
4. Fimmers, C., Schilling, K., Sittig, S., Obdenbusch, M., Brecher, C.: Potentials and challenges of augmented reality for assistance in production technology. In: 7 WGP-Jahreskongress Aachen, 5–6 October, pp. 619–627. Primus Verlag, Aachen (2017)
5. Brown, L., Hua, H.: Magic lenses for augmented virtual environments. IEEE Comput. Graph. Appl. **4**, 64–73 (2006)

Digitization of Industrial Work Environments and the Emerging Challenges of Human-Digitized System Collaborative Work Organization Design

Mohammed-Aminu Sanda[1,2(✉)]

[1] University of Ghana Business School, P. O. Box LG 78, Legon, Accra, Ghana
masanda@ug.edu.gh
[2] Luleå University of Technology, 97187 Luleå, Sweden
mohami@ltu.se

Abstract. This paper discussed how the digitization of human work affect human performance of digitized work entailing both physical and mental activities, as well as the human's social interaction entailing collaboration between the human's and the digitized work system. It is posited that in the digitization process of the industrial work environment, the digitized system designer's lack of knowledge about the external world of the embedded humans may result in the designer's inability to accurately predict the outcomes of his/her decisions on work organization, especially the creation of positive collaboration and social harmony between digitized systems and humans towards enhanced productivity. It is concluded that analytic strategies for understanding the interactive work dynamics between humans and digitized systems, which can serve as key for identifying both the innovative and constraining characteristics of effective and efficient association between humans and digitized work systems towards collaborative work organization design should be developed.

Keywords: Industrial work environment · Digitized system · Human system
Work organization · Work organization design · Social collaboration

1 Introduction

There is an emerging concern among both human work researchers and practitioners (e.g. [1–3]) that despite many promising production concepts, organizational models and development of new technologies during the last decades, the digitized industry continues to face a large gap between expectations and real implementation of realistic organizational practices [4]. Increased global competition has resulted in the production activities of most industrial firms increasingly centered on the lean organisation and the introduction of highly automated systems supported by multi-skilled workers capable of managing multiple productive activities in the firms [3]. In such automated work environment, as argued by [2], technology and work are to be built around 'autonomation', where workers and machines are expected to cooperate. In this regard,

I. L. Nunes (Ed.): AHFE 2018, AISC 781, pp. 384–395, 2019.
https://doi.org/10.1007/978-3-319-94334-3_38

the introduction of process automation and remote operation technologies in an auto-mated work environment is expected to be complemented by an automated work system that is smarter and well-integrated [3]. This kind of work environment fits the type that [2] referred to as the intelligent production system of a "future-automated work system". Yet, there prevails the problem of how to model such intelligent pro-duction system of the "future automated work systems" for them to become enablers for learning and collaboration across organizational borders [2, 3, 5]. Even if the idea of a holistic perspective on production systems is commonplace in most research areas of today [5], there is a true challenge in multidisciplinary research that reconnect the research fields and their theories, methods, ideas and results [5, 6].

As it is observed by [6], there are signs that digitization attempts, such as Industry 4.0 conception, will become more apparent in the workplace. This development, according to [6], requires reflections and considerations so as not to create more problems than can be solved. This observation could be associated with the problem of developing a holistic work organization model to guide the future integration of a firm's digitized work environment and its human and organizational sub-systems [3]. It could also be associated with the challenge of developing specialized knowledge in areas associated with the digitized, organizational and human systems that could contribute towards the attainment of future intelligent-work system [3]. Such challenges reinforce the concerns over the gap between the theories of 'what people do' and 'what people actually do', which according to [7], has given rise to the 'practice' approach in management literature that focuses on the way that actors interact with the social and physical features of context in the everyday activities that constitute practice. For example, [6] has raised the issue of how technology and technological developments are related to working conditions, qualifications, identity, and gender. Thus, a theory of practice, as noted by [7], brings recursiveness and adaptation into a dialectic tension in which the two are inextricably linked. As such, practice does not occur only in macro-contexts which provide commonalities of action, but also in micro-contexts in which action is highly localized [7], and the interaction between these contexts provide an opportunity for adaptive practice [7, 8].

The purpose of this paper, therefore, is to critically reflects on the emerging challenges of designing work organizations in digitized work environments that can lead to positive collaboration and the creation of social harmony between digitized systems and the humans towards enhanced productivity. The objective is to look at how the digitization of human work affect individual's performance of object-oriented activity, with the object of the digitized work activity entailing both physical operations (using digitized tools to perform work) and mental operations (listening to digitized communication models). It also looked at how the digitization of human work affect individuals' subject-oriented activity, which includes social interaction and collabo-ration between the individuals engaged in the activity and the digitized system. In such digitized work activity, the individual is simultaneously engaged in physical activity (the manipulation of digitized technology to programme robotic work tasks) as well as engaged in mental activity (using digitized communication models, for listening to background music during work, information transmission from the control centres, and/or from colleagues approaching or leaving the individuals activity location) [9]. In this respect, it is important to understand how the digitization of the work environment affect human performance of object-oriented activity (if the object of the digitized work activity entails both physical and mental efforts), as well as the human's engagement in subject-oriented activity (if the digitized work activity entails collaboration between the human's activity and the digitized tools). It is also important to understand whether the usage of technology as a mediating object of digitized work activity, and a socializing tool, affects the motivation of humans engaged in the work activity. Finally, it is important to know whether the tacit knowledge of humans in the digitized work environment can be captured and incorporated in future designs of digitized work activities.

2 Industrial Digitization and Workplace Organization

There has been continued systemic structural organizational transformations in several industries for more than a decade now whereby several organizations have moved toward the digitization of their entire value chains in bids to accelerate production activities and organizational productivities. Such moves appear to concentrate on the developments of digitized organizational structures that are more reliant on the use of very efficient and effective technologies with high performance capabilities and less reliant on humans. Yet, there is the argument that the digitization of the work system must not only concern the design of digitized systems that are adapted to humans, but must regard humans as resources in designing better digitized systems (i.e. intelligent automation). The understanding here is that a digitized work system must be seen to consists of a harmonious integration of human and technological components. The system must be seen to consist of practices underscored by processes and infrastructure that enhances human work and enables learning [3]. This is due to the realization that the human-dominant social relations in the work environments are gradually being destroyed and replaced by technology-dominant work environments [4, 9, 10].

Taking into consideration the fact that the human-dominant work environment is supported by well-understood work organization system that has been practiced for so many years, and was continually refined through good understanding of human-human and human-machine work psychology and sociology, the advent of digitized work environment has brought with it significant challenges regarding the development of human-oriented work organization that enhances effective collaboration between the human and the digitized systems he/she uses in work performance. Thus, using the argument of [10] as point of departure, it is important for organizations that have digitized their work environments to develop new mental images for their workers by providing human-technology collaborative work environments with modern work organization that supports high productivity as well as good working and social conditions. The relevance of such human-technology collaborative work environments is underscored by the realization that in the industrial digitization drive, the development of process automation and remote operation technologies has enabled smarter, more integrated production systems.

In the view of [10], such automated systems, should be used as enablers for learning and collaboration across organizational borders. This is due to the observation that increase in the automation of organizations' value chains has resulted in several organizations adopting the lean work organisation system, with few, but multi-skilled workers capable of managing multiple areas of the organizations' digitized activities. Since the technologies being used by the workers and the work whose performance the technologies facilitate are built around 'autonomation', the workers and the technologies should cooperate with one another during the work performance. Thus, since the mode of work activity changes with technological changes, organizations must develop appropriate work methods to associate positively with incorporated technologies, to reduce workers' exposure to risks when interacting with technologies during task performances, especially as the work system transits from mechanization to full automation.

3 Human Factors in Digitized Work Environment

The design of a work organization systems where "hazard preventive planning", "risk reduction/elimination planning", and "technological and behavioral change processes" are key components is a pre-requisite in a digitized industrial firm. So is the development of technology for measurement and process management, organizational design and learning for operators. Development of such work organization system in an organization will require the creation of harmonious relationship between the organization's technical sub-system and its social sub-system. By implication, the development of intelligent automation systems for organizations could be enhanced by creating knowledge on how to harmoniously integrate the technological, organizational and human components of the organization's systems. Such integration stands to provide the basis for the evolution of a community of practice [11] at the workplace, with individual thought considered as essentially social, and developed in interaction with the practical activities of a community, through living and participating in its experiences over time [11].

Based on the notion that a digitized system consists of technical components, information, materials and humans, there is the need for holistic perspectives to be included in the design of work and management systems in organizations. This observation is based on the notion that tools shape the way humans interact with reality [12]. As it is explained by [12], tools usually reflect the experiences of people who have tried to solve problems at an earlier time, and in the process ended up inventing a new tool or modify the original tool to make it more efficient. This experience, according to [12], is embedded in the structural properties of the invented or modified tools, as well as in the knowledge of how the tool should be used. Thus, tools or technologies are created and transformed during the development of their usage-activities, and as such, carry with them specific cultures, which are manifestations of the historical remnants from such development. Therefore, going by the postulation of [12], the use of digitized tools in the industrial work environment is to enable the accumulation and transmission of social knowledge, thereby influencing the nature of individual workers' external behaviour and mental functioning. One of such knowledge derivative is macroergonomics, which has been defined by [13] as a top-down sociotechnical systems approach to the design of the work systems, and the carry-through of the overall work system design to the design of the human-job, human-machine, and human-software interfaces. Thus, to understand practice in a digitized work environment, it is important to move beyond institutional similarities to penetrate the situated and localized nature of practice in specified context [8].

Practice, as argued by [11], is local and situated, arising from the moment-by-moment interactions among actors, as well as between actors and the environments of their action. This implies that in the design process of digitized systems and work environments that is adaptable to humans, the humans themselves should be regarded as useful resources since their inputs could possibly result in the design of better systems. In this regard, the work organization system in a digitized work environment must be seen to consists of not only humans and technology, but also production processes, infrastructure, flows of digital signals, information, material, energy, as well as human work and learning [8]. In such system, as explained by [11], there is always a sense of an ongoing process of social becoming that is realized through a chain of social events or practice, and all these aspects must be in tune. Therefore, overall knowledge on digitized work system design, especially on how to harmoniously integrate the human and the other systems in the value chain is needed. The implication here is that, rather than looking for structural invariants, normative rules of conduct, or preconceived cognitive schema [7], it is important to investigate the processes whereby particular, uniquely constituted circumstances are systematically interpreted to render meaning shared. This represent a paradigm shift from traditional organization system design approach in which more concern is directed towards harnessing human behaviour in the interests of fulfilling organizational goals to a new design approach in which the costs of organization structural constraints on human work is considered.

4 Methodological Consideration

In the light of observations made in the previous sections, the Systemic-structural activity theory (SSAT), a modern synthesis within activity theory which brings together the cultural-historical and systems-structural strands of the tradition with findings and methods from Western human factors/ergonomics and cognitive psychology [14] can be used to guide the development of a conceptual framework for crafting an innovative work organization model for the automated work systems. The SSAT entails the conceptual application of both organizational activity and macroergonomics. The rationale for using these theoretical approaches are that these theories provide dynamic views of strategic practices in organizations and help explain the role of such practices in organizational change [7, 8]. In this regard, an automated work system can firstly be theorized as organizational activity system. It is argued by [15] that the theorization of organizations as activity systems have the tendency to cause bias thinking towards a concern for processes, many of which are goal-directed and boundary-maintaining. These characteristics, according to [15], are central to the open-system or neutral-system model of organizations, which [16] identified as the emerging focus of organizational sociology. Since organizations possess digitized systems for accomplishing work, organizational activity then emphasizes a digitized work system design in which technology affects organizational social relations by structuring transactions between roles that are building blocks of an organization. Therefore, the macroergonomics theory can be used to provide an understanding of the various processes to be entailed in an automated work system as organizational activity. Macroergonomics theory helps in understanding how a work system design can be optimized, in terms of its sociotechnical system characteristics [13]. Based on this understanding, the design characteristics of the digitized work system can be identified and used in the harmonious design of efficient and effective digitized work systems, human work, as well as human-machine and human-software interfaces.

5 Study of Human Work in Digitized Mining Activity

In a study by [9] in a highly digitized underground mine in which both the morphological and functional activities of miners engaged in rock drilling activity was investigated, the external behavioural and internal mental actions and operations of miners engaged in rock drilling activity using the Boomer, which is a highly automated machine with two robotic drilling arms, were observed and video-recorded. The miners' external behavioural actions included various motions they used to transform material tangible objects, while their mental actions were deemed to have either transformed images, concepts or propositions and non-verbal signs in their minds, as suggested by [17].

In the rock drilling activity analysis, the person(s) engaged in the activity, and the way he/she carry out the activity are considered important. In this respect, the individual-psychological aspect of the automated and digitized mining activity is analyzed. Using the systemic analytical approach as the basic paradigm for the analysis of positioning actions [17] in rock drilling activity in deep mines, [9] analysed the miners' activities both morphologically and functionally. In the morphological analysis, [9] described the constructive features of the rock drilling activity, entailing the logical and spatio-temporal organization of the cognitive behavioural actions and operations involved. In the description of the structure of the rock drilling activity, [9] sub-divided the work process into tasks, which were then analyzed individually in terms of mental and motor actions and operations. In the functional analysis, [9] analyzed qualitatively the potential strategies of activity performance associated with the miners' actions and operations, identified as constituting functional blocks, using systemic principles [14]. This analysis allowed [9] to evaluate varieties of performance indicators, such as time and errors, and the selection of the most efficient strategy by the miners. The functional blocks considered by [9] in analyzing the actions and operations in the rock drilling activity included goal, experience, subjective standards of successful result and admissible deviation, decision-making and program formation on explorative performance actions, subjectively relevant task conditions responsible for development of dynamic mental models of task, and assessment of task sense and difficulty.

Using the miners' cognitive and behavioural motor actions and operations as units of analysis. [9] viewed each functional block as a self-regulative system [18, 19] in which the miners use their tacit knowledge in various ways. Based on the functional analysis of data obtained from direct observations and video recordings of the miners' engagement with the production drilling activity, [9] found that a miner's individual object-oriented activity consists of both physical and mental actions and operations whose characteristics are influenced by past experiences. [9] established that the object of a miner's self-regulation of orienting activity could be segregated into physical and mental activities. The object, in terms of the miner's physical activity, was the conduction of production (rock) drilling operations with the miner using informed decisions and programs to enable explorative actions towards performance enhancement, sensemaking in task performance, and determination of motivational level. [9] also established that the object, in terms of the miner's mental activity, is the miner simultaneously observing the production drilling work, assessing task difficulty, and listening to communication models, which activities require the development of stable or dynamic mental models for enhancing the relevant task conditions.

Based on the morphological analysis of the miners' mental assessment of task difficulty, [9] reported the miners' as viewing the use of highly-digitized tractor technology with more than one robotic arm (i.e. boomer) for production drilling tasks as excessive, resulting to task overload for one operator to handle. The reason for this observation, as provided by [9] was that, the miners' viewed the high-level digitization of the work tool adding to the difficulty an operator encounters in his/her ability to effectively manoeuvre the joy-stick handle and its computerized programming command for multi boomers in the automated rock drilling actions and operations. As found by [9], the miners viewed the high-level digitization of the work environment as distracting them from developing the requisite dynamic mental models of task or situation which are required for enhancing the quality of the relevant task conditions. As such, as found by [9], the operators, based on acquired experiences, mostly find ways to guide their highly-digitized work tools for optimum productive performances, as highlighted in the following three miners' commentaries;

- Comment 1: Though I am happy with the digitized work environment, I have also developed enormous knowledge on manipulating the technology to make my work activity easier.
- Comment 2: The technology does not always get it right. Most of the time, I use the experience I have acquired over the years to guide this new technology for optimum performance.
- Comment 3: I have not been writing down the fascinating and intriguing discoveries I made over the years on the best way to use the high-tech equipment in carrying out my tasks… I sometimes share my experiences with colleagues who use the same equipment. I have been imparting some of these knowledge to people I train over here. I explain things to them and hope that they pick it up gradually.

Functional analysis of the commentaries above by [9] showed that the miners possess tacit knowledge developed overtime on various objective activities, which remained unshared, but which is used by the miners to negotiate technology-based standardized task patterns in bids to overcome task repetitiveness, and to increase their productive capacities in terms of waste removal in production time. Additionally, [9] found that the miners use the tacit knowledge to overcome subjective perceptions of technological shortcomings in their task undertakings (based the notion that technologies do not always get it right), such as using their deep tacit knowledge about the rock and self-developed task negotiation skills and abilities of 'reading the rock' to guide their highly-digitized work tools towards effective performances

The findings reported by [9], as highlighted above indicate that despite the miners' use of highly reliable automated mining equipment with improved operating and maintenance procedures, the relevance of their deeply-rooted tacit knowledge about the work environment (i.e. their skills and ability to 'read the rock') moulded by the old mining culture is of relevance for avoiding work interruption when the highly auto-mated tractor used for the work encounters a hard rock. It is important that such tacit knowledge is captured and integrated in the future design of highly digitized work tools, in order to eliminate possible future cognitive challenges on the user. This is because, as people gain experience in their task undertakings, they tend to become more abstract or complex in their conceptualising [9]. It is also due to changes in people's perceptions and interpretations of the world around them, and to which the level of complexity is attributable to the degree to which a given culture or subculture provides an opportunity for exposure to diversity and an active openness towards learning from the newly exposed experiences [9]. This is because, cognitive complexity is the most useful integrating model of psychosocial influences on organisational design [20], with its higher order structural personality dimension underlined by dif-ferent conceptual systems for perceiving reality [21]. Thus, in the theorization of an automated work activity towards the development of an innovative work organization model, the mode of mining activity changes introduced by the technological changes must be understood. For example, a human-technology-work environment design can be enhanced using the model for uncovering and understanding a worker's cognitive and emotional-motivational aspect of task complexity shown in Fig. 1 below.

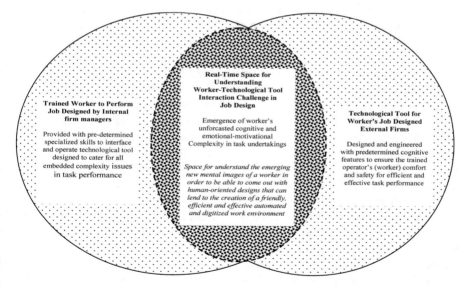

Fig. 1. Model for understanding a worker's emerging cognitive and emotional-motivational aspect of task complexity [8].

As it observable in the Fig. 1 above, operators can simultaneously be engaged in two forms of activities, namely, individual-oriented activity and social interaction.

- The individual-oriented activity entails both physical and mental activities. The object of the physical activity is performance of the digitized/automated task while the object of mental activity is listening to digitized communication models.
- The social interaction entails collaboration between human physical activity and mental activity. The subject is the individual who is simultaneously engaged in a physical activity through the manipulation of digitized computer technology to programme robotic work tasks, and a mental activity through digitized communication models by listening to background music during work, as well as information transmission from the automated system control centers, and/or from colleagues approaching or leaving the individuals activity location inside the automated work system.

6 Discussion

There is a current trend whereby most works on the individual-psychological approaches to organization activity study, considered as basic to the study of human work, are not discussed [12]. The individual-psychological analysis of activity includes the informational (cognitive), the morphological, the functional, and the parametrical methods of activity analyses [22] as highlighted in the previous sections. All these methods, logically organized according to stages and levels of the organizational activity analysis allowed one to tie together the obtained data into a holistic system [14]. As outlined in the sections above, actions, are fundamental components of activities and are subordinate to specific goals. In this context, the goal of an action is deemed a conscious mental representation of the outcome to be achieved [23]. Thus, in an automated mining activity, different actions may be undertaken to meet the same goal. This implies that organizational activities are realized as goal-oriented actions, with operations executed as ways of actions. Organizational activities therefore correspond with the way of goal achievement and are directly determined by the objective conditions in which the goal is given, and in which it also must be achieved [3]. Operations may become routinized and unconscious with practice. Thus, in highly digitized firms, each automated activity may constitute different actions with each action consisting of varieties of operations to be undertaken by different persons or groups, and/or mechanization process.

The interdependent activities to be involved in automated work system can be conceptualized from the learning to be made from existing automated work systems and innovative practices and ideas. Therefore, in the analysis of automated work activity, both the sociocultural and the individual-psychological analyses should be conducted. The main unit of analysis will be the organization. The sub-unit of analysis need to be carried out at two levels. These are the "object oriented" activity level, and the "subject-oriented" activity level. As it is explained by [14], object-oriented activity is performed by a subject using tools on a material object, where the subject of activity is the individual or group of individuals engaged in that activity. In deep-mining activity, this can be ascribed to the various activities entailed in the total mining production process involving humans either operating directly underground with all the associated risks and hazards or operating in control rooms using technology. Similarly, subject-oriented activity (also known as social interaction) involves two or more subjects, and is constituted through information exchange, personal interactions and mutual understanding. During task performance, the object-oriented and subject-oriented aspects of activity continuously transform into one another [14]. Thus, in the analysis of object-oriented activity, inter-subjective relationships also need to be considered [14, 24]. In both sub-analyses, the impact of activity contradictions [25] should be assessed. Contradiction, as a source of tension that arises between elements of the automated activity system identify areas where the system's components no longer match the automated activities they are expected to model (i.e. a signification of misfit between elements within an automated activity, and/or between different automated activities). These contradictions therefore, manifest themselves as problems, ruptures, breakdowns, and clashes [25]. As it is explained by [25], it is important to keep sight of the elements within the automated system. This is because many of the analytic strategies for examining mediated action between the human and the automated system can be made possible by isolating its elements. Such isolation allows various specialized perspectives to bring their insights to bear and serves as key to understanding how change occurs in the mediated action. Through such understanding, an innovative work organization that enhances social collaboration between automated systems and humans can be designed towards increased productivity.

7 Conclusion

The discussions in this paper has established that in the theorization of a highly digitized work activity towards the development of an effective and efficient digitized work organization model, the changes in the organizational activity introduced by the technological changes must be understood. It is therefore, concluded that analytic strategies for examining collaborative work between humans and digitized system should be developed. Such strategies should make it possible to isolate the elements of the collaborative work to allow actors with specialized perspectives and tacit knowledge on the digitized work environment to bring their insights to bear and serves as key to identifying both the innovative and constraining characteristics of the collaborative association to be expected between humans and the digitized work tools. By implication, an innovative work organization that enhances social collaboration between digitized systems and humans can be designed towards increased productivity. Thus, designers of digitized task environment and technologies used by humans need to understand the emerging mental images of humans to enable them to come out with human-oriented designs that can lend to the creation of a friendly, efficient and effective automated and digitized work environment for the production drilling activity. The work environment must also have good work and social conditions that enhance the workers 'emotional-motivational orientation.

References

1. Abrahamsson, L., Johansson, J.: From grounded skills to sky qualifications: a study of workers creating and recreating qualifications, identity and gender at an underground iron ore mine in Sweden. J. Ind. Relat. **48**(5), 657–676 (2006)
2. Bassan, J., Srinivasan, V., Knights, P., Farrelly, C.T.: A day in the life of a mine worker in 2025. In: Proceedings, First International Future Mining Conference, pp. 71–78, Sydney, NSW, 19–21 November 2008
3. Sanda, M.A.: Automation of the work environment and the human-technology collaboration challenge: a critical reflection. Int. Rob. Autom. J. **3**(6), 00074 (2017)
4. Abrahamsson, L., Johansson, B., Johansson, J.: Future of metal mining: sixteen predictions. Int. J. Mining Mineral Eng. **1**(3), 304–312 (2009)
5. Sanda, M.A., Johansson, J., Johansson, B., Abrahamsson, L.: Using systemic approach to identify performance enhancing strategies of rock drilling activity in deep mines. In: Hale, K. S., Stanney, K.M. (eds.) Advances in Neuroergonomics and Cognitive Engineering, Advances in Intelligent Systems and Computing, vol. 488, pp. 135–144. CRC Press, Boca Raton (2012)
6. Johansson, J., Abrahamson, L., Kårebon, B.B., Fältholm, Y., Grane, C., Wykowska, A.: Work and organization in a digital industrial context. Manag. Rev. **28**(3), 281–297 (2017)
7. Jarzabkowski, P.: Strategy as social practice: an activity theory perspective on continuity and change. J. Manag. Stud. **40**(1), 23–55 (2003)
8. Sanda, M.A.: Mediating subjective task complexity in job design: a critical reflection of historicity in self-regulatory activity. In: Carryl, B. (ed.) Advances in Neuroergonomics and Cognitive Engineering, pp. 340–350. Springer, Cham (2017)
9. Sanda, M.A., Johansson, J., Johansson, B., Abrahamsson, L.: Using systemic structural activity approach in identifying strategies enhancing human performance in mining production drilling activity. Theor. Issues Ergon. Sci. **15**(3), 262–282 (2014)
10. Sanda, M.A.: Cognitive and emotional-motivational implications in the job design of digitized production drilling in deep mines. In: Hale, K.S., Stanney, K.M. (eds.) Advances in Neuroergonomics and Cognitive Engineering, Advances in Intelligent Systems and Computing, vol. 488, pp. 211–222. Springer, Switzerland (2016)
11. Lave, J., Wenger, E.: Situated Learning: Legitimate Peripheral Participation. Cambridge University Press, Cambridge (1991)
12. Engeström, Y.: Learning by Expanding: An Activity-Theoretical Approach to Developmental Research. Orienta-Konsultit, Helsinki (1987)
13. Hendrick, H.W., Kleiner, B.M.: Macroergonomics: An Introduction to Work System Design. HFES, Santa Monica (2001)
14. Bedny, G.Z., Karwowski, W.: A Systemic-Structural Theory of Activity: Applications to Human Performance and Work Design. Taylor and Francis, Boca Raton (2007)
15. Aldrich, H.E.: Organizations and Environments. Stanford University Press, Stanford (2008)
16. Thompson, J.: Organizations in Action. McGraw-Hill, New York (1967)
17. Bedny, G.Z., Karwowski, W., Bedny, M.: The principle of cognition and behaviour: implications of activity theory for the study of human work. Int. J. Cogn. Ergon. **5**, 401–420 (2001)
18. Bedny, G.Z., Meister, D.: The factor of significance in the operator activity. In: Proceeding. Human Factors and Ergonomics 41st Annual Meeting, pp. 1075–1078. HFES, Albuquerque (1997)
19. Bedny, G.Z., Karwowski, W.: A systemic structural activity approach to the design of human-computer interaction tasks. Int. J. Hum. Comput. Interact. **2**, 235–260 (2003)

20. Hendrick, W.: Human factors in organizational design and management. Ergonomics **34**, 743–756 (1991)
21. Harvey, O.J., Hunt, D.E., Schroder, H.N.: Conceptual systems and personality organization. Wiley, New York (1961)
22. Bedny, G.Z., Karwowski, W., Bedny, I.S.: Complexity evaluation of computer-based tasks. Int. J. Hum.-Comput. Interact. **28**(4), 236–257 (2012)
23. Leontiev, A.N.: Activity, Consciousness, and Personality. Prentice-Hall, Englewood Cliffs (1978)
24. Sanda, M.A.: Application of systemic structural theory of activity in unearthing employee innovation in mine work. Procedia Manuf. **3C**, 5147–5154 (2015)
25. Engeström, Y.: Expansive learning at work: toward an activity theoretical reconceptualization. J. Educ. Work **14**(1), 133–156 (2001)

Design of a Platform for Sustainable Production Planning and Controlling from an User Centered Perspective

Maximilian Zarte[1]([⊠]), Agnes Pechmann[2], and Isabel L. Nunes[1,3]

[1] Faculty of Science and Technology, Universidade Nova de Lisboa,
Campus Caparica, 2829-516 Caparica, Portugal
m.zarte@campus.fct.unl.pt, imn@fct.unl.pt
[2] Department of Mechanical Engineering, University of Applied Sciences
Emden/Leer, Constantiaplatz 4, 26723 Emden, Germany
agnes.pechmann@hs-emden-leer.de
[3] UNIDEMI, Campus de Caparica, 2829-516 Caparica, Portugal

Abstract. The industrial world is currently being changed by digitalization and computerization of the production with the aid of information and communication technologies (ICT). Existing control and management components and systems are connected to this digitalized production. This industrial development towards Industry 4.0 is an enabler for sustainable development in enterprises. ICT provides opportunities to collect and analyze production data according to sustainable aspects. To visualize and analyze available production data from different sources, human-system interfaces are required. This paper presents a user platform for sustainable production planning and controlling (sPPC). The development of the platform follows the standard DIN EN ISO 9241-210:2010 for user-centered design of human-system interfaces. For the platform for sPPC, the primary and secondary users are identified, user and system requirements are specified, and a prototype of the platform is presented and evaluated.

Keywords: Human-centered design · Sustainable manufacturing
Industry 4.0

1 Introduction

The Brundtland commission defined sustainability as "development that meets the needs of the present without compromising the ability of future generations to meet their own needs" [1]. Moreover, sustainability is bonded by the triple bottom line which balances economic goals, environmental cleanness, and social responsibility [2]. An enabler for sustainable development in enterprises is the digitalization of production systems (so-called Industry 4.0). Existing control and management components and systems are connected to this digitalized production through the establishment of smart factories, smart products, and smart services available through an internet of things and of services [3]. This development provides opportunities for sustainable manufacturing using information and communication technologies (ICT) to collect and analyze

I. L. Nunes (Ed.): AHFE 2018, AISC 781, pp. 396–407, 2019.
https://doi.org/10.1007/978-3-319-94334-3_39

production data. To visualize and analyze available production data from different sources, human-system interfaces are required as connection between the user and the virtual world of industry 4.0. Therefore, sustainable development cannot be achieved without involving users for the design of platforms and services for more sustainable production.

It exists no widely accepted definition for the term human-system interface [4]. In general, human-system interfaces are intuitive and easy-to-use front ends (e.g. platforms, dashboards, applications) for monitoring, analyzing and optimizing specific activities of users to support their decisions [5]. For the design of human-system interfaces, it exists the standard DIN ISO EN 9241-210 which provides requirements and recommendations for human-centered design principles and activities throughout the life cycle of computer-based interactive systems. From a sustainable perspective, the needs and requirements of users must be in the focus of developers for human-system interfaces, to increase the system efficiency, accessibility, and the satisfaction of the users [6]. Obviously, this perspective already supports two dimensions of sustainability: social responsibility (user satisfaction), and economic success (system efficiency). Moreover, a human-centered design considers the complete life cycle of computer-based interactive systems which includes impacts on the users but also on the environment [6].

Due to a lack of tools and instruments for sustainable manufacturing and development, there is often a missing of ability to identify and raise sustainable potentials. This problem was investigated in previous developments of human-system interfaces for sustainable manufacturing and development. Chen and collaborators developed a web-based decision support tool to increase the resources efficiency. Stakeholder groups (e.g. governments, organizations, or industries) can import resources data to highlight the major resource flows and sectoral linkages among complex supply chains to highlight opportunities to improve resource efficiency. The results are presented as Sankey diagram and can be used to develop new strategies for the sustainable development [7]. Steenkamp and collaborators developed a visual resource management system using open-source programs. Based on collected production data, key performance indicators were displayed and used for production planning and controlling [8]. Rackow and collaborators presented the user interface "Green-Cockpit". The free of cost, open source and web-based tool is designed to help companies monitor, interpret, analyze, plan, and report their energy consumption [9]. Travverso and collaborators presented the life cycle sustainability dashboard tool to compare products in terms of their sustainability performance by an overall index, and individual environmental, economic and social indicators. The sustainable performance is displayed through a seven-color code for each indicator, an overall index highlighted by the position of the arrow on the top of the dashboard, and by an overall ranking score [10].

Previous developed human-system interfaces for sustainable development and manufacturing neglect the role of the users. The human-system interfaces are developed as front-end of tools and evaluation systems without considering the requirements, needs, and attributes of the users. Moreover, the developed interfaces are developed for experts of sustainable development and manufacturing. Human-system interfaces supposed be usable for users with less experience by sustainability and sustainable manufacturing.

The paper presents a concept for the design of a platform for sustainable production planning and controlling (sPPC). The design and evaluation of the platform for sPPC follows a user-centered design procedure. The objectives of the platform are to present the planned production, provide indications on potential fields of sustainable improvement, suggest actions for achieving these improvements, permit following their effects, and control the planned production to indicate problems and if the production meets the planned performance. For this approach, the platform integrates data managed by different conventional administrative units (e.g. enterprise resource planning (ERP) system) and production data (e.g. resource consumptions) collected by sensors in the production. Through first usability test with non-experienced users, the platform is evaluated with the aid of a cognitive walkthrough [11]. The feedbacks from the tests-users are discussed and allocated to the Nielsen heuristics [12] to identify potential fields for improvements of the platform for sPPC.

After this introduction, Sect. 2 presents the methodology, which was used for the design of the platform. Section 3 presents the development of the platform including the identification of users and user needs, specification of the system requirements, and description of the platform information and communication architecture according to Industry 4.0. Section 4 presents a prototype of the platform, and the results of the evaluation. Section 5 offers the conclusions and future works, followed by references.

2 Methodology and Methods to Design the Platform for sPPC

The design, prototyping and usability evaluation of the platform for sPPC follows a user-centered design procedure according to the standard DIN EN ISO 9241-210:2010 [6]. The procedure is an iterative process (dash arrows) and consists of five steps (see Fig. 1).

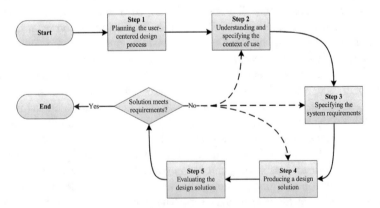

Fig. 1. Iterative procedure to develop the platform for sPPC from a user-centered perspective (adopted from [6]). The procedure consists of five iterative steps (dash arrows).

In Step 1 (planning the user-centered design process), the goals and functions of the platform are defined, and the platform users are identified. In Step 2 (understanding and specifying the context of use), the user needs are specified, which involves the description of the attributes, goals and tasks of the users, and environment where the system will be used. Based on Step 1 and 2, Step 3 (specifying system requirements) specifies the platform requirements. The results of Step 1–3 are presented in Sect. 3.

In Step 4 (producing design solutions) and Step 5 (evaluating the design solutions), design solutions are created and evaluated by experts and test-users. For the development of the platform presented in this paper, two iterative steps were performed. The methods for the design and evaluation are adopted from the study [13]. The method used in this study were discussions with experts, cognitive walkthrough, and Nielsen heuristics.

The first prototype is created on paper and provides examples about the visualization of the production plans and data, arrangement of active panels, buttons and diagrams, and first content of additional information to evaluate the production plans. The paper prototype is evaluated through discussions with experts from the fields production planning and controlling, and system ergonomics. The results of this iterative step are not presented and discussed in this paper.

The second prototype is created with aid of the online tool "justinmind" [14] and provides first functionalities to perform simple tasks. The digital prototype is evaluated with the usability test method cognitive walkthrough [11] with non experienced test-users. For the usability test, a scenario is developed which allows a single interaction session without any flexibility for the test-users. Scenarios combine the limitations of both horizontal prototypes (users cannot interact with real data) and vertical prototypes (users cannot move freely through the human-system interface) [15]. Scenarios have two main advantages [12]: First, scenarios can be used during the design of a human-system interfaces as a way of expressing and understanding the way users eventually will interact with the future system. Second, scenarios can be used during early evaluation of a human-system interface design to get user feedback without the expense of constructing a full running prototype with all required functionalities and features of the target system. The feedbacks from the tests-users are discussed and allocated to the Nielsen heuristics [12] to identify most potential fields for improvements of the platform for sPPC. The Nielsen heuristics represent ten general principles for human-system interface design. They are called "heuristics" because they are broad rules of thumb and not specific usability guidelines. An overview of the prototype and the results from the evaluation are presented in Sect. 4.

3 User and System Requirements of the Platform for sPPC

3.1 User and User Needs

The first step in the human-system interface design process is to study the intended users and use of the human-system interface. It is necessary to know the type of people who will be using the system [12]. There are several types of users that are relevant to system development [16]. For this paper, primary and secondary users are considered.

The primary users are the main user of the system and needs all functionalities of the human-system interface. The secondary users need only partial functionalities of the human-system interface.

In general, methods for the identification of the users and the description attributes, tasks, and goals of the users can be categorized in two groups [15]: data-based identification and assumption-based identification of users. Because of missing information about users for sustainable management systems, the users are identified and described based interviews with experts from the fields of production planning and controlling and system ergonomics.

For the platform for sPPC, three types of users are identified in enterprises:

1. Production scheduler
2. Production manager
3. Manager for environmental and sustainable topics.

User 1 (*production scheduler*) is a primary user and uses the platform for sPPC in periodical time intervals (daily, weekly) to improve the degree of sustainability of the planned production. The main attributes of this user are experience in conventional production planning and controlling (PPC), and knowledge about the economic, social and environmental consequences of actions to improve the sustainability of the production plan. To schedule the production according to sustainable aspects, the user must be able to consider sustainable impacts of the planned production, control the planned production performance according to sustainable goals, and modify production plans to meet target production performances and sustainable goals.

User 2 (*production manager*) is a secondary user and uses the platform for sPPC in periodical time intervals (weekly, monthly, quarterly) to control the previous production performance with target performances. The main attributes of this user are experience in conventional production controlling. To control the production, the user must be able to compare the performance previous production inputs and outputs with target inputs and outputs.

User 3 (*manager for environmental and sustainable topics*) is also a secondary user and uses the platform for sPPC in periodical time intervals (monthly, quarterly, yearly) to report and analyze sustainable impacts of the production, and to control the status of sustainable goals. The main attributes of this user are experience in conventional production controlling, and knowledge about sustainable impacts from competitors using similar processes (e.g. through sustainable reports from the competitors). To control the production according to sustainable aspects, the user must be able to analyze and to present production inputs and outputs for sustainable reporting, to compare resource consumptions and emissions of a single consumer (present values with historical values for specific timeframes), and to compare resource consumptions and sustainable impacts of two or more consumers (internal and external benchmarks).

3.2   Specification of the System Requirements

Based on the user needs, system requirements and functionalities relevant for the users are specified. The following system requirements and functionalities represent the basis for the development of the platform for sPPC.

Graphical Presentation of Production Plans: With the aid of time-diagrams, the considered production plan (historical and planned production) of specific production resources (machines, employees, external services), products, and customer orders must be presented for specific timeframes (day, week, month, year). Moreover, relevant information, such as start and end time of productions steps, required resources (e.g. energy, materials), and deadlines of costumer orders must be available for the users.

Graphical Presentation of Production Data and Sustainable Indicator: With the aid of time-diagrams and related to the considered production plan (historical and planned production), production data (e.g. energy demand, renewable energy production), and sustainable indicator (e.g. self-consumption ratio) must be presented for specific timeframes (day, week, month, year).

Presentation of the Sustainable Score: With the aid graphic elements and related to the considered production plan, the sustainable average score must be presented. Moreover, detailed information to best-praxis sustainable score, target sustainable score, and scores of single sustainable indicators must be available for the users.

Adjustment of Production Steps: With the aid of data inputs, start, end, and process times of production tasks and orders must be changeable to improve the sustainable score of the considered production plan. Moreover, the effects of possible adjustments in the production plan on resource consumptions, emissions, sustainable indicators, and sustainable score must be presented to the users.

Security of the System: With the aid of user rights, functionalities and information contents of the platform for sPPC, such as adjustment of production steps, and presentation of specific historical production plans, must be allocated to specific users.

3.3   Functions, Goals, and Architecture of the Platform for sPPC

Assessing the degree of sustainability is a very complex task which needs expert knowledge and experience. A detailed description of models for the evaluation of sustainability and sustainable development is out of scope of this paper. The goals of the platform are to act as front-end system for the users to present the results of models for the evaluation sustainability and sustainable development, modify the planned production to increase the sustainability, and analyse the production according to sustainable aspects, and to control production and sustainable goals. For these functions, additional production management systems are required, such as an ERP system (for e.g. conventional production planning and controlling) and simulations (for e.g. forecast of renewable energy generation), and databases (for e.g. recording of production data and management data).

These systems are structured in a system architecture in Fig. 2 as parts of a cyber physical system (CPS). A CPS is an integration of computation with physical processes. Embedded computers and networks monitor and control the physical processes, usually with upper and lower feedback loops where physical processes affect computations [17]. According to the reference architecture model industry 4.0 (RAMI4.0) [18], the architecture of the platform for sPPC is structured in layers, which represent physical parts (red background) and virtual parts (green background) of the CPS. The feedback loops are presented as lower and upper arrows.

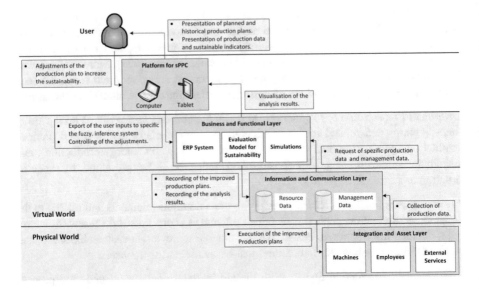

Fig. 2. The architecture of the platform for sPPC as part of a cyber physical system (CPS).

The *integration and asset layer* contain the production system, including machines, employees, and external services, which are required to produce the production output. With the aid of ICT, production data are collected (upper arrow) and production task are received (lower arrow).

The *information and communication layer* retrieve the production data from the asset layer on a regular basis and records it into databases. With the aid of database queries, specific information are requests (upper arrow) and management data (e.g. production plans, evaluation results) are recorded (lower arrow).

The *business and functional layer* contain production management systems, models for the sustainable evaluation, and simulations, which are used to analysis the collected production data, to plan, optimize the production. With the aid of the platform, the results of the business and functional layer are visualized to the user (upper arrow) and adjustments in production plans are made by the user to improve the degree of sustainability (lower arrow).

Due to the open architecture, the platform for sPPC can be operated as web-based application via computers and tablets independent of the location of the production system and users.

4   Digital Design Solution of the Platform for Sustainable Planning

4.1   Overview of the Platform for sPPC

Figure 3 presents the digital prototype of the platform for sPPC. To access the platform, a normal user log-in is required, which is not presented in this paper.

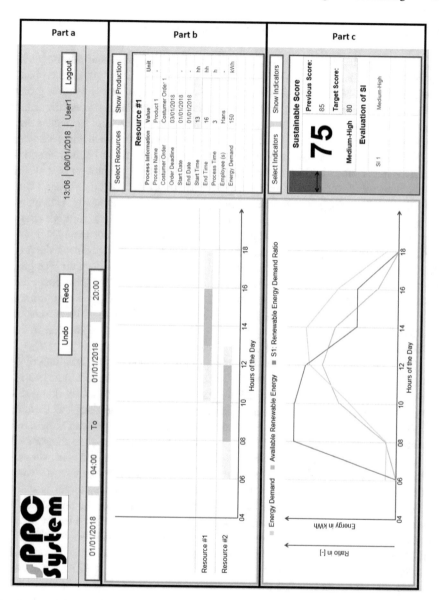

Fig. 3. Overview of the Interface separated in three parts (a, b, and c) which contain buttons (red marks), active panels (blue marks) and time diagrams (green marks).

Through the user log-in, different user rights can be allocated to users limiting the functionality of the platform. The platform for sPPC can be separated in three parts (Part a, b, and c) which contain buttons (red marks), active panels (blue marks), and time diagrams (green marks).

Part a provides basic functionalities ("Undo", "Redo" and "Logout" buttons) and information (current time, date and logged in user) to the user. Moreover, through the data input fields, the user can specify the timeframe for production planning and controlling.

Part b presents the production plan as Gantt-time diagram and additional information in the active panel. With the button "Select Resources", the active panel is changed to a tree-node menu which contains lists of the involved resources, products, and costumer orders for the specified timeframe. Due to check boxes, the user can choose which resources, products, and/or costumer orders are presented in the Gantt-time diagram. With the button "Show Production", the Gantt-time diagram is refreshed and shows the selection of the user. With a click on one of the colored bars, the active panel shows additional information to the selected production task. The user can see the process name, costumer order ID, deadline of the costumer order, start and end time of productions steps, process time, and required resources (e.g. employees, energy). The user can change the start time of the production task through drag and drop of the colored bar or via an input field in the active panel for the start time. For production steps with changeable process time (e.g. idle time of machines), the process time can be changed through an input field in the active panel, which changes the length of the colored bar. Because of the limited functionality of the tool "Justinmind", it was not possible to connect production tasks with each other. In case that production tasks can only perform if previous tasks are finished, start times must be automatically changed in the whole production plan. This feature must be implemented in future prototypes of the platform for sPPC.

Part c presents the production data and sustainable indicator as line-time diagram and additional information of the status of the sustainable score and indicator in the active panel. With the button "Select Indicators", the active panel is changed to a tree-node menu which contains lists of the available production data and sustainable indicator for the specified timeframe. Due to check boxes, the user can choose which production data and sustainable indicator are presented in the line-time diagram. With the button "Show Indicators", the line-time diagram is refreshed and shows the selection of the user. In the active panel, the user can see the sustainable score (as value, linguistic value, bar chart), best-practice sustainable score, target sustainable score, and a list of sustainable indicators with linguistic values for the considered time frame. The list of sustainable indicators is sorted according to the linguistic values and shows first the worst sustainable indicator. Because of the limited functionality of the tool "Justinmind", it was not possible to connect production tasks with the production data and sustainable indicator. In case that the start time of a production task is changed, the values in the line-time diagram must be automatically changed and the new sustainable score must be presented. This feature must be implemented in future prototypes of the platform for sPPC.

4.2 Evaluation of the Platform for sPPC

To test usability of the platform a scenario was developed in which the test-users must perform tasks with aid of the platform. For the scenario, six tasks have been defined:

1. Log-in in the sPPC system with the username "User1" and the password "1234".

2. Choose the start and end date 01/01/2018, and the start time 04:00 and end time 20:00.
3. Select all available resources, products, and costumer orders and consider the production plan.
4. Select all available production data and sustainable indicator and consider only the historical information in the diagram.
5. Change the start time of the production task "Product 2" (Resource #2) to 6am.
6. Change the process time of the production task "Idle Mode" (Resource #1) to 0,5 h.

For the usability test, four test-users are selected. The test-users were students from the industrial engineering faculty and have no practical experience in production management and sustainable manufacturing. The procedure for the usability tests was as follow: First, the test-users were introduced to the tasks and goals of the primary user "Production scheduler" of the platform for sPPC. Second, the test-user performs the defined scenario. During the tests, the test user got no supports to perform the tasks and the time is stopped how long the test-users need to perform the tasks. After performing the defined tasks, third, the test-users gave feedback about problems which occurred during the test and gave recommendations for improvements. The feedbacks and recommendations from the tests-users are discussed and allocated to the Nielsen heuristics [12] to identify potential fields for improvements of the platform for sPPC.

The Nielsen heuristics *"Help and documentation"*, *"Help users recognize, diagnose, and recover from errors"*, and *"Error prevention"* consider the support and feedback for the user in case of errors and problems with functionalities. The prototype contains no support and error feedbacks systems. To meet these heuristics, support and error feedbacks systems must be implemented in future prototypes.

The Nielsen heuristic *"User control and freedom"* consider functionalities to leave unwanted states without having to go through an extended dialogue (e.g. undo and redo buttons). The prototype contains buttons for undo and redo but without functionality. The functionality must be implemented in future prototypes.

According to the Nielsen heuristic *"Match between system and real world"*, the system should speak the users' language, with words, phrases and concepts familiar to the user, rather than system-oriented terms. This heuristic can only be considered limited, because non-experienced users were selected for the usability tests. Therefore, the test-users were not familiar with the concept of sustainability and PPC. Despite this limitation, the test-users criticized, that the system language is not the native language. Specially, for non-experience users who are not familiar with English words and phrases for PPC have problems to understand the meanings of e.g. buttons and diagrams for the presentation of the production plan. In future prototypes, optional languages and a glossary for important words should be provided to the users to increase the usability for non-experienced users. Moreover, future prototypes must be tested with experienced users to consider this heuristic from another perspective.

The Nielsen heuristic *"Recognition rather than recall"* considers the user's memory load by making objects, actions, and options visible. The user should not have to remember information from one part of the dialogue to another. The test-users miss the functionality to compare various timeframes with each other. Moreover, is was discussed to implement a second screen for the specification of timeframes, and for selection of resources, products, costumer orders, production data, and sustainable indicator to decrease the information content with only one screen. In future prototypes, a sequence must be implemented to set the Gant-time and line-time diagrams.

The Nielsen heuristic *"Flexibility and efficiency of use"* considers if accelerators e.g. shortcuts, are used to speed up the interaction with the human-system interface. The test-users suggest labelling and marking the production tasks in the Gant-time diagram to indicate possible interactions with the bars in future prototypes. Moreover, useful shortcuts can be implemented to increase the speed to interact with the platform.

The Nielsen heuristics *"Aesthetic and minimalist design"*, *"Consistency and standards"*, and *"Visibility of system status"* consider the design and information content of the prototype. The tests users were satisfied with design of the prototype. The information contents were sufficient to perform the user tasks without help in an acceptable timeframe.

Table 1 presents the times which the users needed to perform the tasks. These times were compared with the references time of an experienced user (developer of the system) and show that there is still space to improve the intuitive operation of the platform to decrease the time performing specific tasks. These results can be used as basis for future usability tests of new prototypes.

Table 1. Results for the times, which the test-users needed to perform the user-tasks. The times are compared with reference time of experienced user (developer of the system).

User	Time [mm:ss]	Difference to the Reference User [mm:ss]
Reference-User	01:57	± 00:00
Test-User 1	7:15	+ 05:18
Test-User 2	5:57	+ 04:00
Test-User 3	8:26	+ 06:29
Test-User 4	4:02	+ 02:05

5   Conclusion and Future Work

The development towards Industry 4.0 is an enabler for sustainable development in enterprises. This development provides opportunities for sustainable manufacturing using ICT to collect and analyze production data. To visualize and analyze available production data from different sources and tools, a digital design solution is presented. The development of the platform for sPPC follows the standard DIN EN ISO 9241-210:2010 for user-centered design of computer systems. For the platform for sPPC, primary and secondary users are identified, user needs and system requirements are specified, and design solutions are created and evaluated. Moreover, the system architecture of the platform according to RAMI4.0 is presented. To test the usability of the platform for sPPC a cognitive walkthrough has been performed, in which the users have to perform tasks with aid of the platform for sPPC. The feedbacks from the tests-users are discussed and allocated to the Nielsen heuristics. Based on the results of the discussion, the platform must be improved in future works. Through an implementation in a case study for sPPC, more features can be adapted in the proto-typed and tested with experienced user in PPC and real data.

References

1. WCED: Report of the World Commission on Environment and Development: Our Common Future (1987)
2. Elkington, J.: Cannibals with Forks: The Triple Bottom Line of 21st Century Business. Capstone, Oxford (1997)
3. Stock, T., Seliger, G.: Opportunities of sustainable manufacturing in industry 4.0. Procedia CIRP **40**, 536–541 (2016)
4. Adam, F., Pomerol, J.-C.: Developing practical decision support tools using dashboards of information. In: Burstein, F., Holsapple, C. (eds.) International Handbooks on Information Systems, Handbook on Decision Support Systems 2: Variations, 1st edn, s. l, pp. 151–173. Springer (2008)
5. Gröger, C., Hillmann, M., Hahn, F., Mitschang, B., Westkämper, E.: The operational process dashboard for manufacturing. In: Forty Sixth CIRP Conference on Manufacturing Systems, vol. 7, pp. 205–210 (2013)
6. Ergonomie der Mensch-System-Interaktion – Teil 210: Prozess zur Gestaltung gebrauch stauglicher interaktiver Systeme, 9241-210:2010 (2011)
7. Chen, P.-C., Liu, K.-H., Ma, H.-W.: Resource and waste-stream modeling and visualization as decision support tools for sustainable materials management. J. Clean. Prod. **150**, 16–25 (2017)
8. Steenkamp, L.P., Hagedorn-Hansen, D., Oosthuizen, G.A.: Visual management system to manage manufacturing resources. Procedia Manuf. **8**, 455–462 (2017)
9. Rackow, T., et al.: Green cockpit: transparency on energy consumption in manufacturing companies. Procedia CIRP **26**, 498–503 (2015)
10. Traverso, M., Finkbeiner, M., Jørgensen, A., Schneider, L.: Life cycle sustainability dashboard. J. Ind. Ecol. **16**(5), 680–688 (2012)
11. Wharton, C., Rieman, J., Clayton, L., Polson, P.: The cognitive walkthrough method: a practitioner′s guide. In: Nielsen, J. (ed.) Usability Inspection Methods, pp. 105–140. Wiley, New York (1994)
12. Nielsen, J.: Usability Engineering. Kaufmann, San Francisco (1993)
13. Nunes, I.L.: Teaching usability to industrial engineering students. In: Nunes, I.L. (ed.) Advances in Human Factors and System Interactions: Proceedings of the AHFE 2016 International Conference on Human Factors and System Interactions, 27–31 July 2016, Walt Disney World®, Florida, USA. Advances in Intelligent Systems and Computing, vol. 497, s. l., pp. 155–162. Springer International Publishing, Cham (2017)
14. Justinmind: Justinmind. https://www.justinmind.com/. Accessed 11 December 2017
15. Nielsen, L.: Personas - User Focused Design. Springer, London (2013)
16. Crowston, K.: "Personas" to support development of cyberinfrastructure for scientific data sharing. JESLIB **4**(2), e1082 (2015)
17. Lee, E.A.: Cyber physical systems: design challenges. In: Center for Hybrid and Embedded Software Systems (2008)
18. Reference Architecture Model Industrie 4.0 (RAMI4.0), DIN SPEC 91345:2016-04 (2016)

Author Index

A

Acosta-Vargas, Patricia, 197, 210, 233, 246, 265
Afzaal, Maryam, 40
Ahmed, Muhammad, 40
Akin, N. Tugbagul Altan, 257
Ali, Asad, 40
Arezes, Pedro, 134

B

Babić, Snježana, 76
Baldeon, Jonathan, 210
Bendzioch, Sven, 332
Benter, Martin, 343
Bibi, Zarina, 40
Bliss, James P., 184
Bouazzaoui, Sarah, 101
Brkic, Vesna Spasojevic, 56

C

Calle-Jimenez, Tania, 197, 210, 221, 233, 246, 265
Campos, Pedro, 69
Cardoso, Heitor, 274
Carruth, Daniel W., 33
Chen, Chien-Liang, 156
Costa, Nelson, 134
Costa, Susana, 134

D

Dar, Mahnoor, 40
Deb, Shuchisnigdha, 33
Deuse, Jochen, 299
Ding, Lu, 26
Doeltgen, Martin, 19

Dombrowski, Uwe, 353
Dworschak, Bernd, 312

E

Ebert, Achim, 109
Eguez-Sarzosa, Adrián, 221
Esparza, Danilo, 246
Esparza, Wilmer, 197, 210, 233, 265
Essdai, Ahmed, 56
Etinger, Darko, 76

F

Falkowska, Julia, 164
Ferreira, Nuno, 274
Figueira, José R., 122
Fonseca, Micaela, 274
Frey, Darren, 33
Fu, Shan, 26

G

Gamas, Filipa, 285
Ganz, Walter, 312
Gokturk, Mehmet, 257
Gomes, Madalena, 274
González, Mario, 210, 246, 265
Guevara, César, 197, 210, 233, 265

H

Hamann, Bernd, 109
Heck, Bo, 109
Herfs, Werner, 374
Hering, Friederike, 19
Hinrichsen, Sven, 332
Hu, Huimin, 88
Hudson, Christopher R., 33

© Springer International Publishing AG, part of Springer Nature 2019
I. L. Nunes (Ed.): AHFE 2018, AISC 781, pp. 409–410, 2019.
https://doi.org/10.1007/978-3-319-94334-3

J

Jadán, Janio, 197, 210, 233, 246
Jadán-Guerrero, Janio, 265
Jan, Yih-Kuen, 156
Jeong, Heejin, 3, 9
Jeske, Tim, 321

K

Kabir, Imran, 40
Kilijańska, Barbara, 164
Kuhlang, Peter, 343

L

Lee, Byung Cheol, 3, 9
Lennings, Frank, 321
Li, Wei, 88
Liau, Ben-Yi, 156
Lin, Na, 88
Lin, Yung-Sheng, 156
Liu, Zhongqi, 146
Lo, Yu-Chou, 156
Lopes, Arminda Guerra, 69
Lu, Yanyu, 26
Luján-Mora, Sergio, 221
Lung, Chi-Wen, 156

M

Maettig, Benedikt, 19
Masood, Fatima, 40
McGinley, John, 33
Moniz, Sara, 69
Morais, Pedro, 285

N

Noehring, Fabian, 299
Nunes, Isabel L., 197, 210, 233, 246, 265, 363, 396

O

Orehovački, Tihomir, 76

P

Park, Jaehyun, 3, 9
Park, Jangwoon, 3, 9
Pechmann, Agnes, 363, 396
Pham, Thanh, 9
Pilco, Hennry, 197

Q

Quaresma, Claudia, 274
Quintão, Carla, 274, 285

R

Reimer, Anne, 353
Renzi, Adriano B., 46
Ribeiro, Nuno, 134
Rupprecht, Franca, 109
Rybarczyk, Yves, 197, 210, 233, 246, 265

S

Sanchez-Gordon, Sandra, 197, 210, 233, 246, 265
Sanda, Mohammed-Aminu, 384
Schilling, Katrin, 374
Schnalzer, Kathrin, 312
Shi, Bo, 175
Simões, Paulo, 134
Simões-Marques, Mário, 122
Soares, Luisa, 69
Sobecki, Janusz, 164
Stefanak, Tobias, 353
Storms, Simon, 374
Stowasser, Sascha, 321

T

Taylor, Aysen K., 101
Tiller, Lauren N., 184

V

Veljkovic, Zorica, 56
Vigário, Ricardo, 274, 285
Villarreal, Santiago, 197, 210, 233, 246, 265

W

Weber, Marc-André, 321
Wienzek, Tobias, 299
Woestmann, René, 299
Wu, Haimei, 88
Würfels, Marlene, 321

Y

Yin, Qingsong, 146

Z

Zarte, Maximilian, 363, 396
Zerka, Katarzyna, 164
Zhang, Liwei, 146
Zhou, Qianxiang, 146
Zia, Hafiz Usman, 40

Printed in the United States
By Bookmasters